AF595013

Novel Applications and Technologies for the Industrial Exploitation of Algal Derived Marine Bioactives as Nutraceuticals or Pharmaceuticals

Novel Applications and Technologies for the Industrial Exploitation of Algal Derived Marine Bioactives as Nutraceuticals or Pharmaceuticals

Editors

Marco García-Vaquero
Brijesh K. Tiwari

MDPI • Basel • Beijing • Wuhan • Barcelona • Belgrade • Manchester • Tokyo • Cluj • Tianjin

Editors
Marco García-Vaquero
University College Dublin,
Dublin, Ireland

Brijesh K. Tiwari
Teagasc Food Research Centre,
Dublin, Ireland

Editorial Office
MDPI
St. Alban-Anlage 66
4052 Basel, Switzerland

This is a reprint of articles from the Special Issue published online in the open access journal *Marine Drugs* (ISSN 1660-3397) (available at: https://www.mdpi.com/journal/marinedrugs/special_issues/Novel_Applications_Technologies_Industrial_Exploitation_Algal_Derived).

For citation purposes, cite each article independently as indicated on the article page online and as indicated below:

LastName, A.A.; LastName, B.B.; LastName, C.C. Article Title. *Journal Name* **Year**, *Volume Number*, Page Range.

ISBN 978-3-0365-7008-2 (Hbk)
ISBN 978-3-0365-7009-9 (PDF)

Contents

About the Editors

Marco García-Vaquero

Marco Garcia-Vaquero, DVM, Ph.D., MIFST is Assistant Professor at University College Dublin, Ireland. Previously, he was research officer at TEAGASC, Food Research Center, Ireland. Dr. Garcia-Vaquero's primary research interest focuses on the utilization of algae and other agri-food waste or by-products for the generation of novel food ingredients. His research explores the use and optimization of innovative technologies for the extraction of high-value compounds from agri-food by-products as nutraceuticals. Additionally, he is interested in chemical analyses of these compounds and their biological activities, and the safety and bio-accessibility of these compounds in vitro. Throughout his career, Dr. Garcia-Vaquero has written over 50 peer-reviewed research publications and over 30 book chapters. He has also co-edited multiple books in different aspects of food technology and food chemistry. He is currently an editorial board member of *Marine Drugs*, and a management committee member in the European research networks of researchers.

Brijesh K. Tiwari

Brijesh K Tiwari, M.Sc., Ph.D., FIFST, FRSC is a principal research officer at TEAGASC and adjunct Professor at University College Dublin, Ireland. Previously, he was a Research Professor at the Food Research Center; Manchester Metropolitan University, UK. He has been an external research project evaluator for Research Council of Norway and External evaluator for the Norwegian Strategic Research Programme, Norway; Executive Agency for Higher Education and Research, Development and Innovation. Dr. Tiwari performs extensive research all around the world, has won many awards, is an editorial board member and a keynote speaker at international conferences, and has published over 175 peer-reviewed research publications and over 100 book chapters. He has also co-edited 10 books and is a book series editor for the IFST Advances in Food Science book series.

Editorial

Nutraceuticals from Algae: Current View and Prospects from a Research Perspective

Brijesh K. Tiwari [1] and Marco Garcia-Vaquero [2,*]

[1] Teagasc Food Research Centre, Ashtown, Dublin 15, D15 DY05 Dublin, Ireland
[2] Section of Food and Nutrition, School of Agriculture and Food Science, University College Dublin, Belfield, Dublin 4, D04 V1W8 Dublin, Ireland
* Correspondence: marco.garciavaquero@ucd.ie

Citation: Tiwari, B.K.; Garcia-Vaquero, M. Nutraceuticals from Algae: Current View and Prospects from a Research Perspective. *Mar. Drugs* **2022**, *20*, 671. https://doi.org/10.3390/md20110671

Received: 3 June 2022
Accepted: 22 October 2022
Published: 26 October 2022

Introduction

In recent years, algae, both microalgae and macroalgae, have attracted the attention of the scientific community as a source of multiple active molecules or bioactives, including polysaccharides, fatty acids, proteins and peptides, polyphenols, diterpenes, steroids, and alkaloids. Compared with other natural sources, marine organisms have taken the lead in the discovery of new drugs. Oceans cover more than 70% of the Earth's surface and represent a challenging environment for the growth of marine organisms, with extreme fluctuations in water level, solar radiation, and temperature; an environment in which algae are able to thrive by producing unique metabolites, different in composition from those of terrestrial plants, which play a major role in the protection of biomass [1]. The unique chemical features of algal compounds and their reported health benefits have contributed to expanding the interest in these molecules far beyond their pharmaceutical and cosmeceutical applications, with the food industry gaining interest in the incorporation of these molecules as functional foods or nutraceuticals [2]. Innovative technologies and processes are currently being explored to improve the use of algae as a source of nutraceuticals. The future use of algal biomass for food and nutraceutical purposes needs a dynamic research environment to understand the knowledge gaps and develop new strategies for the optimum exploitation of these resources, including the improved production/cultivation and understanding of the biomass, the sustainable and green extraction of high-value compounds, and finally, exploring the biological properties of these compounds using in vitro, ex vivo, and in vivo models. Further multidisciplinary research is needed to establish new or improve current methods for algae production, including the application of stressors or manipulations of the biomass targeting the overproduction of high-value compounds; as well as analyzing the changes in composition of the wild biomass to establish a relationship between the stressors and compositional changes in algae that can guide their future culture and industrial exploitation. Moreover, the use of innovative and emerging technologies, including ultrasound, microwaves, electric fields, and high-pressure and supercritical fluids, have also shown promising results for the development of efficient and green extraction, isolation, purification, and preservation processes of algal compounds for their future use as nutraceuticals. Finally, the health benefits of these compounds will also have to be demonstrated using in vitro, ex vivo, and/or in vivo model systems to enable the future commercialization of marine compounds as nutraceuticals and to establish health claims. This Special Issue contains nine articles, including five research articles and four reviews, covering multiple innovative aspects of the exploitation of algae for the development of nutraceuticals. Here, we provide a brief overview of what the reader will find in this Special Issue.

Wang et al. [3] reviewed and proposed improvements to adaptive laboratory evolution (ALE), an innovative method to explore strain improvements for microalgal production in developing new biological and phenotypic functions and improving the performance of strains in microalgal biotechnology. The authors identified several challenges and proposed solutions for improvements of ALE, including the generation of a high-quality mutant library to identify the genetic diversity of ALE, the need for multi-omics to promote the efficient data mining and implementation of ALE experiments, and the need to incorporate novel cultivation strategies, such as red LED light and phytohormones, to accelerate the ALE process, amongst other strategies that will need upgraded and developed software for more effective data interpretation [3].

Bolaños-Martínez et al. [4] provided an update on microalgal biotechnology strategies for pharmaceutical applications, including techniques for the generation of recombinant proteins and genetic engineering processes, including viral-based vector constructions. The use of viral vectors is relatively new in algae, and they are gaining momentum for the production of biopharmaceuticals because they have higher yields and shorter production times compared with chloroplast and nuclear-stable transformation methods. The authors also emphasized that in February 2022, one company produced the first COVID-19 vaccine in plants (COVIFENZ®) that was approved by the Health Agency in Canada, fact that can be a stepping-stone for the green production of these products for human use [4].

Čagalj et al. [5] generated extracts from *Cystoseira compressa* collected in the Central Adriatic Sea using microwave-assisted extraction technology from algae collected during the seasonal growth period (May–September). The authors analyzed the total phenolic content, total tannin content, antioxidant activities (measured as ferric reducing antioxidant power (FRAP), 2,2-diphenyl-1-picrylhydrazyl radical scavenging activity (DPPH) and oxygen radical absorbance capacity (ORAC)), and the antimicrobial effect of these compounds against *Listeria monocytogenes*, *Staphylococcus aureus*, and *Salmonella enteritidis*. The authors reported the highest antibacterial activity in extracts generated in June, July, and August, and associated those results with compounds produced when the sea temperature was at its highest. Moreover, the authors emphasized the usefulness of the *C. compressa* biomass as a source of nutraceuticals [5].

Meinita et al. [6] reviewed the biological activities, as well as the safety and toxicity levels of fucosterol from marine algae, relevant for the use of these compounds in the nutraceutical and pharmaceutical industries. The literature search focused on the period 2002–2020, and following the Preferred Reporting Items for Systematic Reviews and Meta-Analyses method, the authors identified 43 studies that could help to fill certain research gaps. Overall, the review concluded that fucosterol exhibited low toxicity in animal cell lines, human cell lines, and animals. However, the authors emphasized the need for further safety and toxicity reports of this compound under clinical settings [6].

O'Connor et al. [7] reviewed the current and emerging processes for the generation of algal bioactive peptides, including pre-treatments for the extraction of protein from algae and methods for the generation of hydrolysates and purification of these compounds. The authors outlined the main biological properties attributed to determining bioactive peptide sequences isolated from algae, including anti-hypertensive, antioxidant, and anti-proliferative/cytotoxic effects assayed in vitro and/or in vivo, and emphasized the use of in silico tools, such as quantitative structural activity relationships (QSARs) and molecular docking, as powerful tools to accelerate the discovery of these promising compounds [7].

Ahmed et al. [8] reviewed the potential of marine bioactive peptides for human health due to their unique chemical structures, physicochemical, and biological activities. The authors focused on marine microorganisms, including microalgae, bacteria, and fungi, considered as good sources of amino acids and peptides, and emphasized the relevance of the marine biome and the opportunities it offers for the valorization of marine-biome-based bioactive peptides. The review also summarizes FDA-approved marine bioactive peptides, as well as the legislation, challenges, and future perspectives for the increased use of these compounds [8].

Garcia-Vaquero et al. [9] focused on experimentally exploring the application of innovative extraction technologies (ultrasound-assisted extraction (UAE), microwave-assisted extraction (MAE), ultrasound–microwave-assisted extraction (UMAE), hydrothermal-assisted extraction (HAE), and high-pressure-assisted extraction (HPAE)). The authors used fixed extraction conditions (solvent: 50% ethanol; extraction time: 10 min; algae/solvent ratio: 1/10) when using all the selected innovative technologies, and they explored the application of a time post-treatment (0 and 24 h) for the recovery of phytochemicals (total phenolic content, total phlorotannin content, total flavonoid content, total tannin content, and total sugar content) and associated antioxidant properties (DPPH and FRAP) from *Fucus vesiculosus* and *Pelvetia canaliculata*. Overall, UAE generated extracts with the highest phytochemical contents from both macroalgae, with the highest yields of compounds generated from *F. vesiculosus* which included the total phenolic content (445.0 $\pm$ 4.6 mg gallic acid equivalents/g), total phlorotannin content (362.9 $\pm$ 3.7 mg phloroglucinol equivalents/g), total flavonoid content (286.3 $\pm$ 7.8 mg quercetin equivalents/g), and total tannin content (189.1 $\pm$ 4.4 mg catechin equivalents/g). The DPPH antioxidant activities were at the highest levels in extracts generated by UAE and UMAE from both macroalgae, whereas no clear pattern was appreciated for FRAP. Moreover, the authors determined that after the application of these innovative technological treatments, additional storage post-extraction did not improve the yields of phytochemicals or antioxidant properties of the extracts [9].

Elhady et al. [10] isolated three fatty acids (docosanoic acid 4, hexadecenoic acid 5, and alpha hydroxy octadecanoic acid 6) and three ceramides (A (1), B (2), and C (3)) from the macroalga *Hypnea musciformis*. The authors analyzed the biological properties of these isolated compounds and determined that ceramides A (1) and B (2) had in vitro cytotoxic activity against human breast adenocarcinoma (MCF-7) cell lines. Furthermore, when assayed in vivo using a mouse model of Ehrlich ascites carcinoma (EAC), both compounds reduced the size of the tumors in inoculated mice: in the case of ceramide A (1) at a dose of 1 mg/kg; and in the case of B (2) at a dose of 2 mg/kg. Overall, the authors' findings demonstrated the cytotoxic, apoptotic, and antiangiogenic effects of ceramides from *Hypnea musciformis* [10].

Wei et al. [11] researched the effects of polysaccharides extracted from the macroalgae *Sargassum fusiforme* by water extraction (SfW) and acid extraction (SfA) on the cecal and fecal microbiota of mice fed high-fat diets (HFDs) by 16S rRNA gene sequencing. The authors determined that 16 weeks of HFD administration dramatically impaired the homeostasis of both the cecal and fecal microbiota, without affecting the relative abundance of *Firmicutes*, *Clostridiales*, *Oscillospira*, and *Ruminococcaceae* in cecal microbiota and the Simpson's index of fecal microbiota. Co-treatments with SfW and SfA exacerbated body weight gain and altered the abundance of genes encoding monosaccharide-transporting ATPase, α-galactosidase, β-fructofuranosidase, and β-glucosidase with the latter showing more significant potency. Overall, the authors concluded that SfW and SfA could regulate the cecal microbiota, pointing out the relevance of further studies on the influence of macroalgal polysaccharides as gut microbiota regulators [11].

Funding: The Editors acknowledge the funding received by BiOrbic SFI Bioeconomy Research Centre funded by Ireland's European Structural and Investment Programmes, Science Foundation Ireland (16/RC/3889) and the Department of Agriculture Food and the Marine (DAFM) under the umbrella of the European Joint Programming Initiative "A Healthy Diet for a Healthy Life" (JPI-HDHL) and of the ERA-NET Cofund ERA HDHL (GA No 696295 of the EU Horizon 2020 Research and Innovation Programme).

Acknowledgments: The Editors would like to thank all authors that participated in this Research Topic entitled "Novel Applications and Technologies for the Industrial Exploitation of Algal Derived Marine Bioactives as Nutraceuticals or Pharmaceuticals". Special acknowledgment is given to the team of *Marine Drugs* for all the invaluable help received to make this Special Issue a success; and to all the reviewers who contributed with their recommendations to the success of this Special Issue.

Conflicts of Interest: The authors declare no conflict of interest.

References

1. Garcia-Vaquero, M.; Rajauria, G.; Miranda, M.; Sweeney, T.; Lopez-Alonso, M.; O'Doherty, J. Seasonal Variation of the Proximate Composition, Mineral Content, Fatty Acid Profiles and Other Phytochemical Constituents of Selected Brown Macroalgae. *Mar. Drugs* **2021**, *19*, 204. [CrossRef] [PubMed]
2. Biris-Dorhoi, E.-S.; Michiu, D.; Pop, C.R.; Rotar, A.M.; Tofana, M.; Pop, O.L.; Socaci, S.A.; Farcas, A.C. Macroalgae—A Sustainable Source of Chemical Compounds with Biological Activities. *Nutrients* **2020**, *12*, 3085. [CrossRef] [PubMed]
3. Wang, J.; Wang, Y.; Wu, Y.; Fan, Y.; Zhu, C.; Fu, X.; Chu, Y.; Chen, F.; Sun, H.; Mou, H. Application of Microalgal Stress Responses in Industrial Microalgal Production Systems. *Mar. Drugs* **2022**, *20*, 30. [CrossRef] [PubMed]
4. Bolaños-Martínez, O.C.; Mahendran, G.; Rosales-Mendoza, S.; Vimolmangkang, S. Current Status and Perspective on the Use of Viral-Based Vectors in Eukaryotic Microalgae. *Mar. Drugs* **2022**, *20*, 434. [CrossRef]
5. Čagalj, M.; Skroza, D.; Razola-Díaz, M.D.C.; Verardo, V.; Bassi, D.; Frleta, R.; Generalić Mekinić, I.; Tabanelli, G.; Šimat, V. Variations in the Composition, Antioxidant and Antimicrobial Activities of Cystoseira compressa during Seasonal Growth. *Mar. Drugs* **2022**, *20*, 64. [CrossRef] [PubMed]
6. Meinita, M.D.N.; Harwanto, D.; Tirtawijaya, G.; Negara, B.F.S.P.; Sohn, J.-H.; Kim, J.-S.; Choi, J.-S. Fucosterol of Marine Macroalgae: Bioactivity, Safety and Toxicity on Organism. *Mar. Drugs* **2021**, *19*, 545. [CrossRef]
7. O'Connor, J.; Garcia-Vaquero, M.; Meaney, S.; Tiwari, B.K. Bioactive Peptides from Algae: Traditional and Novel Generation Strategies, Structure-Function Relationships, and Bioinformatics as Predictive Tools for Bioactivity. *Mar. Drugs* **2022**, *20*, 317. [CrossRef]
8. Ahmed, I.; Asgher, M.; Sher, F.; Hussain, S.M.; Nazish, N.; Joshi, N.; Sharma, A.; Parra-Saldívar, R.; Bilal, M.; Iqbal, H.M.N. Exploring Marine as a Rich Source of Bioactive Peptides: Challenges and Opportunities from Marine Pharmacology. *Mar. Drugs* **2022**, *20*, 208. [CrossRef] [PubMed]
9. Garcia-Vaquero, M.; Ravindran, R.; Walsh, O.; O'Doherty, J.; Jaiswal, A.K.; Tiwari, B.K.; Rajauria, G. Evaluation of Ultrasound, Microwave, Ultrasound–Microwave, Hydrothermal and High Pressure Assisted Extraction Technologies for the Recovery of Phytochemicals and Antioxidants from Brown Macroalgae. *Mar. Drugs* **2021**, *19*, 309. [CrossRef] [PubMed]
10. Elhady, S.S.; Habib, E.S.; Abdelhameed, R.F.A.; Goda, M.S.; Hazem, R.M.; Mehanna, E.T.; Helal, M.A.; Hosny, K.M.; Diri, R.M.; Hassanean, H.A.; et al. Anticancer Effects of New Ceramides Isolated from the Red Sea Red Algae Hypnea musciformis in a Model of Ehrlich Ascites Carcinoma: LC-HRMS Analysis Profile and Molecular Modeling. *Mar. Drugs* **2022**, *20*, 63. [CrossRef] [PubMed]
11. Wei, B.; Xu, Q.-L.; Zhang, B.; Zhou, T.-S.; Ke, S.-Z.; Wang, S.-J.; Wu, B.; Xu, X.-W.; Wang, H. Comparative Study of Sargassum fusiforme Polysaccharides in Regulating Cecal and Fecal Microbiota of High-Fat Diet-Fed Mice. *Mar. Drugs* **2021**, *19*, 364. [CrossRef] [PubMed]

MDPI

Review

Application of Microalgal Stress Responses in Industrial Microalgal Production Systems

Jia Wang [1], Yuxin Wang [1], Yijian Wu [2], Yuwei Fan [1], Changliang Zhu [1], Xiaodan Fu [3], Yawen Chu [4], Feng Chen [5], Han Sun [1,]* and Haijin Mou [1,]*

[1] College of Food Science and Engineering, Ocean University of China, Qingdao 266003, China; wj7154@stu.ouc.edu.cn (J.W.); wangyuxin7553@stu.ouc.edu.cn (Y.W.); fyw@stu.ouc.edu.cn (Y.F.); zhuchangliang@ouc.edu.cn (C.Z.)

[2] School of Foreign Languages, Lianyungang Technical College, Lianyungang 222000, China; wuyj5131@163.com

[3] State Key Laboratory of Food Science and Technology, Nanchang University, Nanchang 330047, China; luna_9303@163.com

[4] Heze Zonghoo Jianyuan Biotech Co., Ltd, Heze 274000, China; sunhanias@163.com

[5] Institute for Advanced Study, Shenzhen University, Shenzhen 518060, China; sfchen@szu.edu.cn

* Correspondence: sunhan@ouc.edu.cn (H.S.); mousun@ouc.edu.cn (H.M.)

Abstract: Adaptive laboratory evolution (ALE) has been widely utilized as a tool for developing new biological and phenotypic functions to explore strain improvement for microalgal production. Specifically, ALE has been utilized to evolve strains to better adapt to defined conditions. It has become a new solution to improve the performance of strains in microalgae biotechnology. This review mainly summarizes the key results from recent microalgal ALE studies in industrial production. ALE designed for improving cell growth rate, product yield, environmental tolerance and wastewater treatment is discussed to exploit microalgae in various applications. Further development of ALE is proposed, to provide theoretical support for producing the high value-added products from microalgal production.

Keywords: adaptive laboratory evolution; microalgal production; environmental tolerance

Citation: Wang, J.; Wang, Y.; Wu, Y.; Fan, Y.; Zhu, C.; Fu, X.; Chu, Y.; Chen, F.; Sun, H.; Mou, H. Application of Microalgal Stress Responses in Industrial Microalgal Production Systems. *Mar. Drugs* **2022**, *20*, 30. https://doi.org/10.3390/md20010030

Academic Editors: Marco García-Vaquero and Brijesh K. Tiwari

Received: 29 November 2021
Accepted: 23 December 2021
Published: 26 December 2021

1. Introduction

Adaptive laboratory evolution (ALE) refers to obtaining the expected biological evolution under given laboratory conditions. As an innovative method, it is to make up for the neglect of molecular genetic mechanisms in Darwin's theory of evolution and its development. As an experimental method, it uses high-throughput sequencing of DNA as a tool to effectively simulate the evolutionary process of the selection (Figure 1).

Compared with natural selection, ALE is the process of implementing the "rules" of natural evolution for specific populations (mainly microorganisms) in the laboratory under controlled conditions, exerting pressure on them to obtain the required characteristics until the new strains with favorable mutations are developed [1]. Over the past few decades, ALE has been successfully used to develop microorganisms with the required phenotypes.

Microalgae have attracted extensive attention from all walks of life and penetrated various fields. For example, they are used for the production of bioactive compounds [2] and bioenergy [3,4], as well as in wastewater treatment facilities [5]. Therefore, microalgae are being intensively developed and utilized for various applications. To better utilize and control microalgal biomass and product yield, microalgae are suitable for ALE research with their advantages of fast growth rate, short generation time, easy to control in different cultivation systems and convenience preservation. Altering the environment for microalgal culture through ALE, we can use low-cost investment in exchange for the higher biomass concentration and product yield.

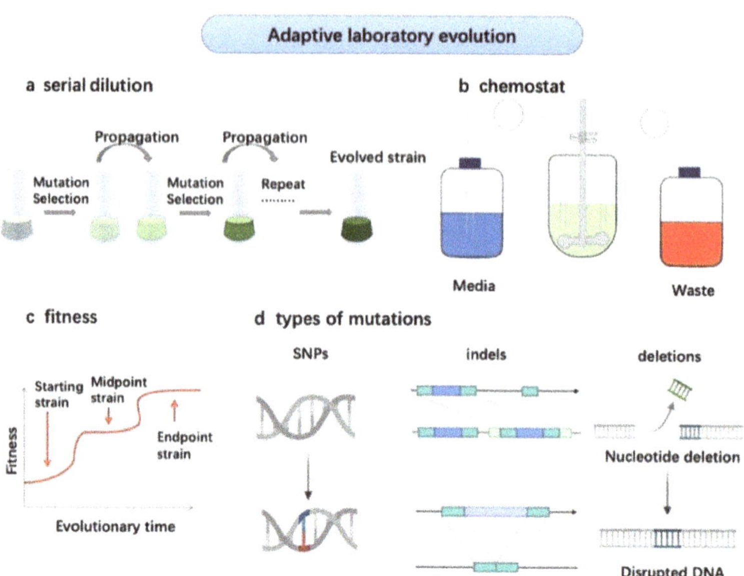

Figure 1. Adaptive laboratory evolution for strains improvement. The serial dilution is applied for mutation selection during the propagation until the evolved strains obtained (**a**). The improved strains can be cultured for high-density cultivation and wastewater treatment (**b**). The strains undergo starting, midpoint and endpoint periods (**c**) with types of nucleotide deletion and disrupted DNA (**d**).

ALE is less common in microalgal experiments at the early stage. Recently, numerous studies explored the effects of environment conditions or chemical compounds on cell growth and product accumulation; the genetic materials are not changed in a short time [6]. The microalgal production still needs condition configurations for high productivity in each batch cultivation. However, ALE induces the accumulation of beneficial mutant genes, resulting in new genotypes more suitable for the stressful environment [7]. Later experiments revealed the adaptation mechanism of planktonic alga, *Skeletonema costatuni*, under strong light and a high temperature, which provided a foundation for ALE in microalgal engineering [8].

The improvement of microalgal strains is one of the major applications of ALE in microalgae (Figure 1). It can exert pressure on the process of microalgae growth and metabolism through batch or continuous culture, to make microalgae constantly adapt to the new environment, and evolve towards the beneficial mutation, including high biomass concentration and product yield. The serial dilution is applied to obtain the evolved strain, which usually continues for 3 months to 2 years. The fitness of strain undergoes starting, midpoint and endpoint periods. Then, the improved strains can be cultured in chemostat for high-cell density growth or wastewater treatment. Unlike genetic engineering, ALE does not need to know the genetic basis of the target phenotype in advance [9], the generated mutants are generally recognized as safe (GRAS) [10]. ALE with the above advantages, has been initially developed mainly for bacterial and fungal models, allowing cells containing beneficial random gene mutations to reproduce more rapidly under environmental stress. When this strategy is applied to microalgae, it is mainly used to improve the growth rate, product yield, stress tolerance and the ability of nitrogen and phosphorus removal in wastewater [7]. These can meet the industrial demand of microalgal production.

2. Adaptive Laboratory Evolution Experimental Design

In an ALE experiment, strains are cultured in a unique pressure under artificial environmental conditions for a long time. Therefore, the formation of evolved strains is promoted, and populations best adapted to the growth environment outperform the residual ones [11]. It is necessary for ALE to take various factors into consideration simultaneously in microalgae, such as the strain, stressful condition, cultivation condition and cultivation strategy.

2.1. Cultivation Modes

A range of culture methods have been successfully used for ALE, including continuous culture, batch culture and staged culture.

2.1.1. Continuous Culture

The continuous culture can maintain process conditions with constant nutrient supply and cell densities [11]. Similar to "Bioreactor Batch Cultivations", the same bioreactor method can be used for continuous evolution experiments. The process of this training is that the addition of fresh nutrients into the medium during exponential growth at a suitable rate would allow the biomass to increase at a given rate indefinitely. A steady state can be achieved so that the microbial population grows at a constant rate in a constant environment [12]. However, there are also some disadvantages, such as the high expense and difficult control.

2.1.2. Batch and Fed-Batch Culture

Batch culture refers to a method of culturing strains using a certain amount of medium in a closed reactor. The characteristic of this mode is to load culture medium and inoculate bacteria at the beginning of culture. The volume of culture medium and culture temperature during the process are maintained.

Fed-batch culture is a variation on batch culture, which is fed continuously or sequentially with substrate without removing any of the biomass. Compared with conventional batch culture, fed-batch culture has several advantages, including sufficient nutrients, decreased fermentation time and higher productivity [13]. However, the operation of the fed-batch culture is more complicated, which requires an appropriate feeding strategy in detail. Erythritol production by fed-batch culture of *Trichosporon* sp. resulted in a high constant productivity [14]. According to the general expression patterns, there were outgoing differences in gene expression profiles between the batch and fed-batch cultures that can be attributed to the fed-batch process [15]. Fed-batch culture has been applied widely to increase biomass and lipid productivity [16].

2.1.3. Staged Cultivation

Conditions for product accumulation are usually different from those for biomass accumulation [17], the strategy of two-stage process has been found to conquer this paradox. According to this method, there is a new ALE approach named as "chemical modulators based adaptive laboratory evolution" (CM-ALE). The first step used acetyl-CoA carboxylase (ACCase), as a pressure to increase the lipid and docosahexaenoic acid (DHA) productivity of strains by 50% and 90%, respectively. Then, the second step used a sesamol based on the stress of ACCase, to increase the cell growth rate and make the productivity of lipid and DHA up to 100% and 130%, respectively [18]. This demonstrated that the two-step CM-ALE can achieve mutual improvement between desired products and cell growth. The two-stage process was considered as the better approach for productivity improvement. However, the choice of pressure is indispensable with the strategy applied [19]. The understanding of the relationship between carbon metabolism and the ROS quenching mechanism can provide strategies for microalgal production [18].

In addition, heterotrophic cultivation has become a tempting option to increase cell density by overcoming microalgal dependency on light [20]. Therefore, ALE strategy domesticating autotrophic strains to heterotrophic types is a promising approach for high density cultivation of microalgae.

2.2. Choice of Stress Conditions and Equipment

During microbial ALE, a strain is cultivated under clearly defined conditions for prolonged periods of time. The selective stress serves as the foremost step for the success of ALE, which can be classified into environment stress and the nutrient stress [21]. Microbial characteristics should be considered for selecting pressure. The special properties of microalgal strains are anticipated to amplify to improve the productivity in industrial production. For example, *Chlorella vulgaris*, *Neochloris oleeoabundans* and *Scenedesmus obliquus* [22] are known as producing lipids, so pressure promoting lipid accumulation is preferentially selected, such as nitrogen starvation and high light.

The culture equipment is another important factor. The approaches for ALE usually include serial transfer, colony transfer or chemostat culture [17]. The chemostat is commonly used for the continuous addition of medium and simultaneous removal of culture broth [23]. In addition, ALE is suitable for automation in photobioreactor (PBR) to achieve continuous monitoring, improve experimental robustness, increase throughput and minimize manual labor [24,25].

3. Increased Cell Growth Rate

The microalgae are a group of prospective resources. Their cultivation is simple, and requires less fresh water and fertile land compared to the other terrestrial plants [26]. Microalgae can be used in the pharmaceuticals, nutraceuticals, biofertilizers, bioplastics, biofuels, cosmetics and feeds for aquaculture and poultry [27–29], consequently addressing environment pollution [30]. The productivity and yield are still the key indicators for the biotechnological and economic feasibility of microalgae [7]. Recently, genetic engineering is applied to improve microalgal growth for large-scale production, which can eventually realize the commercial utilization of microalgae [31,32]. However, the use of transgenic microalgae in outdoor aquaculture systems is still limited for security reasons. Therefore, increasing ideas focus on ALE training, which can avoid risks with high growth rate of microalgae. The existing stress methods for increasing cell growth rate are shown in Table 1.

Table 1. Summary of targeting increased growth rate.

Stress Type	Strain	Stress Effect	Reference
Light intensity	*Chlamydomonas reinhardtii*, CC-124, CC-124H, CC-124 L	Fast growth rate cultivated on 120 μmol photons m^{-2} s^{-1}	[33]
	Microcoleus vaginatus	The biomass can arrive in 546.0 mg L^{-1}	[34]
	Chlorella vulgaris	Biomass density rose to approximately 20 g L^{-1} under 680 nm LEDs	[35]
Carbon	*Eubacterium limosum* ATCC 8486	Significant increased optical density (600 nm) and growth rate by 2.14 and 1.44 folds, respectively, under syngas conditions with 44% CO over 150 generations	[36]
	Haematococcus pluvialis	Biomass and astaxanthin yields t in an atmosphere comprising 15% CO_2 were 1.3 times and 6 times higher than in normal air	[37]

Recently, the method to improve the growth rate is mainly performed through controlling light factors by using ALE. In the early process of exploring natural selection, to improve the performance of strains, *C. reinhardtii* was evolved for 1880 generations in liquid medium under continuous light. At the end of experiment, evolved cells had a growth rate that was 35% higher than the progenitor population [38]. This significant growth enhancement was largely due to the improvement of acetic acid metabolism, which showed that the utilization route of organic carbon in algae can provide direction for strain improvement. The process completely transformed the strain at the gene level. In other words, continuous light can enhance the utilization of acetic acid for fundamental processes, such as DNA replication and protein translation. Although whether the genetic modification was stable or permanent at that time was a subject for debate, it has clearly demonstrated the ability of algae genomes to adapt to environmental changes, and the potential of this strategy for future microalgae engineering, which is later known as ALE. Since then, there have been many studies on regulating light to increase cell growth rates. *C. reinhardtii* was cultivated in TAP media with a light intensity of 50 µmol photons m^{-2} s^{-1}, the final biomass concentration can reach 1.48 times of the starting strains [39].

Light-emitting diode (LED) is a novel light source, which has the advantages of high efficiency, reliability, long life and low power consumption [40]. For microalgal cultivation, LED allows for artificial control of the spectral output, light intensity and light frequency for light configuration [35]. Biological productivity and light capture efficiency are crucial indicators to evaluate the economic feasibility of production mode by LED technology. Although LED is a little expensive, LED-based PBRs will become practical for producing algal biomass. Using LED-based PBRs for *C. vulgaris* can provide a biomass productivity of up to 2.11 g L^{-1} d^{-1}, with a light yield of 0.81 g DCW/Einstein. This demonstrated that LED-based PBRs, combined with microalgae biotechnology, can efficiently convert carbon dioxide into biomass and valuable products.

4. Improved Product Yield

Microalgae provide an abundance of value-added products that can accumulate up to 10–70% of specific biochemical substances (such as lipids and carbohydrates). These components have different functions because of various features. Carotenoids from microalgae can be used in medicine, cosmetics and food; lipid from microalgae can be used as a raw material to produce biodiesel to replace fossil fuels; carbohydrates produced by microalgae can be processed into bioethanol; and astaxanthin from microalgae has strong antioxidant activity and is used in health products, food and feed industries. However, the yield of these active products is not large enough to meet the industrial demand without external interference, so ALE can be applied to obtain high-yield strains for commercial utilization. The existing stress methods for improving product yield are shown in Table 2.

Table 2. Summary of targeting increased product yield.

Stress Type	Strain	Stress Effect	Reference
Carbon	*Haematococcus pluvialis*	Oil content increased to 35.2% under 15% CO_2	[41]
	Crypthecodinium cohnii	DHA-rich lipids accumulation in the strain can increase by 15.49% at 45 g L^{-1} glucose concentrations	[42]
	Chlorella pyrenoidosa G32	Starch content in the first few days under high glucose stress was eight times higher than that under low glucose stress	[43]
	Zymomonas mobilis ATCC ZW658	Maximum ethanol productivity attaches to 3.3 g L^{-1} h^{-1} in dual substrate mixture containing 5% (*w/v*) of glucose and 5% (*w/v*) xylose	[44]
Salt	*Synechocystis* sp. CCNM 2501	β-carotene produced at 1 M salinity is three times higher than the control	[45]
	Marine *Phaeodactylum tricornutum*	The addition of 20 g L^{-1} NaCl increased the total FA productivity to 219.0 ± 10.7 mg L^{-1} d^{-1}, and the biological yield reached 80% of the salt-free culture	[46]

Table 2. *Cont.*

Stress Type		Strain	Stress Effect	Reference
		Marine *Schizochytrium* sp.	Showed a maximal cell dry weight (CDW) of 134.5 g L^{-1} and lipid yield of 80.14 g L^{-1} under 30 g L^{-1} NaCl medium	[47]
		Chlamydomonas reinhardtii	Lipid content (73.4%) and lipid productivity (10.9 mg L^{-1} d^{-1})	[48]
		Marine *Dunaliella salina*	When salt concentration was increased from 4 to 9%, β-carotene yield was increased by 30-fold	[49]
Light	Light quality	Marine *Dunaliella salina*	The all-trans β-carotene and lutein content was increased to 3.3 times and 2.3 times of initial levels combining red LED (75%) with blue LED (25%)	[50]
		Marine *Stauroneis* sp.	The highest EPA proportions and yields were obtained under blue LED in f/2 medium (16.5% and 4.8 mg g^{-1}) and the fucoxanthin yield was the highest when cells were subjected to blue LEDs (5.9 mg g^{-1})	[51]
	Light intensity	*Haematococcus pluvialis*	The highest astaxanthin accumulation with 15.76 mg g^{-1} in the experimental group with light intensity of 350 μmol photons m^{-2} s^{-1}	[52]
		Marine *Phacodactylum tricornutum*	Biomass production and fucoxanthin accumulation enhanced under combined red and blue light	[53]
		Marine *Dunaliella salina*	The β-carotene production of 30 pg $cell^{-1}$ d^{-1} under high light intensity	[54]
		Desmodesmus sp.	The light intensity resulted in an enhanced lutein productivity of 3.6 mg L^{-1} d^{-1}	[55]
		Haematococcus pluvialis	Through a two-stage cultivation system in conjunction with light stress, a final astaxanthin productivity of 11.5 mg L^{-1} d^{-1} was obtained	[56]
Temperature		*Haematococcus pluvialis*	The net biomass and astaxanthin yields increased 5 and 2.9-fold under the culture temperature was 28 °C (daytime) and < 28 °C (night)	[57]
Oxygen		Marine *Schizochytrium* sp.	Observed 84.34 g/L of cell dry weight and 26.40 g L^{-1} of DHA yield with high oxygen	[58]
Nitrogen		*Chlamydomonas reinhardtii*	Total lipid content of the strain increased suddenly from 24.27% to 44.67% after nitrogen deficiency for 6 h	[39]
		Synechococcus elongatus cscB	The production of polyhydroxyalkanoates (PHA) of about 23.8 mg L^{-1} d^{-1} and a maximal titer of 156 mg L^{-1}	[59]
		Marine *Synechococcus* sp. NKBG 15041c	Under nitrogen ambient (3 mM $NaNO_3$) conditions also gave a higher yield of glycogen (404 μg mL^{-1} $OD_{730}{}^{-1}$)	[60]
		Chromochloris zofingiensis	Increase lipid and astaxanthin productivity to 457.1 and 2.0 mg L^{-1} d^{-1}	[61]

Table 2. *Cont.*

Stress Type	Strain		Stress Effect	Reference
Sulfur	*Chlamydomonas reinhardtii*		Lipid accumulation in sulfur-free medium was 66% higher than usual	[62]
Phosphorus	*Chlorella vulgaris*		Oil content in medium without KH_2PO_4 was 1.02 times higher than that in control group	[63]
Chemical regulator	*Crypthecodinium cohnii*		Adding sethoxydim to 60 µM doubles lipid production	[18]
Combined	Light and CO_2	*Haematococcus pluvialismutant*	Yield of astaxanthin under 15% CO_2 and strong light was 6 times higher than that of control group	[37]
	Temperatures and salinities	Marine *Schizochytrium* sp.	A maximal cell dry weight of 126.4 g L^{-1} and DHA yield of 38.12 g L^{-1} under concomitant low temperature and high salinity	[47]
	Light and nitrogen	*Limnothrix* sp. CACIAM25	Produced a high lipid content at a low level of $NaNO_3$ concentration (1 g L^{-1}) and a high level of light intensity (100 µmol photons m^{-2} s^{-1})	[64]

Because of the requirement to supply the global market, people are interested in extracting from algae and higher plants to gain β- carotenoids and lutein. Therefore, ALE has been applied to microalgae to produce a high content of carotenoids. It has been suggested that light plays a key role in the biosynthesis of carotenoids, through light signal sensing and downstream regulation [65]. It is well known that β-carotene can be overproduced in the marine microalga *Dunaliella salina,* in response to stressful light conditions [50]. Previous study explored the effect of red light-emitting diode (LED) lighting on growth rate and biomass yield, which identified the optimal photon flux for marine *D. salina* growth. The red-light photon stress alone at a high level was not capable of upregulating carotenoid accumulation. Therefore, combining red LED (75%) with blue LED (25%) allowed growth at a higher total photon flux, and increased β-carotene and lutein accumulation [50]. An efficient culture system with increased light energy efficiency and economy of operation can be developed in combination with genetically based methods, such as ALE for strain development.

Haematococcus pluvialis is considered as the best natural source to product astaxanthin. The antioxidant capacity of astaxanthin is the most outstanding than other carotenoids, which endows astaxanthin in suppressing tumor growth, improving body immunity and the scavenging of active oxygen and free radicals [66]. Many efforts are devoted to increase astaxanthin yield from *H. pluvialis.* More astaxanthin was obtained at the melatonin concentration of 5–15 mu mol/L, at 27–29 °C and light intensity of 198–216 µmol photons m^{-2} s^{-1} [67]. The application of nitrogen stress and excess light based on fed-batch culture can improve lipid and astaxanthin productivity to 457.1 and 2.0 mg L^{-1} d^{-1} [61]. The study revealed excess light can lead to more available carbon molecules to synthesize astaxanthin.

Microbial lipids often contain abundant polyunsaturated fatty acids, including eicosapenteanoic acid (EPA), DHA and arachidonic acid (AA), which can be used as the source of functional food and raw materials for green biofuels. ALE strategy has greatly improved the ability of microbial oil production, aiming to culture special microalgae strains with convenient, cost-saving and high-yield oil. The salt stress and nutrient osmotic pressure are widely used to increase lipid content in ALE [46].

These stressful conditions have largely changed the carbon flux to the expected products. ^{13}C-MFA is used to be an accurate tool for describing the central carbon metabolism [68]. There is a conflict between lipid accumulation and cell growth; the massive accumulation of lipid is against for cell division. To resolve the conflict between cell growth and lipid accumulation, a general countermeasure is two-stage cultivation strategy. The first stage is for a rapid growth to gain the maximum biomass production, subsequently the second process for the lipid accumulation under various stress conditions at the expense of cell growth [17,69]. Studies have shown that *Parachlorella* sp. can be cultured in two stages, increasing the total FA productivity to 219.0 ± 10.7 mg L^{-1} d^{-1}, and the biomass reached by 80% [29]. Similarly, in the first stage, microalgae grew under red LED to obtain the maximum biomass; in the second phase, green LED was used for stress to produce a large amount of lipids [70].

However, two-stage culture strategy can also lead to large labor expenditure and economic consumption, so it is necessary to culture algae strains that can grow rapidly under stress for a long time. Therefore, transcriptome analysis of strains produced by ALE can be carried out to obtain two different metabolic responses to stress and reveal the different gene expression patterns among the strains. The understanding of microalgal evolution under stress is beneficial for the development of the strains with rapid growth under stress [71]. Based on gene expression patterns of different metabolic reactions under stress, the ALE can be applied to all lipid-rich microalgae.

5. Enhanced Environmental Tolerance

ALE, to enhance the environmental tolerance of strains (Table 3), is profitable for the industrial production of microalgae. For example, cyanobacteria, a group of Gram-negative prokaryotes capable of performing oxygenic photosynthesis [72], is a promising cell factory to convert carbon dioxide into useful chemicals by autotrophic mode. Since toxicity produced is a key challenge, when cyanobacteria is the host for photosynthetic production of chemicals, tolerance of cyanobacteria to solvents is required to improve by ALE. *Synechococcus elongatus* PCC 11801 was cultured in ALE stage, and a new strain resistant to n-butanol and 2, 3-butanediol was obtained, which also showed high tolerance to other alcohols. The evolved strain had high tolerance without obvious growth lag phenomenon compared with original strain. The mutation mechanism was revealed by whole genome sequencing [73].

Table 3. Summary of targeting increased stress tolerance.

Tolerance Type	Strain	Stress Effect	Reference
Butanol	*Synechocystis* sp. PCC 6803	A 150% increase of the butanol (0.2–0.5% *v/v*) tolerance	[71]
	Synechococcus elongatus PCC 11801	A 100% improvement in concentrations tolerated (2–5 g L^{-1} n-butanol and 15–30 g L^{-1} 2,3-butanediol)	[73]
Temperature	*Symbiodinium* spp.	Tolerance to 31 °C	[74]
	Marine *Thalassiosira pseudonana* CCMP 1335	Tolerance to 32 °C	[75]
Light	*Synechocystis* sp. PCC 6803	Tolerance to 2000 μmol photons m^{-2} s^{-1}	[76]
Cadmium	*Synechocystis* sp. PCC 6803	Tolerated $CdSO_4$ with a concentration up to 9.0 μM	[77]
Acid	*Synechocystis* sp. PCC 6803	Tolerance to pH 5.5	[78]
Salt	*Chlorella* sp.	Tolerance to 30 g L^{-1} NaCl	[79]
Carbon dioxide	*Chlorella* sp.	They grew rapidly in 30% CO_2	[80]
Oxygen	Marine *Schizochytrium* sp.	A 32.4% increase in dry weight	[58]
Flue gas	*Desmodesmus* spp.	Tolerance to 100% unfiltered flue gas	[81]
	Chlorella sp.	1.2 g L^{-1} d^{-1} CO_2 fixation rate 2.7 g L^{-1} biomass concentration 68.4% carbohydrate content	[82]

Light factors of light source, light quality, light intensity and light cycle affect microalgal growth. Microalgae have different appropriate light intensities. The light compensation points of microalgae can only maintain cell basic metabolism without growth. The higher light intensity promotes microalgal growth. However, when the light intensity reaches the saturation point of photosynthesis, microalgal growth rate no longer increases. A further increase of light intensity will lead microalgae to suffer a photoinhibition. At present, studies indicated that enhanced light tolerance of the cyanobacterium *Synechocystis* by ALE could increase the biomass. By combining repeated mutagenesis and exposure to increasing light intensities, the modified strains can grow rapidly under extremely high light intensities [75]. The ALE maximized photosynthesis and thus increased the accumulation of photosynthetic products, converting carbon dioxide into useful chemicals.

In addition, large-scale cultivation of microalgae is mostly used for environmental governance. The atmospheric CO_2 levels reached the "dangerously high" levels of 400 ppm. One of the reasons for the large emission of greenhouse gases is the combustion of fuel in power plants. Efforts are underway to scrub CO_2 from flue gas emitted from pulverized coal power plants using carbon sequestration and storage technology [83], which will make the production costly. Subsequently, microalgae are rapid converters of solar energy to biomass by assimilating atmospheric CO_2 [84]. There has been demonstrated to use industrial flue for culturing microalgae [85], but flue gas supplementation at higher flow rates leads to the acidification of the growth medium, which typically cannot be tolerated by microalgae. Therefore, the microalgal tolerance is required to enhance to offset carbon emissions from fossil fuel combustion. A study showed that mixed biodiverse microalgal communities can be selected and adapted to tolerate growth in 100% flue gas from an unfiltered coal-fired power plant that contained 11% CO_2 [80]. Cheng et al. have reported that the adaptive evolution against simulated flue gas containing 10% CO_2, 200 ppm NOx and 100 ppm SOx can obtain a new strain, *Chlorella* sp. CV, which can tolerate simulated flue gas conditions and the maximum CO_2 fixation rate was 1.2 g L^{-1} d^{-1} [81]. It can be helpful to establish a new process for CO_2 capture directly from industrial flue gas.

6. Promoted Nitrogen and Phosphorus Removal in Wastewater

Wastewater, resulting from various farming, domestic and industrial water operations, has been a key pollution of the environment for a long time in many countries or regions [86]. However, current methods for wastewater treatment have many problems, such as high energy consumption and cost, and heavy secondary pollution [87]. It is reported that some microalgae can utilize external organic carbon sources to remove nitrogen and phosphorus efficiently, and absorb heavy metals by functional groups on its cell surface [88–90]. Moreover, microalgae can utilize the nutrients from wastewater discharge to form beneficial products, including methane and hydrogen [22,91,92]. The application of removing nitrogen and phosphorus from wastewater by microalgae is developing rapidly [93]. The existing stress methods for increasing the ability of nitrogen and phosphorus removal in wastewater are shown in Table 4.

Table 4. Summary examples of increasing the ability of nitrogen and phosphorus removal in wastewater [1].

Stress type	Types of Wastewater	Strain	Removal Rate	Reference
Temperature	Municipal wastewater	*Chlorella vulgaris*	TN (96.5%) TP (99.2%) COD (83.0%)	[94]
Light	Artificial wastewater	*Chlorella kessleri*	NH_3-N (97.8%) NO_3^--N (88.1%)	[95]
	Municipal wastewater	Marine *Spirulina platensis*	PO_4^{3-}-P (93%) NH_4^+-N (83%)	[96]
Salt	Municipal wastewater	Marine *Dunaliella salina*	NO_3^--N (100%) NH_4^+-N (75.5%) PO_4^{3-}-P (63.5%)	[97]
	Sludge liquor	*Chlorella vulgaris*	COD (85.3%) TN (99.6%)	[98]
Phosphorus	Municipal wastewater	*Chlorella vulgaris*	PO_4^{3-}-P (>99%)	[99]
		Desmodesmus communis, Tetradesmus obliquus, Chlorella protothecoides	DIP (>99.9%) DIP (>99.9%)	[100]
Sodium acetate	Municipal wastewater	*Scenedesmus obliquus*	TN (82.20%) TP (76.35%)	[101]
Phenol	Phenolic wastewater	*Chlorella* sp.	Phenol (100%)	[102]

[1] TN: total nitrogen; TP: total phosphorus; COD: chemical oxygen demand; DIN: dissolved inorganic nitrogen and DIP: dissolved inorganic phosphorus.

As mentioned above, light stress is used to improve the biomass and biodiesel production of microalgae [103]. Therefore, in wastewater treatment, light stress increases microalgal biomass to improve the removal rate of nitrogen and phosphate. Mahsa et al. have reported that the total biomass and protein concentrations of marine *Spirulina platensis* were observed under blue light, at around 100 μmol photons m^{-2} s^{-1} of 13.4 and 9.0 g L^{-1}, respectively. However, the highest phosphate and ammonium removal were about 145 and 218 mg L^{-1} under purple light, at around 100 μmol photons m^{-2} s^{-1}, respectively [97]. Results showed that light intensity and wavelength, combined with semi-batch cultivation, can be designed to achieve the highest biomass and production, as well as to maximize the removal of phosphorous and ammonium.

Obviously, when microalgae absorb nitrogen, phosphorus and heavy metals in wastewater, their biomass and by-product yield are correspondingly increased. Previous research was to develop large-scale production to produce oil-rich algal biomass from wastewater. Almost all nutrients in sewage were consumed, and microalgal biomass and oil content could reach 2.0 g^{-1} L^{-1} d^{-1} and 25.25% (*w/w*) [97]; *Chlorella* cultivated in pig manure wastewater could also obtain 0.23 g m^{-2} d^{-1} fatty acid production capacity [98]. In addition, bacteriostatic compounds against antibiotic-resistant bacteria have been discovered from microalgae during phytoremediation of swine wastewater [99]. Aigars et al. have reported phosphorus removal from municipal wastewater and the biomass were enhanced after four microalgal species were exposed to a phosphorus starvation medium, which indicated that the species, N/P molar ratio in the wastewater and P content of biomass could control the efficiency of phosphorus uptake [99]. *C. vulgaris* was continuously cultured for 15 days in municipal wastewater, in 3 temperature regimes. In this study, the analysis revealed that in alternating high–low temperature conditions, biomass production had the potential for biofuel production, with the highest lipid content (26.4% of total dry biomass) [95]. Therefore, some microalgae have the great potential to remove nutrient from wastewater and produce valuable compounds, simultaneously.

The technologies aim at promoting nitrogen and phosphorus metabolism in microalgae, by converting the nutrients into biomass [91]. There are two main pathways of nitrogen and phosphorus removal with microalgae for efficient wastewater treatment. One is performed through biochemical pathways for the uptake of nutrient components into the biomass for nucleic acids and proteins production. Additionally, the other can ingest phosphate for storing as an acid-insoluble polyphosphate granule [104].

Nowadays, microalgae technology is promising for wastewater treatment. However, adaptability metabolism of microalgae to extreme conditions in industrial wastewater is poorly understood. The microalgae harvesting in wastewater requires equipment investment and large amounts of energy, as the biomass concentration is usually below 0.6 g L^{-1} after the treatment. Therefore, efficient microalgal systems to treat wastewater at a large-scale production are in urgent needs.

7. Industrial Application of ALE in Microalgae

There are two mainly aspects about the industrial application of ALE in microalgae. Firstly, ALE is used as an optimal tool to improve the photosynthetic efficiency of microalgae in PBRs. The low and high light intensities limited the microalgal productivity at a large-scale production. ALE can develop an evolved microalgal strain with a new fixed trait, which can later be used as industrial strain for the enhanced production of the target metabolite. With the growing greenhouse effect, it is very exciting to produce biofuels, such as lipid, bioethanol, methane and hydrogen from microalgae via ALE.

Secondly, ALE has been applied to improve microalgal strains for industrial wastewater treatment. Since phosphorus and nitrogen in wastewater are the major nutrients causing eutrophication of aquatic ecosystems [99], it is necessary to search for a more effective means of nutrient removal. ALE can obtain microalgal strains for rapid nutrient loading, high tolerance to heavy metal ions and low nutrient effluent concentration [105]. At present, *Chlorella* sp. and *S. obliquus* are the most promising unicellular algae in wastewater treatment.

8. Challenges and Prospects

ALE is an emerging approach in artificial conditions, imitating the process of natural evolution to improve organisms, and being performed for various purposes at the laboratory level. The industrial microorganisms have become a key producer in various fields, including food, pharmaceuticals and other value-added chemical production [106]. ALE is one of the most effective approaches to eliminate obstacles and maximize productivity in bio-based processes, having advantages, such as the easy control of culture conditions, short generation time, easy manufacturing and storage of living fossil records for each period [107].

Meanwhile, based on these known influencing conditions, suitable combinations can make the effect more significant. Recent research showed that ALE can be combined with genetic engineering to improve the efficiency of gene transformation [108]. ALE was applied to streamline fitness recovery of genomically recoded *Escherichia coli*, for industrial-scale protein production [108]. High salinity accelerated the synthesis of phospholipids to restore lipid, and ALE culture was therefore carried out to obtain an evolutionary strain of *Nitzschia* sp. to increase lipid content to 51.2% [109]. Therefore, future research can combine several factors to obtain higher returns.

However, ALE currently still faces many challenges. First is the formation of large mutant libraries and the need for large-scale screening of the required evolutionary bodies. Because the generation of genetic diversity has become a core technology for accelerating ALE, a high-quality mutant library is crucial to its success [6]. Second, with cell division, effective genes can disappear. The application of multi-omics analyses can promote the efficient data mining for the implementation of ALE experiments. The third is that the time span of ALE experiments is very long, and the end point is entirely up to the researcher's decision. The novel cultivation strategy, such as red LED light and phytohormone, can potentially accelerate ALE progress based on microalgal cell physiology. Last, the strains produced by ALE have multiple mutations, and the interpretation of genotypes requires tedious omics analysis. The upgrade and development of related software will provide more effective data for ALE performance.

Whatever the purpose, its essence is to continuously culture microalgae under certain pressure, screen out the most adaptable strains and constantly culture to meet our expectations. It is in response to "natural selection and survival of the fittest". With in-depth research to conquer those difficulties, ALE combined with systems biology and synthetic biology tools is a prominent strategy. It is acknowledged that ALE is bound to become a popular technology in microalgae for the future.

Author Contributions: Conceptualization, resources and supervision, H.S. and H.M.; writing—original draft preparation, J.W., Y.W. (Yuxin Wang), Y.W. (Yijian Wu), X.F. and Y.F.; writing—review and editing, J.W., C.Z., F.C., Y.C., H.S. and H.M. All authors have read and agreed to the published version of the manuscript.

Funding: China Postdoctoral Science Foundation (No. 2021M703039) and the National Key Research and Development (R&D) Program of China (2019YFD0900201).

Institutional Review Board Statement: Not applicable.

Informed Consent Statement: Not applicable.

Data Availability Statement: Not applicable.

Conflicts of Interest: The authors declare no conflict of interest.

References

1. Maria, M.; Alexandra, D.; Seraphim, P.; George, A. Adaptive laboratory evolution principles and applications in industrial biotechnology. *Biotechnol. Adv.* **2021**, *56*, 1–16.
2. Lee, H.-S.; Kim, Z.-H.; Park, H.; Lee, C.-G. Specific light uptake rates can enhance astaxanthin productivity in Haematococcus lacustris. *Bioprocess Biosyst. Eng.* **2016**, *39*, 815–823. [CrossRef] [PubMed]
3. Kim, Z.-H.; Park, Y.-S.; Ryu, Y.-J.; Lee, C.-G. Enhancing biomass and fatty acid productivity of *Tetraselmis* sp. in bubble column photobioreactors by modifying light quality using light filters. *Biotechnol. Bioprocess Eng.* **2017**, *22*, 397–404. [CrossRef]
4. Zhang, T.-Y.; Hu, H.-Y.; Wu, Y.-H.; Zhuang, L.-L.; Xu, X.-Q.; Wang, X.-X.; Dao, G.-H. Promising solutions to solve the bottlenecks in the large-scale cultivation of microalgae for biomass/bioenergy production. *Renew. Sustain. Energy Rev.* **2016**, *60*, 1602–1614. [CrossRef]
5. Salama, E.-S.; Kurade, M.B.; Abou-Shanab, R.A.I.; El-Dalatony, M.M.; Yang, I.-S.; Min, B.; Jeon, B.-H. Recent progress in microalgal biomass production coupled with wastewater treatment for biofuel generation. *Renew. Sustain. Energy Rev.* **2017**, *79*, 1189–1211. [CrossRef]
6. Zheng, Y.Y.; Hong, K.Q.; Wang, B.W.; Liu, D.Y.; Chen, T.; Wang, Z.W. Genetic Diversity for Accelerating Microbial Adaptive Laboratory Evolution. *ACS Synth. Biol.* **2021**, *10*, 1574–1586. [CrossRef] [PubMed]
7. Alaina, J.L.; Anagha, K.; Matthew, C.P. Adaptive Laboratory Evolution for algal strain improvement: Methodologies and applications. *Algal Res.* **2020**, *53*, 102122.
8. Kuebler, J.E.; Davison, I.R.; Yarish, C. Photosynthetic adaptation to temperature in the red algae *Lomentaria baileyana* and *Lomentaria orcadensis*. *Eur. J. Phycol.* **1991**, *26*, 9–19. [CrossRef]
9. Yinan, W.; Aysha, J.; XinHui, X.; Chong, Z. Advanced strategies, and tools to facilitate and streamline microbial adaptive laboratory evolution. *Trends Biotechnol.* **2021**, *1036*, 66–72.
10. Neha, A.; George, P.P. Microalgae strain improvement strategies: Random mutagenesis and adaptive laboratory evolution. *Trends Plant Sci.* **2021**, *26*, 1199–1200.
11. Michaela, G.; Thorsten, H.; Felix, M.; Anina, B.; Julia, H.-B.; Andreas, F.; Alexander, N.; Marcus, P.; Jörn, K.; Bastian, B.; et al. Continuous Adaptive Evolution of a Fast-Growing *Corynebacterium glutamicum* Strain Independent of Protocatechuate. *Front. Microbiol.* **2019**, *10*, 1648.
12. Hoskisson, P.A.; Hobbs, G. Continuous culture—Making a comeback? *Microbiology* **2005**, *151*, 3153–3159. [CrossRef] [PubMed]
13. Youssef, F.; Roukas, T.; Biliaderis, C.G. Pullulan production by a non-pigmented strain of *Aureobasidium pullulans* using batch and fed-batch culture. *Process Biochem.* **1999**, *34*, 355–366. [CrossRef]
14. Jinbyung, P.; Byungcheol, S.; Jungryul, K.; Yongkun, P. Production of erythritol in fed-batch cultures of *Trichosporon* sp. *J. Ferment. Bioeng.* **1998**, *86*, 577–580.
15. Lee, Y.Y.; Wong, K.T.; Nissom, P.M.; Wong, D.C.; Yap, M.G. Transcriptional profiling of batch and fed-batch protein-free 293-HEK cultures. *Metab. Eng.* **2007**, *9*, 52–67. [CrossRef]
16. Henley, W.J. The past, present and future of algal continuous cultures in basic research and commercial applications. *Algal Res.* **2019**, *43*, 101636. [CrossRef]
17. Sun, X.-M.; Ren, L.-J.; Zhao, Q.-Y.; Ji, X.-J.; Huang, H. Microalgae to produce lipid and carotenoids: A review with focus on stress regulation and adaptation. *BioMed Cent.* **2018**, *11*, 227.
18. Diao, J.; Song, X.; Cui, J.; Liu, L.; Shi, M.; Wang, F.; Zhang, W. Rewiring metabolic network by chemical modulator-based laboratory evolution doubles lipid production in *Crypthecodinium cohnii*. *Metab. Eng.* **2019**, *51*, 88–98. [CrossRef] [PubMed]
19. Bin, L.; Jin, L.; Peipei, S.; Xiaonian, M.; Yue, J.; Feng, C. Sesamol Enhances Cell Growth and the Biosynthesis and Accumulation of Docosahexaenoic Acid in the Microalga *Crypthecodinium cohnii*. *J. Agric. Food Chem.* **2015**, *63*, 5640–5645.
20. Perez-Garcia, O.; Escalante, F.M.; De-Bashan, L.E.; Bashan, Y. Heterotrophic cultures of microalgae: Metabolism and potential products. *Water Res.* **2011**, *45*, 11–36.
21. Nam, H.; Conrad, T.M.; Lewis, N.E. The role of cellular objectives and selective pressures in metabolic pathway evolution. *Curr. Opin. Biotechnol.* **2011**, *22*, 595–600. [CrossRef] [PubMed]
22. Chamberlin, J.; Harrison, K.; Zhang, W. Impact of Nutrient Availability on Tertiary Wastewater Treatment by *Chlorella vulgaris*. *Water Environ. Res.* **2018**, *90*, 2008–2016. [CrossRef] [PubMed]
23. Narang, A. The Steady States of Microbial Growth on Mixtures of Substitutable Substrates in a Chemostat. *J. Theor. Biol.* **1998**, *190*, 241–261. [CrossRef]
24. Sandberg, T.E.; Salazar, M.J.; Weng, L.L.; Palsson, B.O.; Feist, A.M. The emergence of adaptive laboratory evolution as an efficient tool for biological discovery and industrial biotechnology. *Metab. Eng.* **2019**, *56*, 1–16. [CrossRef] [PubMed]
25. Winkler, J.D.; Kao, K.C. Recent advances in the evolutionary engineering of industrial biocatalysts. *Genomics* **2014**, *104*, 406–411. [CrossRef] [PubMed]
26. Sen, T.J.; Ying, L.S.; Wayne, C.K.; Kee, L.M.; Wei, L.J.; ShihHsin, H.; Loke, S.P. A review on microalgae cultivation and harvesting, and their biomass extraction processing using ionic liquids. *Bioengineered* **2020**, *11*, 116–129.
27. Baldev, E.; MubarakAli, D.; Ilavarasi, A.; Pandiaraj, D.; Ishack, K.A.S.S.; Thajuddin, N. Degradation of synthetic dye, Rhodamine B to environmentally non-toxic products using microalgae. *Colloids Surf. B Biointerfaces* **2013**, *105*, 207–214. [CrossRef]
28. Davoodbasha, M.; Edachery, B.; Nooruddin, T.; Lee, S.-Y.; Kim, J.-W. Evidence of C16 fatty acid methyl esters extracted from microalga for effective antimicrobial and antioxidant property. *Microb. Pathog.* **2018**, *115*, 233–238. [CrossRef]

29. Deepika, P.; MubarakAli, D. Production and assessment of microalgal liquid fertilizer for the enhanced growth of four crop plants. *Biocatal. Agric. Biotechnol.* **2020**, *28*, 101701. [CrossRef]
30. Pathak, J.; Rajneesh; Maurya, P.K.; Singh, S.P.; Hader, D.-P.; Sinha, R.P. Cyanobacterial Farming for Environment Friendly Sustainable Agriculture Practices: Innovations and Perspectives. *Front. Environ. Sci.* **2018**, *6*, 7. [CrossRef]
31. Schoepp, G.N.; Stewart, L.R.; Vincent, S.; Quigley, J.A.; Dominick, M.; Mayfield, P.S.; Burkart, D.M. System and method for research-scale outdoor production of microalgae and cyanobacteria. *Bioresour. Technol.* **2014**, *166*, 273–281. [CrossRef]
32. Griffiths, M.J.; Harrison, S.T.L. Lipid productivity as a key characteristic for choosing algal species for biodiesel production. *J. Appl. Phycol.* **2009**, *21*, 493–507. [CrossRef]
33. Shin, S.-E.; Koh, H.G.; Kang, N.K.; Suh, W.I.; Jeong, B.-R.; Lee, B.; Chang, Y.K. Isolation, phenotypic characterization, and genome wide analysis of a *Chlamydomonas reinhardtii* strain naturally modified under laboratory conditions: Towards enhanced microalgal biomass and lipid production for biofuels. *Biotechnol. Biofuels* **2017**, *10*, 308. [CrossRef]
34. Lan, S.; Wu, L.; Zhang, D.; Hu, C. Effects of light and temperature on open cultivation of desert cyanobacterium *Microcoleus vaginatus*. *Bioresour. Technol.* **2015**, *182*, 144–150. [CrossRef]
35. Fu, W.; Gudmundsson, O.; Feist, A.M.; Herjolfsson, G.; Brynjolfsson, S.; Palsson, B.Ø. Maximizing biomass productivity and cell density of Chlorella vulgaris by using light-emitting diode-based photobioreactor. *J. Biotechnol.* **2012**, *161*, 242–249. [CrossRef] [PubMed]
36. Kang, S.; Song, Y.; Jin, S.; Shin, J.; Bae, J.; Kim, D.R.; Lee, J.-K.; Kim, S.C.; Cho, S.; Cho, B.-K. Adaptive Laboratory Evolution of *Eubacterium limosum* ATCC 8486 on Carbon Monoxide. *Front. Microbiol.* **2020**, *11*, 402. [CrossRef]
37. Cheng, J.; Li, K.; Yang, Z.; Lu, H.; Zhou, J.; Cen, K. Gradient domestication of *Haematococcus pluvialis* mutant with 15% CO_2 to promote biomass growth and astaxanthin yield. *Bioresour. Technol.* **2016**, *216*, 340–344. [CrossRef] [PubMed]
38. Perrineau, M.-M.; Gross, J.; Zelzion, E.; Price, D.C.; Levitan, O.; Boyd, J.; Bhattacharya, D. Using natural selection to explore the adaptive potential of *Chlamydomonas reinhardtii*. *PLoS ONE* **2017**, *9*, e92533. [CrossRef]
39. Yu, S.; Zhao, Q.; Miao, X.; Shi, J. Enhancement of lipid production in low-starch mutants *Chlamydomonas reinhardtii* by adaptive laboratory evolution. *Bioresour. Technol.* **2013**, *147*, 499–507. [CrossRef] [PubMed]
40. Yam, F.K.; Hassan, Z. Innovative advances in LED technology. *Microelectron. J.* **2004**, *36*, 129–137. [CrossRef]
41. Li, K.; Cheng, J.; Qiu, Y.; Tian, J.; Zhou, J.; Cen, K. Potential use of accumulation lipid as a biodiesel feedstock *Haematococcus pluvialis* induced by 15% CO_2. *Acta Energ. Sol. Sin.* **2020**, *41*, 351–355.
42. Li, X.; Pei, G.; Liu, L.; Chen, L.; Zhang, W. Metabolomic analysis and lipid accumulation in a glucose tolerant *Crypthecodinium cohnii* strain obtained by adaptive laboratory evolution. *Bioresour. Technol.* **2017**, *235*, 87–95. [CrossRef] [PubMed]
43. Liu, B. Starch and Lipid Contents and Transcriptome Analysis of Chlorella G32 Based on Glucose Concentration. Zhejiang University: Hangzhou, China, 2018.
44. Sarkar, P.; Mukherjee, M.; Goswami, G.; Das, D. Adaptive laboratory evolution induced novel mutations in *Zymomonas mobilis* ATCC ZW658: A potential platform for co-utilization of glucose and xylose. *J. Ind. Microbiol. Biotechnol.* **2020**, *47*, 329–341. [CrossRef] [PubMed]
45. Chetan, P.; Imran, P.; Tonmoy, G.; Rahulkumar, M.; Kaumeel, C.; Vamsi Bharadwaj, S.V.; Shristi, R.; Sandhya, M. Selective carotenoid accumulation by varying nutrient media and salinity in *Synechocystis* sp. CCNM 2501. *Bioresour. Technol.* **2015**, *197*, 363–368.
46. Hun, K.Z.; Kwangmin, K.; Hanwool, P.; Soo, L.C.; Won, N.S.; June, Y.K.; Young, J.J.; Joo, H.S.; Gyun, L.C. Enhanced Fatty Acid Productivity by *ParaChlorella* sp., a Freshwater Microalga, via Adaptive Laboratory Evolution Under Salt Stress. *Biotechnol. Bioprocess Eng.* **2021**, *26*, 223–231.
47. Sun, X.-M.; Ren, L.-J.; Bi, Z.-Q.; Ji, X.-J.; Zhao, Q.-Y.; Huang, H. Adaptive evolution of microalgae *Schizochytrium* sp. under high salinity stress to alleviate oxidative damage and improve lipid biosynthesis. *Bioresour. Technol.* **2018**, *267*, 438–444. [CrossRef]
48. Hang, L.T.; Mori, K.; Tanaka, Y.; Morikawa, M.; Toyama, T. Enhanced lipid productivity of *Chlamydomonas reinhardtii* with combination of NaCl and $CaCl_2$ stresses. *Bioprocess Biosyst. Eng.* **2020**, *43*, 971–980. [CrossRef]
49. Takagi, M.; Karseno; Yoshida, T. Effect of salt concentration on intracellular accumulation of lipids and triacylglyceride in marine microalgae Dunaliella cells. *J. Biosci. Bioeng.* **2006**, *101*, 223–226. [CrossRef]
50. Fu, W.; Guðmundsson, Ó.; Paglia, G.; Herjólfsson, G.; Andrésson, Ó.S.; Palsson, B.Ø.; Brynjólfsson, S. Enhancement of carotenoid biosynthesis in the green microalga *Dunaliella salina* with light-emitting diodes and adaptive laboratory evolution. *Appl. Microbiol. Biotechnol.* **2013**, *97*, 2395–2403. [CrossRef]
51. Rachel, P.; Lorraine, A.; Dónal, M.G.; Thomas, J.S.; Eoin, G.; Nicolas, T. Differential responses in EPA and fucoxanthin production by the marine diatom *Stauroneis* sp. under varying cultivation conditions. *Biotechnol. Prog.* **2021**, *37*, 3197.
52. Telli, M.; Sahin, G. Effects of gradual and sudden changes of salinity and light supply for astaxanthin production in *Haematococcus pluvialis* (Chlorophyceae). *Fundam. Appl. Limnol.* **2020**, *194*, 11–17. [CrossRef]
53. Zhiqian, Y.; Maonian, X.; Manuela, M.; Yuetuan, Z.; Sigurdur, B.; Weiqi, F. Photo-Oxidative Stress-Driven Mutagenesis and Adaptive Evolution on the Marine Diatom *Phaeodactylum tricornutum* for Enhanced Carotenoid Accumulation. *Mar. Drugs* **2015**, *13*, 6138–6151.
54. Lamers, P.P.; van de Laak, C.C.W.; Kaasenbrood, P.S.; Lorier, J.; Janssen, M.; Vos, R.C.H.D.; Bino, R.J.; Wijffels, R.H. Carotenoid and fatty acid metabolism in light stressed *Dunaliella salina*. *Biotechnol. Bioeng.* **2010**, *106*, 638–648. [CrossRef]

55. Xie, Y.; Ho, S.-H.; Chen, C.-N.N.; Chen, C.-Y.; Ng, I.-S.; Jing, K.-J.; Chang, J.-S.; Lu, Y. Phototrophic cultivation of a thermo-tolerant *Desmodesmus* sp. for lutein production: Effects of nitrate concentration, light intensity, and fed-batch operation. *Bioresour. Technol.* **2013**, *144*, 435–444. [CrossRef] [PubMed]
56. Claude, A.; Yuval, M.; Aliza, Z.; Sammy, B. On the relative efficiency of two- vs. one-stage production of astaxanthin by the green alga *Haematococcus pluvialis*. *Biotechnol. Bioeng.* **2007**, *98*, 300–305.
57. Wan, M.; Zhang, J.; Hou, D.; Fan, J.; Li, Y.; Huang, J.; Wang, J. The effect of temperature on cell growth and astaxanthin accumulation of *Haematococcus pluvialis* during a light–dark cyclic cultivation. *Bioresour. Technol.* **2014**, *167*, 276–283. [CrossRef]
58. Sun, X.-M.; Ren, L.-J.; Ji, X.-J.; Chen, S.-L.; Guo, D.-S.; Huang, H. Adaptive evolution of *Schizochytrium* sp. by continuous high oxygen stimulations to enhance docosahexaenoic acid synthesis. *Bioresour. Technol.* **2016**, *211*, 374–381. [CrossRef]
59. Löwe, H.; Hobmeier, K.; Moos, M.; Kremling, A.; Pflüger-Grau, K. Photoautotrophic production of polyhydroxyalkanoates in a synthetic mixed culture of *Synechococcus elongatus* cscB and *Pseudomonas putida* cscAB. *Biotechnol. Biofuels* **2017**, *10*, 190. [CrossRef]
60. Amr, B.; Shouhei, T.; Mitsuharu, N.; Stefano, F.; Peter, L.; Koji, S. Glycogen Production in Marine Cyanobacterial Strain *Synechococcus* sp. NKBG 15041c. *Mar. Biotechnol.* **2018**, *20*, 109–117.
61. Sun, H.; Ren, Y.; Lao, Y.; Li, X.; Chen, F. A novel fed-batch strategy enhances lipid and astaxanthin productivity without compromising biomass of *Chromochloris zofingiensis*. *Bioresour. Technol.* **2020**, *308*, 123306. [CrossRef]
62. Atsushi, S.; Rie, M.; Naomi, H.; Mikio, T.; Norihiro, S. Responsibility of regulatory gene expression and repressed protein synthesis for triacylglycerol accumulation on sulfur-starvation in *Chlamydomonas reinhardtii*. *Front. Plant Sci.* **2014**, *5*, 444.
63. Chu, F.-F.; Chu, P.-N.; Cai, P.-J.; Li, W.-W.; Lam, P.K.S.; Zeng, R.J. Phosphorus plays an important role in enhancing biodiesel productivity of *Chlorella vulgaris* under nitrogen deficiency. *Bioresour. Technol.* **2013**, *134*, 341–346. [CrossRef]
64. Aboim, J.B.; de Oliveira, D.T.; de Mescouto, V.A.; dos Reis, A.S.; da Rocha Filho, G.N.; Santos, A.V.; Xavier, L.P.; Santos, A.S.; Gonçalves, E.C.; do Nascimento, L.A.S. Optimization of Light Intensity and $NaNO_3$ Concentration in Amazon Cyanobacteria Cultivation to Produce Biodiesel. *Molecules* **2019**, *24*, 2326. [CrossRef] [PubMed]
65. Lamers, P.P.; Janssen, M.; De Vos, R.C.H.; Bino, R.J.; Wijffels, R.H. Exploring and exploiting carotenoid accumulation in *Dunaliella salina* for cell-factory applications. *Trends Biotechnol.* **2008**, *26*, 631–638. [CrossRef] [PubMed]
66. Seabra, L.M.J.; Pedrosa, L.F.C. Ataxanthin: Structural and functional aspects. *Rev. Nutr. Braz. J. Nutr.* **2010**, *23*, 1041–1050.
67. Yu, X.; Yue, C.; Ding, W. Application of melatonin for improving astaxanthin content in Haematococcus pluvialis. CN107418993-A; CN107418993-B, 1 December 2017.
68. Sun, H.; Li, X.; Ren, Y.; Zhang, H.; Mao, X.; Lao, Y.; Wang, X.; Chen, F. Boost carbon availability and value in algal cell for economic deployment of biomass. *Bioresour. Technol.* **2020**, *300*, 122640. [CrossRef] [PubMed]
69. Chen, G.; Zhao, L.; Qi, Y. Enhancing the productivity of microalgae cultivated in wastewater toward biofuel production: A critical review. *Appl. Energy* **2015**, *137*, 282–291. [CrossRef]
70. Ra, C.-H.; Kang, C.-H.; Jung, J.-H.; Jeong, G.-T.; Kim, S.-K. Effects of light-emitting diodes (LEDs) on the accumulation of lipid content using a two-phase culture process with three microalgae. *Bioresour. Technol.* **2016**, *212*, 254–261. [CrossRef]
71. Wang, Y.; Shi, M.; Niu, X.; Zhang, X.; Gao, L.; Chen, L.; Wang, J.; Zhang, W. Metabolomic basis of laboratory evolution of butanol tolerance in photosynthetic *Synechocystis* sp. PCC 6803. *BioMed Cent.* **2014**, *13*, 151. [CrossRef]
72. Srivastava, V.; Amanna, R.; Rowden, S.J.; Sengupta, S.; Madhu, S.; Howe, C.J.; Wangikar, P.P. Adaptive laboratory evolution of the fast-growing cyanobacterium *Synechococcus elongatus* PCC 11801 for improved solvent tolerance. *J. Biosci. Bioeng.* **2021**, *131*, 491–500. [CrossRef]
73. Supekar, S.D.; Skerlos, S.J. Reassessing the Efficiency Penalty from Carbon Capture in Coal-Fired Power Plants. *Environ. Sci. Technol.* **2015**, *49*, 12576–12584. [CrossRef]
74. Chakravarti, L.J.; Beltran, V.H.; van Oppen, M.J. Rapid thermal adaptation in photosymbionts of reef-building corals. *Glob. Change Biol.* **2017**, *23*, 4675–4688. [CrossRef]
75. Schaum, C.-E.; Buckling, A.; Smirnoff, N.; Studholme, D.J.; Yvon-Durocher, G. Publisher Correction: Environmental fluctuations accelerate molecular evolution of thermal tolerance in a marine diatom. *Nat. Commun.* **2018**, *9*, 1719. [CrossRef] [PubMed]
76. Dann, M.; Ortiz, E.M.; Thomas, M.; Guljamow, A.; Lehmann, M.; Schaefer, H.; Leister, D. Enhancing photosynthesis at high light levels by adaptive laboratory evolution. *Nat. Plants* **2021**, *7*, 681–695. [CrossRef] [PubMed]
77. Xu, C.; Sun, T.; Li, S.; Chen, L.; Zhang, W. Adaptive laboratory evolution of cadmium tolerance in *Synechocystis* sp. PCC 6803. *BioMed Cent.* **2018**, *11*, 205. [CrossRef]
78. Uchiyama, J.; Kanesaki, Y.; Iwata, N.; Asakura, R.; Funamizu, K.; Tasaki, R.; Agatsuma, M.; Tahara, H.; Matsuhashi, A.; Yoshikawa, H.; et al. Genomic analysis of parallel-evolved cyanobacterium *Synechocystis* sp. PCC 6803 under acid stress. *Photosynth. Res.* **2015**, *125*, 243–254. [CrossRef]
79. Li, X.; Yuan, Y.; Cheng, D.; Gao, J.; Kong, L.; Zhao, Q.; Wei, W.; Sun, Y. Exploring stress tolerance mechanism of evolved freshwater strain *Chlorella* sp. S30 under 30 g/L salt. *Bioresour. Technol.* **2018**, *250*, 495–504. [CrossRef]
80. Li, D.; Wang, L.; Zhao, Q.; Wei, W.; Sun, Y. Improving high carbon dioxide tolerance and carbon dioxide fixation capability of *Chlorella* sp. by adaptive laboratory evolution. *Bioresour. Technol.* **2015**, *185*, 269–275. [CrossRef] [PubMed]
81. Aslam, A.; Thomas-Hall, S.R.; Mughal, T.A.; Schenk, P.M. Selection, and adaptation of microalgae to growth in 100% unfiltered coal-fired flue gas. *Bioresour. Technol.* **2017**, *233*, 271–283. [CrossRef]
82. Cheng, D.; Li, X.; Yuan, Y.; Yang, C.; Tang, T.; Zhao, Q.; Sun, Y. Adaptive evolution and carbon dioxide fixation of *Chlorella* sp. in simulated flue gas. *Sci. Total Environ.* **2019**, *650*, 2931–2938. [CrossRef]

83. Andreeva, N.A.; Melnikov, V.V.; Snarskaya, D.D. The Role of Cyanobacteria in Marine Ecosystems. *Russ. J. Mar. Biol.* **2020**, *46*, 154–165. [CrossRef]
84. Hsueh, H.T.; Chu, H.; Yu, S.T. A batch study on the bio-fixation of carbon dioxide in the absorbed solution from a chemical wet scrubber by hot spring and marine algae. *Chemosphere* **2007**, *66*, 878–886. [CrossRef] [PubMed]
85. Moreira, D.; Pires, J.C.M. Atmospheric CO_2 capture by algae: Negative carbon dioxide emission path. *Bioresour. Technol.* **2016**, *215*, 371–379. [CrossRef] [PubMed]
86. Mofihur, M.; Rizwanul Fattah, I.M.; Senthil, K.P.; Arafat, S.S.Y.; Ashrafur, R.S.M.; Ahmed, S.F.; Chyuan, O.H.; Shiung, L.S.; Anjum, B.I.; Yunus, K.T.M.; et al. Bioenergy recovery potential through the treatment of the meat processing industry waste in Australia. *J. Environ. Chem. Eng.* **2021**, *9*, 105657. [CrossRef]
87. Xu, J.W.; Liu, C.; Hsu, P.C.; Zhao, J.; Wu, T.; Tang, J.; Liu, K.; Cui, Y. Remediation of heavy metal contaminated soil by asymmetrical alternating current electrochemistry. *Nat. Commun.* **2019**, *10*, 2440. [CrossRef]
88. Kumar, D.; Pandey, L.K.; Gaur, J.P. Metal sorption by algal biomass: From batch to continuous system. *Algal Res.* **2016**, *18*, 95–109. [CrossRef]
89. De Wilt, A.; Butkovskyi, A.; Tuantet, K.; Leal, L.H.; Fernandes, T.V.; Langenhoff, A.; Zeeman, G. Micropollutant removal in an algal treatment system fed with source separated wastewater streams. *J. Hazard. Mater.* **2016**, *304*, 84–92. [CrossRef] [PubMed]
90. Gouveia, L.; Graça, S.; Sousa, C.; Ambrosano, L.; Ribeiro, B.; Botrel, E.P.; Neto, P.C.; Ferreira, A.F.; Silva, C.M. Microalgae biomass production using wastewater: Treatment and costs. *Algal Res.* **2016**, *16*, 167–176. [CrossRef]
91. Pacheco, D.; Rocha, A.C.; Pereira, L.; Verdelhos, T. Microalgae Water Bioremediation: Trends and Hot Topics. *Appl. Sci.* **2020**, *10*, 1886. [CrossRef]
92. Razzak, S.A.; Ali, S.A.M.; Hossain, M.M.; de Lasa, H. Biological CO_2 fixation with production of microalgae in wastewater—A review. *Renew. Sustain. Energy Rev.* **2017**, *76*, 379–390. [CrossRef]
93. Cai, T.; Park, S.Y.; Li, Y.B. Nutrient recovery from wastewater streams by microalgae: Status and prospects. *Renew. Sustain. Energy Rev.* **2013**, *19*, 360–369. [CrossRef]
94. Xu, K.; Zou, X.; Wen, H.; Xue, Y.; Qu, Y.; Li, Y. Effects of multi-temperature regimes on cultivation of microalgae in municipal wastewater to simultaneously remove nutrients and produce biomass. *Appl. Microbiol. Biotechnol.* **2019**, *103*, 8255–8265. [CrossRef]
95. Lee, K.; Lee, C.-G. Effect of light/dark cycles on wastewater treatments by microalgae. *Biotechnol. Bioprocess Eng.* **2001**, *41*, 351–355. [CrossRef]
96. Mahsa, B.; Marjan, A.; Hasan, J.; Ali, B.; Soroosh, D.; Esmaeili, B.M.; Abdeltif, A. Effect of light intensity and wavelength on nitrogen and phosphate removal from municipal wastewater by microalgae under semi-batch cultivation. *Environ. Technol.* **2020**. [CrossRef]
97. Liu, Y.; Yildiz, I. The effect of salinity concentration on algal biomass production and nutrient removal from municipal wastewater by *Dunaliella salina*. *Int. J. Energy Res.* **2018**, *42*, 2997–3006. [CrossRef]
98. Yang, C.; Wang, S.; Yang, J.; Xu, D.; Li, Y.; Li, J.; Zhang, Y. Hydrothermal liquefaction and gasification of biomass and model compounds: A review. *Green Chem.* **2020**, *22*, 8210–8232. [CrossRef]
99. Aigars, L.; Fredrika, M.; Elīna, Z.; Linda, M.; Tālis, J. Increasing Phosphorus Uptake Efficiency by Phosphorus-Starved Microalgae for Municipal Wastewater Post-Treatment. *Microorganisms* **2021**, *9*, 1598.
100. Aigars, L.; Linda, M.; Tālis, J. Microalgae starvation for enhanced phosphorus uptake from municipal wastewater. *Algal Res.* **2020**, *52*, 102090.
101. Junzhi, L.; Jinye, Y.; Yaming, G.; Houfeng, H.; Mei, L.; Feng, G. Improved lipid productivity of Scenedesmus obliquus with high nutrient removal efficiency by mixotrophic cultivation in actual municipal wastewater. *Chemosphere* **2021**, *285*, 131475.
102. Wang, L.; Xue, C.; Wang, L.; Zhao, Q.; Wei, W.; Sun, Y. Strain improvement of *Chlorella* sp. for phenol biodegradation by adaptive laboratory evolution. *Bioresour. Technol.* **2016**, *205*, 264–268. [CrossRef]
103. Gude, V.G.; Blair, M.F.; Kokabian, B. Light and growth medium effect on *Chlorellla vulgaris* biomass production. *Abstr. Pap. Am. Chem. Soc.* **2014**, *2*, 674.
104. Whitton, R.; Ometto, F.; Pidou, M.; Jarvis, P.; Villa, R.; Jefferson, B. Influence of light regime on the performance of an immpbilised microalgae reactor for wastewater nutrient removal. *Algal Res. Biomass Biofuels Bioprod.* **2019**, *4*, 148.
105. Levasseur, W.; Perre, P.; Pozzobon, V. A review of high value-added molecules production by microalgae in light of the classification. *Biotechnol. Adv.* **2020**, *41*, 107545. [CrossRef] [PubMed]
106. Lee, S.; Kim, P. Current Status and Applications of Adaptive Laboratory Evolution in Industrial Microorganisms. *J. Microbiol. Biotechnol.* **2020**, *30*, 793–803. [CrossRef] [PubMed]
107. Elena, S.F.; Lenski, R.E. Evolution experiments with microorganisms: The dynamics and genetic bases of adaptation. *Nat. Rev. Genet.* **2003**, *4*, 457–469. [CrossRef] [PubMed]
108. Wannier, T.M.; Kunjapur, A.M.; Rice, D.P.; McDonald, M.J.; Desai, M.M.; Church, G.M. Adaptive evolution of genomically recoded *Escherichia coli*. *Proc. Natl. Acad. Sci. USA* **2018**, *115*, 3090–3095. [CrossRef]
109. Cheng, J.; Feng, J.; Sun, J.; Huang, Y.; Zhou, J.; Cen, K. Enhancing the lipid content of the diatom *Nitzschia* sp. by 60 Co-γ irradiation mutation and high-salinity domestication. *Energy* **2014**, *78*, 9–15. [CrossRef]

MDPI

Review

Current Status and Perspective on the Use of Viral-Based Vectors in Eukaryotic Microalgae

Omayra C. Bolaños-Martínez [1,2], Ganesan Mahendran [1,2], Sergio Rosales-Mendoza [3,4] and Sornkanok Vimolmangkang [1,2,*]

1 Department of Pharmacognosy and Pharmaceutical Botany, Faculty of Pharmaceutical Sciences, Chulalongkorn University, Bangkok 10330, Thailand; omayra.cbm@hotmail.com (O.C.B.-M.); mahendran0007@gmail.com (G.M.)
2 Center of Excellence in Plant-Produced Pharmaceuticals, Chulalongkorn University, Bangkok 10330, Thailand
3 Laboratorio de Biofarmacéuticos Recombinantes, Facultad de Ciencias Químicas, Universidad Autónoma de San Luis Potosí, Av. Dr. Manuel Nava 6, San Luis Potosí 78210, Mexico; rosales.s@uaslp.mx
4 Sección de Biotecnología, Centro de Investigación en Ciencias de la Salud y Biomedicina, Universidad Autónoma de San Luis Potosí, Av. Sierra Leona 550, Lomas 2a Sección, San Luis Potosí 78210, Mexico
* Correspondence: sornkanok.v@pharm.chula.ac.th; Tel.: +662-218-8358

Abstract: During the last two decades, microalgae have attracted increasing interest, both commercially and scientifically. Commercial potential involves utilizing valuable natural compounds, including carotenoids, polysaccharides, and polyunsaturated fatty acids, which are widely applicable in food, biofuel, and pharmaceutical industries. Conversely, scientific potential focuses on bioreactors for producing recombinant proteins and developing viable technologies to significantly increase the yield and harvest periods. Here, viral-based vectors and transient expression strategies have significantly contributed to improving plant biotechnology. We present an updated outlook covering microalgal biotechnology for pharmaceutical application, transformation techniques for generating recombinant proteins, and genetic engineering tactics for viral-based vector construction. Challenges in industrial application are also discussed.

Keywords: biopharmaceuticals; recombinant proteins; transient expression; viral vectors

Citation: Bolaños-Martínez, O.C.; Mahendran, G.; Rosales-Mendoza, S.; Vimolmangkang, S. Current Status and Perspective on the Use of Viral-Based Vectors in Eukaryotic Microalgae. *Mar. Drugs* **2022**, *20*, 434. https://doi.org/10.3390/md20070434

Academic Editors: Marco García-Vaquero and Brijesh K. Tiwari

Received: 5 May 2022
Accepted: 27 June 2022
Published: 29 June 2022

1. Introduction

Microalgae are unicellular microorganisms found in marine and freshwater ecosystems over a wide range, from very small (a few micron) to large (a few hundreds of microns). They can rapidly produce biomass from solar energy, CO_2, and nutrients, such as nitrogen, sulfur, and phosphorous. Simple maintenance and cultivation in artificial environments offer a profitable platform to produce and extract bioactive compounds compared with other bioresources. Here, microalgae produce various metabolites with applications in pharmaceutical, cosmetic, bioenergy, and food/feed industries [1,2]. Various microalgae-derived products for food and feed have already been commercialized by different companies worldwide, including A4F-Algae 4 Future (Portugal), Blue Biotech (Germany), DIC Lifetec (Japan), E.I.D Parry (India), Ocean Nutrition (Canada), Phycom (Netherlands), Chlorella Co. (Taiwan), and Solazyme, Inc. (San Francisco), all of which used their bioactive compounds as colorants, additives, or supplements [3].

Biopharmaceuticals are complex molecules of biological origin used to diagnose, prevent, treat, and cure diseases or conditions in human beings and animals. According to their biological structure, biopharmaceuticals can be classified into amino acids, nucleic acids, and vaccines. In biopharmaceutical terms, these molecules are specifically produced under biotechnological processes based on genetically engineered organisms used as an expression host [4]. The main organisms used here are bacteria, yeast, mammalian cells, and insect cells, with each system having their own advantages as well as limitations [5–8].

Recently, microalgae have attracted increasing scientific interest due to their versatile growth and functional metabolic properties, as well as their biopharmaceutical production. Microalgae possess distinct attributes that have attracted the attention of biotechnologists, who have developed advanced genetic and molecular tools to leverage microalgae as green bioreactors to produce biopharmaceuticals. These attributes include their ability to grow and culture under heterotrophic, autotrophic, and mixotrophic conditions, the capacity to realize post-translational modifications and proper protein maturation, and the distinction of some microalgae species as "Generally Recognized as Safe" by the Food and Drug Administration (FDA). This status is conferred to any substance, chemical, or a whole organism that is safe for human consumption, owing to the absence of pathogens, microorganism, or related endotoxins. Mostly heterotrophic microalgae are FDA-approved for biotechnological applications due to their large-scale growing capacity and high cell density compared with other organisms [9,10]

To increase the yield and accelerate time to obtain and improve biopharmaceutical quality, microalgae biotechnology uses various expression methods and genetic and molecular biology strategies. These methods include stable nuclear and chloroplast expression and, in recent years, transient expression using viral-based vectors that allow high protein accumulation in a short period of time. However, the method using *Agrobacterium tumefaciens* transformation makes oral formulations of algal biomass unusable due to residual bacteria. Conversely, viral vectors for this purpose are limited and are mainly designed using elements derived from plant viruses. We present an updated outlook covering microalgal biotechnology for pharmaceutical applications, transformation techniques for obtaining recombinant proteins, and genetic engineering tactics for viral-based vectors construction (Figure 1). Finally, we discuss the potential challenges in industrial application.

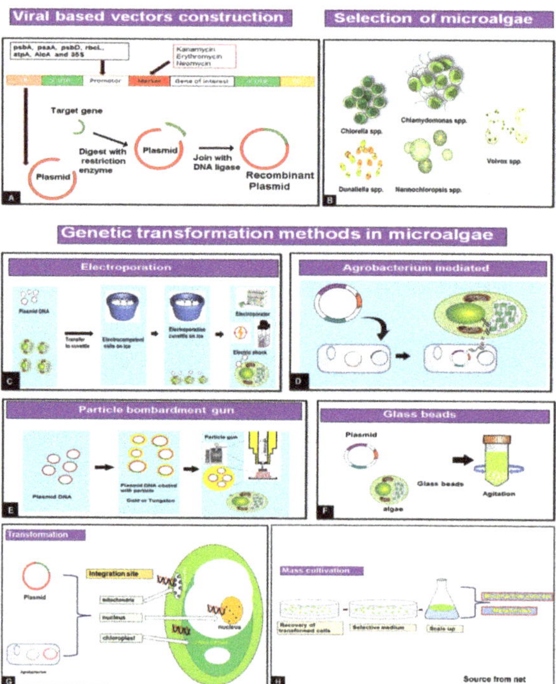

Figure 1. Overview of microalgal biotechnology for biopharmaceutical application. The essential components are from the vector design and selection of gene interests, microalgal hosts, and methods of transformation to finally obtain either bioactive metabolites or biopharmaceuticals. (**A**). Plasmid construction and transfer to *Agrobacterium*. (**B**). Selection of microalgae for genetic transformation.

(**C**). Method to transfer plasmid DNA using electroporation. (**D**). Introduction to target gene through the *Agrobacterium*-mediated method. (**E**). Stepwise protocol for the transfer of genes. (**F**). Traditional algae transformation method (glass beads). (**G**). Transformation methods (direct or Agrobacterium mediated) and integration into algae cell. (**H**). Stages of development for large-scale production of valuable biopharmaceuticals.

2. Genetic Engineering Transformation Methods for Biopharmaceutical Production of Microalgae

During the last 20 years, 40 different microalgae species genetic engineering methods have been developed [11,12]. *Chlamydomonas reinhardtii*, *Dunaliella salina*, *Volvox carteri*, *Haematococcus pluvialis*, and *Phaeodactylum tricornutum* are widely used microalgae for transforming foreign transgene expression studies, as well as biopharmaceutical protein production [13–17]. Here, microalgae genomes, such as nuclear, chloroplast, and mitochondrial transformation protocols, have been explored. In microalgae, four traditional methods are widely used to deliver foreign genes into microalgal genomes, including agitation with glass beads [18], particle bombardment [19], electroporation [20], and *Agrobacterium*-mediated transformation [21–25]. Of these methods, glass beads and *Agrobacterium* do not require any specialized apparatus, are less labor-intensive, and are relatively fast [18,26]. Additionally, bacterial conjugation as well as natural and liposome-mediated transformation also have been employed, each of these exhibiting their own advantages and drawbacks. The most notable disadvantages presented for some methods include the need for optimizing the transformation conditions, the low efficiency, and the high cost of the equipment or interface used [27].

Using the agitation method, transformation involves agitating wall-deficient cells/protoplasts of microalgae with foreign genes, glass beads, and polyethylene glycol (surfactant) [28–30]. This method can be used for both nuclear and chloroplast transformation. Furthermore, studies show cell wall-removed protoplasts are sufficient for gene transformation in *Chlorella ellipsoidea* [31]. Glass bead agitation has also been reported in chloroplast genetic engineering in *C. reinhardtii* using agitation of DNA/cell suspensions with glass beads [32]. The glass bead method also includes low transformation efficiencies due to thick cell walls, agitation duration, velocity, and surfactant concentration [12,33–35]. Table 1 presents and compares the limitations of different transformation methods.

Table 1. Comparison and limitations of genetic transformation methods in microalgae.

Method	Advantage	Disadvantage	Integration Site	Transformation Efficiency	Microalgae Species	Reference
Glass beads	Simple, controlability, high cell-survival rate, affordable, and minimal physical damage to cells	Cell wall removal and low transformation efficiency	Nucleus	~10^3 µg DNA^{-1}	*Chlamydomonas reinhardtii*	[18]
				NR	*Dunaliella salina*	[36]
				NR	*Platymonas subcordiformis*	[37]
Electroporation	Simple, affordable equipment, and high transformation efficiency	Transformation frequency affected by higher pulse strength and length, medium composition, temperature and membrane characteristics	Nucleus	6×10^3 per µg of DNA 2.5×10^4 per µg of DNA	*Chlamydomonas reinhardtii*	[38] [39]
				1.1×10^7 per µg of DNA	*Nannochloropsis limnetica*	[40]
				NR	*Chlamydomonas reinhardtii*	[20]
				NR	*Dunaliella salina*	[41]
				NR	*Scenedesmus obliquus*	[42]
				NR	*Monoraphidium neglectum*	[43]
				NR	*Chlorella pyrenoidosa*	[44]
				NR	*Nannochloropsis oculata*	[45]

Table 1. *Cont.*

Method	Advantage	Disadvantage	Integration Site	Transformation Efficiency	Microalgae Species	Reference
Agrobacterium-mediated	Transformation of large DNA fragments, simple, stable, and efficient	Variation of transformation efficiencies and transformation frequency affected by physical and chemical factors, silenced transformants, lower number of multiple insertions	Nucleus	311–355 × 10^{-6}	*Chlamydomonas reinhardtii*	[23]
				NR	*Haematococcus pluvialis*	[22]
				41.0 ± 4 CFU per 10^6 cells	*Dunaliella bardawil*	[46,47]
Biobalistic	Most effective method for the transformation of chloroplasts/nuclear, multiple copies insertion. More DNA integration and copy number	Cost effective, required specialized equipment, and recovery low	Nuclear/ Chloroplast Genome	~2.5 × 10^{-5} DNA	*Volvox carteri*	[48]
				1.9 × 10^{-6} to 4.2 × 10^{-6} per µg of DNA 10^8	*Chlamydomonas reinhardtii*	[49,50]
				NR		[51]
				NR	*Phaeodactylum tricornutum*	[52]
				NR	*Cyclotella cryptica* and *Navicula saprophila*	[53]
				NR	*Cylindrotheca fusiformis*	[54]
Silicon carbide whiskers	Similar protocol	Low transformation frequency	Nucleus	NR	*Chlamydomonas reinhardtii*	[55,56]
Lithium acetate/ polyethylene	Simple operation, low cost, less damage to the host cells and high transformation efficiency	Growth rate transformation temperature and plasmid concentration	Nucleus	113 colonies µg^{-1} DNA	*Dunaliella salina*	[57]

NR: Not Reported.

The *Agrobacterium*-based transformation method has previously been applied to *C. reinhardtii* [23,50,58], *H. pluvialis* [22,59], *Chlorella vulgaris* [60], *Parachlorella kessleri* [61], *Dunaliella bardawil* [46,47], *D. salina* [62], *Euglena gracilis* [63], *Cenedesmus almeriensis* [64], and *Dictyosphaerium pulchellum* [65]. According to Bashir et al. (2018), efficiency transformation using the Agrobacterium-based method was 50-fold higher that the glass bead method. However, different transformation efficiencies have been reported with *Agrobacterium*-based protocols [23,50,58]. Factors such as co-cultivation temperature, optical density, infection time, pre-culture duration, and acetosyringone concentration can substantially affect transformation efficiency [47,60]. In a study by Kumar et al. (2004), the *Agrobacterium*-based method performed equally as well as electroporation for stable integration into *Parachlorella kessleri* [61].

Electroporation is the most common and effective method for performing high-intensity electric pulses across the microalgae cell membranes to allow exogenous DNA to pass through cells [66–68]. This method has been reported in *C. reinhardtii* [20,38,39], *Nannochloropsis limnetica* [40], *D. salina* [41], *Scenedesmus obliquus* [42], *Monoraphidium neglectum* [43], *Chlorella pyrenoidosa* [44], *C. vulgaris* [69], *Chlorella zofingiensis* [70], and *Nannochloropsis oculata* [45]. Advantages include a rapid protocol, low labor, and high speed. Electroporation has been also been reported with transformation efficiencies up to 100-fold over agitation [12]. However, transformation efficiencies may be affected by electric-strength, pulse, and cell wall complexity [20,71,72].

Particle bombardment is an early and highly reproducible transformation method due to its ability to deliver genes into the nucleus, mitochondria, and chloroplast genomes without disturbing the cell walls [19,49,73,74]. This method is based on a DNA-coated ejection device with tungsten or gold metal particles that can detect target cells. Successful transformation using particle bombardment have previously been reported for *C. reinhardtii* [9,49–51,75,76], *D. salina* [77], *Haematoccucs pluvialis* [59], *V. carteri* [48], *P. tri-*

cornutum [52], *Cyclotella cryptica* and *Navicula saprophila* [53], *Cylindrotheca fusiformis* [54], and *Schizochytrium* sp. ATCC 20888 [78,79].

Among these techniques, the particle gun method is the most efficient for direct DNA delivery into cells. Generally, the gene gun method shows high transformation efficiency; however, this method is costly. Both particle bombardment and electroporation can be applied to transfer not only endogenous DNA but also proteins into microalgae cells. The most important application introduced Cas9 protein-gRNA ribonucleoproteins (RNPs) into microalgae, namely, into *C. reinhardtii*, *P. tricornutum*, and *Tetraselmis* sp. cells, for DNA-free genome editing [80–83].

In addition to the aforementioned methods used to introduce foreign DNA into microalgae cells, other transformation methods are also available. Hawkins and Nakamura (1999) showed *Chlorella* sp. protoplast cells and plasmids can be generated by mixing with polyethylene glycol and dimethyl sulfoxide for human growth hormone gene transformation [84]. Similarly, Liu et al. (2013) described novel, simple, reliable, and cost-effective transformation of *C. ellipsoidea* protoplast cells by mixing foreign DNA with PNC solution (NaCl, $CaCl_2$, and 40% PEG 4000) [71]. Other methods include stable nuclear transformation systems for *Pleurochrysis carterae* using polyethylene glycol (PEG)-mediated transfer of hygromycin B-resistance genes [85]. Recent reports present genetic transformation of microalgae by bacterial conjugation [86,87] and gene injection [88]. In addition to these techniques, other emerging methods, such as cell-penetrating peptides, nanoparticles, metal–organic frameworks, and liposomes, have not yet been demonstrated in microalgae [12,89–91].

3. Microalgae Nuclear and Chloroplast-Based Expression

Microalgae contain nuclear, mitochondrial, and chloroplast genomes, each of which have their own transcription, translation, and post-translation properties [92]. Nuclear expression in microalgae offers numerous benefits, such as targeting recombinant protein expression in specific organelles, protein glycosylation, post-translational modification, and secretion [93]. In nuclear-based expression, the position of an exogenous gene into a microalgal genome occurs as a random insertion and usually transgenic cells are selected via phenotypic variation or antibiotic resistance. Generally, this approach results in low yields. Although the reasons for this phenomenon are not completely understood, possible explanations could be attributed to the RNA-silencing process, transcript instability, positional effects of transgenes, and an inaccessible chromatin structure [94].

Using chloroplasts to express foreign genes has become a promising alternative to the nuclear genome. Microalgae chloroplasts serve as the main cell factory for synthesizing several metabolic pathway enzymes and appropriate transformation objects for producing isoprenoids, carbohydrates, lipid, carotenoids, pigments, fatty acids, and proteins [95,96]. Further, this organelle lacks a gene-silencing mechanism and may be used to protect proteins from degradation and involve some post-translation modifications, such as phosphorylation. These multiple functions in a single cell organelle are the most important traits for its heterologous gene expression in microalgae [97,98]. For delivery, the foreign gene must pass through several membranes, which represent a greater challenge. The preferred method to achieve this goal is particle bombardment. In particular, *C. reinhardtii* has been descried in numerous transformation studies for producing foreign proteins due to the chloroplast genome being fully sequenced and offering a unique advantage in the transformation system [99]. Further, various transformation methods have been reported for *C. reinhardtii* chloroplasts, among which are the marker-free chloroplast transformation system [100] and glass bead agitation using cell wall-deficient cells [28–30]. Finally, a chloroplast transformation system based on electroporation has also been developed for *Phaeodactylum tricornutum* [101].

4. Algal Biotechnology in Pharmaceutical Applications

In biochemistry, metabolites are defined as small molecules of <1.5 kilodaltons (KDa) that act as intermediates or end products in cellular metabolism and are classified as primary and secondary. Primary metabolites are directly involved in growth, development, and reproduction, whereas secondary are not implicated in these processes but offer an important ecological function and are typically linked to specific environmental conditions or developmental stages [102]. In microalgae, diverse bioactive metabolites have been studied for their antifungal, anticancer, antibacterial, and immunosuppressive properties [103–107].

Further, bioactive compounds obtained from microalgae, such as β-carotene, polyunsaturated fatty acids (Omega-3), clionasterol, phycocyanin, lutein, astaxanthin, canthaxanthin, fucoxanthin, zeaxanthin, docosahexaenoic acid (DHA), and eicosapentaenoic acid (EPA), can be applied as nutraceuticals, food additives, or in the cosmetics industry. Amino acids, such as tryptophan, lysine, leucine and arginine, vitamins B and E, essential minerals, and carbohydrates, are used in human and animal nutrition. Further, metabolites obtained from microalgae can be used in biofertilizer production as a source of nitrogen- and phosphorous-rich biomass residues as feedstock and in the bioenergy industry as bulk oil and biomass residue feedstock for jet fuel, biodiesel, bioethanol, biogas, biochar, and biohydrogen production. Furthermore, some microalgae strains can be used in wastewater treatment by reducing the amount of nitrogen, phosphate, and chemical oxygen demand, as well as removing heavy metals (copper (Cu), iron (Fe), manganese (Mn), and zinc (Zn)) and pharmaceutical pollutants (triclosan and hormones (17β-estradiol and 17α-ethinylestradiol) [108–113]. Interestingly, potential industrial applications and commercialization of microalgae-derived biomass and bioactive compounds in the food industry has recently been explored by Camacho et al. (2019). This analysis introduced the potential for formulation as prebiotics or as part of functional food/feed for human and animal consumption. Further, various industries can commercialize products, including phycocyanin, lutein, β-carotene, astaxanthin, eicosapentaenoic acid (EPA), and docosahexaenoic acid (DHA) (ω–3), derived from microalgae to be used as food colorants or supplements [3]. Currently, different species of microalgae have been used in the food/feed industry, such as *Porphyridium cruentum*, *Pavlova salina*, *Tisochrysis lutea*, *Chaetoceros muelleri*, *Nannochloropsis* spp., *Skeletonema* spp., *Thalassiosira pseudonana*, *Schizochytrium* sp., and *Crypthecodinium cohnii*, and wastewater bioremediation, including *Scenedesmus obliquus*, *Franceia* sp., *Ankistrodesmus* sp., *Tetraedron* sp., *Chlorella* sp., and *Mesotaenium* sp. [114].

Conversely, using single-cell engineering microalgae as a green factory to produce biopharmaceuticals includes recombinant expression of numerous antigenic proteins that act as human and animal vaccine candidates against viral or bacterial diseases and parasitic infections. Among these candidates, expression of viral epitopes from Zika virus [115], avian influenza [116], human papillomavirus [117], hepatitis B [41], and human immunodeficiency virus (HIV) [118], as well bacterial proteins from *Staphylococcus aureus* [76] and *Histophilus somni* [119], are well studied. Regarding parasitic infections, proteins from *Plasmodium falciparum* that cause malaria are also expressed in microalgae [120,121]. Furthermore, microalgae are also used to produce monoclonal antibodies, hormones, cytokines, growth factors, immunotoxins, and proteins to prevent non-communicable diseases [122–125]. A detailed recompilation of biopharmaceuticals produced in microalgae are summarized in Table 2. In addition, these recombinant microalgae cells can be utilized as an effective oral drug delivery platform formulated as pills, tablets, or freeze-dried cells [9]. A study by Kwon et al. (2019) demonstrated that the green fluorescent protein (GFP) expressed in chloroplasts of *C. reinhardtii* remained intact after biomass lyophilization [126].

Table 2. Production of recombinant biopharmaceuticals proteins in microalgae.

Microalgae Strain	Transformation Method	Integration Site	Protein Expressed	Yields Obtained	Application	Reference
Dunaliella salina	Lithium acetate/PEG	Nucleus	SKTI	0.68% TSP	Antivirus and anticancer	[57]
	Agrobacterium-mediated transformation		H5HA	225 µg TSP/2g	Avian influenza	[127]
	Electroporation		HBsAg	3.11 ± 0.50 ng/mg	Hepatitis B	[41]
	Glass beads		VP28	3.04 ± 0.26 ng/mg and 78 µg/100 mL culture	White spot syndrome in crayfish	[128]
	Biolistics	Chloroplast	sTRAIL	0.67% TSP	Tumor cells and virus-infected cells	[129]
Chlamydomonas reinhardtii	*Agrobacterium*-mediated transformation	Nucleus	HBcAgII	0.05% TSP	Hypertension	[130]
			IFN-α2a	NA	Immunity	[131]
			RBD	1.61 µg/g FWB	COVID-19	[35]
			bFGF	1.025 ng/g FWB	Growth factor	
	Glass beads	Chloroplast	HPV16 E7 mutated form r E7GGG-His6, E7GGG and E7GGG-FLAG	E7GGG-His6 (0.02%), E7GGG (0.1%) and E7GGG-FLAG (0.12%) TSP	Cancer	[117]
			WSSV VP28	NA	White spot disease in shrimp	[132]
			hGH	0.5 mg hGH/L	Growth Hormone	[32]
			dsRNA	NA	Yellow head virus infection in shrimp	[133]
	Biolistics		ctxB-pfs25	0.09% TSP and 20 µg/mL	Malaria	[51]
			pfs25 and pfs28	Pfs25 (0.5%) and Pfs28 (0.2%) TSP		[134]
			c.r.pfs48/45	NR		[75]
	Glass beads	Nucleus	AMA1/MSP1-GBSS	0.2 to 1.0 mg of protein/mg		[120]
			hVEGF-165, hPDGF-B, and hSDF-1	0.06% TSP, 0.003% TSP, 0.0006% TSP	Tissue hypoxia, wound healing	[135]
			P24, CpP24, CrP24, P24w	0.25% TSP	AISD	[118]
			hEGF	0.2%–0.25% TSP (40 mg/L)	hEGF deficiency	[136]
			Endolysin (Cpl-1 and Pal)	~1.3 mg/g ADW	*Streptococcus pneumoniae* infection	[137]
			ALFPm3	0.35% TSP	Anti-bacteria, anticancer, and antiviral activity	[138]
			IF	NA	Autoimmune disease pernicious anemia	[139]

Table 2. *Cont.*

Microalgae Strain	Transformation Method	Integration Site	Protein Expressed	Yields Obtained	Application	Reference
Chlamydomonas reinhardtii	Biolistics	Chloroplast	αCD22	0.7% TSP	Cancer	[124]
			83K7C	100 mg/1 g of DAB	Anthrax	[140]
			HSV8 scfv	0.5% TSP	Herpes simplex virus	[141]
			HSV8-lsc	>1% TSP	Herpes simplex virus	[142]
			M-SAA	0.25% TSP	Protection against intestinal bacterial and viral infections in newborns	[143]
			apcA and apcB	2–3% TSP	Inhibit the S-180 carcinoma in mice	[144]
			hMT-2	NA	UV-B effects	[145]
			CTB:p210	60 μg/g of FWB	Atherosclerosis	[146]
			Ara h 1 and Ara h 2	NA	Peanut allergy	[147]
			Bet v 1.0101	0.01 and 0.04% TSP	Allergy	[148]
			IL-2 and PfCelTOS	1.5% TSP	Malaria	[121]
			IFN-β1	NA	Multiple sclerosis	[149]
			VEGF	0.1% TSP	Depression and pulmonary arteries	[149]
			HMGB1	1% TSP	Response of the brain to neural injury and wound healing	[149]
			CelK1	0.003% TSP	Bioethanol and biogas production	[150]
	Biolistics	Nucleus	huBuChE	0.4% TSP	Pesticide poisonings	[151]
	Electroporation	Nucleus	Mytichitin-A	0.28% TSP	Growth inhibition of fungi, viruses, parasites, and bacteria	[152]
			ToAMP4	0.32% TSP	Antimicrobial	[153]
			hLF	1.82% TSP	Antibacterial	[154]
Schizochytrium sp	*Agrobacterium*-mediated transformation	Nucleus	HER-2, MUC1, MAM-A, and WT1	637 μg/g FWB	Breast cancer	[155]
			ZK1, ZK2, ZK3, and LTB	365 μg/g FWB	Zika disease	[115]
			LTB:RAGE	380 μg/g FWB	Alzheimer disease	[156]
			GP1 and LTB	1.25 mg/g FWB (6 mg/L of culture)	Ebola	[79]
Schizochytrium sp. ATCC 20888	Biobalistic	Nucleus	HA	5–20 mg/l	Influenza	[78]

Table 2. *Cont.*

Microalgae Strain	Transformation Method	Integration Site	Protein Expressed	Yields Obtained	Application	Reference
Chlorella vulgaris	*Agrobacterium*-mediated transformation	Nucleus	RBD	1.14 μg/g FWB	COVID-19	[35]
			bFGF	1.61 ng/g FWB	Growth factor	
Chlorella sp	Electroporation	Nucleus	Scygonadin and hepcidin	NA	Antibacterial	[157]
Chlorella sorokiniana ATCC-22521) or *Chlorella vulgaris* C-27	PEG	Nucleus	hGH	200–600 ng/mL	Cell regeneration/hGH deficiency	[84]
Chlorella ellipsoidea	Biobalistic	Chloroplast	fGH	420 μg fGH protein/L	Growth hormone	[123]
Dunaliella tertiolecta and *C. reinhardtii*	Biobalistic	Plastids	Xylanase, α-galactosidas, Phytase, phosphate anhydrolase, and β-mannanase	NA	Animal feeds and biofuel production	[158]
Haematococcus pluvialis	Biobalistic	Chloroplast	Piscidi-4	NA	Antimicrobial	[159]

PEG: Polyethylene glycol; SKTI: Soybean Kunitz trypsin inhibitor; TSP: Total soluble protein; H5HA: Hemagglutinin-Influenza A virus; TSP: Total soluble protein, HBsAg: Hepatitis B surface antigen; HBcAgII: Angiotensin II fusion to hepatitis B virus (HBcAg); HPV16 E7: Human papillomavirus 16 E7 protein; ctxB-pfs25: Plasmodium falciparum surface protein (Pfs25) fused to cholera toxin (CtxB); pfs25 and pfs28: Plasmodium falciparum surface protein 25 and 28; c.r.pfs48/45: Plasmodium falciparum surface protein 48/45; AMA1/MSP1-GBSS: Apical major antigen or major surface protein fused to granule bound starch synthase; CTB-D2: fibronectin-binding domain D2, fused to the cholera toxin B subunit protein; hGAD65: Human glutamic acid decarboxylase; CSFV E2: classical swine fever virus structural protein E2; αCD22: Immunotoxin protein; 83K7C: Human IgG1 monoclonal antibody 83K7C against the PA83 anthrax antigen; DAB: dry algal biomass; HSV8 scfv: single-chain variable regions antibody against Herpes simplex virus glycoprotein D; HSV8-lsc: Large single-chain antibody directed against Herpes simplex virus glycoprotein D; huBuChE: A fusion protein containing luciferase and the human butyrylcholinesterase; AISD: Acquired immunodeficiency syndrome: FWB: Fresh weight biomass; IL-2 and PfCelTOS: PfCelTOS fused to human interleukin-2; sTRAIL: Tumor factor-related apoptosis inducing ligand; IFN-β1: Human interferon β1; VEGF: Human vascular endothelial growth factor; HMGB1: High mobility group protein B1; hEGF: Human epidermal growth factor; ALFPm3: Anti-Lipopolysaccharide factor isoform 3; CelK1: Bacterial endoglucanase (CelK1, Glycohydrolase, family 5) enzyme; hGH: human growth hormone, M-SAA: Bovine mammary-associated amyloid; hMT-2: Metallothionein-2; IFN-α2a: Human interferon-α; IF: Human protein intrinsic factor; WSSV VP28: White spot syndrome virus protein; ToAMP4: Taraxacum officinale antimicrobial peptide 4; hLF: Human lactoferrin; HER-2 Human Epidermal Growth Factor Receptor-2; MUC1: Mucin-like glycoprotein 1; WT1: Wilms' Tumor Antigen; MAM-A: Mammaglobin-A; LTB:RAGE: Receptor of Advanced Glycation End products fused to *E. coli* heat-labile enterotoxin B subunit; GP1: Complex viral proteins from Zaire ebolavirus; HA: Recombinant hemagglutinin from A/Puerto Rico/8/34 (H1N1) influenza virus; fGH: flounder growth hormone.

5. Viral-Based Expression Vectors for Recombinant Protein, Vaccine, and Biopharmaceutical Production

Currently, biotechnology and genetic engineering is harnessing numerous viruses or their component parts to produce heterologous proteins for human and animal use. Given the expression of epitopes from influenza A virus can be fused with the hepatitis B core antigen in *Nicotiana benthamiana* plants, generation of virus-like particles (VLPs) in insect cells for the human papilloma virus as a vaccine-delivery vehicle for genetic material can generate an immune response in the human body, as recently developed for a COVID-19 vaccine [160–162]. Furthermore, polymerases and reverse-transcriptases from viral origins, in addition to elements such as transcriptional promoters, terminators, silencing suppressors, and internal ribosomal entry sites, form part of a molecular toolbox for genetic engineers, biologists, and biotechnologists.

The common approach for generating viral-based expression vectors involves inserting a determinate viral genome sequence into an expression vector downstream of a cell-type-specific promoter. The coding sequence of a heterologous gene is then inserted into the viral genome sequence as part of a viral polyprotein or downstream to a subgenomic promoter. The construct is then transferred to host cells for transcription and subsequent translation processes by host molecular machinery [163]. During the last decade, the design, generation, and use of viral-based expression vectors for producing heterologous proteins have gained increasing scientific interest, mainly in the plant biotechnology field. To achieve this goal, expression strategies have focused on RNA and DNA plant viruses, of which tobamovirus, comovirus, potexvirus, and geminivirus are the most exploited genera.

Developing and applying this approach has followed an interesting path. First, by creating first-generation expression vectors or full virus strategies based on expression of the gene of interest (GO), this approach has also produced its own viral genes and subsequent translation as an individual antigenic or fusion protein on the C-terminal of the capsid protein (CP). Using these vectors, several immunogens have been produced, reaching up to 10% of the total soluble protein (TSP) in *Nicotiana benthamiana* plants. However, stability is negatively related to insert size, hence the proteins larger than 30 KDa are poorly expressed in a chimeric CP form and epitopes should be 25 amino acids at maximum length [164–166]. These drawbacks slowed the development of second-generation viral vectors, whereby using a full virus was replaced with a deconstructed virus genome containing essential elements for replication and non-viral sequence integration to accomplish other functions, such as replicon formation using T-DNA delivered via *A. tumefaciens*. Using *Agrobacterium* for DNA delivery offers considerable advantages given the efficient transfer capacity by infiltration of plant leaves. Plants species using this approach include spinach, sunflower, red beetroot, and *N. benthamiana*, presenting maximum yields up to 50% of TSP in a 4–5 day period where the size of the GO can be up to 2 Kb and proteins of 80 KDa can be produced [167–169].

Special attention should be directed toward DNA virus-based vectors, specifically those applying elements from geminivirus, a twinned icosahedral virus with a single-strand DNA (ssDNA) arranged in one (monopartite) or two components (bipartite) encoding proteins essential for the replication process, pathogenicity, suppression of plant gene silencing, and intercellular and long-distance movement of the virus [170,171]. In general, these vectors are based on a transient expression system, the advantages of which include rapid product expression, high production rate, flexibility, and scalability. A geminivirus engineered for biopharmaceuticals is *Bean Yellow Dwarf Virus* (BeYDV), which has been modified to leverage its Rep protein under independent promoter control. With this strategy, diverse BeYDV-based expression vectors have been engineered and an assortment of antigens and monoclonal antibodies have been generated [172–174]. For microalgae, the geminiviral vector pBYR2e was used for expression of the receptor-binding domain (RBD) from SARS-CoV-2 and fibroblast growth factor (bFGF) in two freshwater microalgal species. Yields reached up to 1.61 µg/g and 1.14 µg/g for RBD when expressed in *C. reinhardtii* and *C. vulgaris*, respectively [35]. Conversely, Berndt et al. (2021) reported expression of RBD-fused GFP in *C. reinhardtii*. Interestingly, the protein targeted three different cellular localizations: (i) in the endoplasmic reticulum–Golgi pathway; (ii) secreted out of the cell into the culture media; and (iii) directed to the chloroplasts. In the latter, although under higher expression, the protein appeared to be truncated by ~5 kDa at the amine end, whereas the end targeted to the ER was produced with the expected size and correct amino acid sequence. For obtaining proteins, the transgene was placed into the pBR9 and pOpt vectors; in particular, the pBR9 vector containing the *sh ble* zeocin resistance selection marker with a food and mouth disease virus (FMDV) 2A self-cleaving sequence placed between the coding sequences, resulted in accumulation of two separate proteins [175].

Another geminivirus-based vector, named Algevir, has been developed with diverse antigenic proteins and epitopes expressed in the marine microalgae *Schizochytrium* sp., which was engineered using the Rep protein and origin of replication (Ori) from the begomovirus *Ageratum enation* to produce and replicate circular DNA carrying the GO and AlcR gene, as well as the AlcA promoter from *Aspergillus nidulans* to obtain ethanol-induced expression. This innovative system has produced viral and bacterial proteins at a maximum level production of 1.25 mg/g fresh biomass for GP1 from *Zaire ebolavirus* [79]. Table 3 shows the viral-based vectors used for biopharmaceutical production in microalgae. However, yields produced in microalgae with a nuclear approach and using viral-based vectors do not fully outcompete those produced in chloroplasts whereby targets allow production of 3.28 mg/L of culture medium [176]. The strategy based on protein production in this organelle requires a long time and construction of detailed vectors containing specific sequences for integration by homologous recombination. Here, optimizing viral-based vectors is needed to increase the protein yield and improve stability, which requires transient expression as a primary approach given that some transgene products may become toxic for host cells, leading to very low yields under stably transformed lines. Alternatively, microalgae viruses can be naturally used to drive gene expression at different infection stages and viral elements can be explored throughout the design process of novel viral-based vectors or when improving current models. Updating the functions of viral genes and the genome composition is an important requirement for executing a rational design in which regulatory elements, such as promoters, terminators, or replication proteins, help reach strong GO expression. Finally, exploring the possibility of directly purifying recombinant proteins using elements from lytic viruses presents an alternative approach [177].

Table 3. Virus-based vectors used for biopharmaceutical production.

<table>
<tr><th>Microalgae Host</th><th>Type of Transformation</th><th>Name</th><th>Viral Elements</th><th>Protein Expressed</th><th>References</th></tr>
<tr><td>Schizochytrium sp.</td><td>Transient nuclear/Inducible expression</td><td>Algevir</td><td>Cauliflower mosaic virus:
35S promoter
35S terminator
Ageratum enation virus:
Replication protein "Rep"
Origin of replication "Ori"</td><td>The GP1 from Zaire ebolavirus and LTB
RAGE (23–54 amino acids)
The ZK1, ZK2, ZK3 from the E protein from Zika virus fused to LTB
The multiepitope protein BCB comprised epitopes from HER-2, MUC1, WT1, MAM-A fused to LTB</td><td>[79,115,155,156]</td></tr>
<tr><td>Chlamydomonas reinhardtii</td><td rowspan="2">Transient nuclear</td><td rowspan="2">pBYR2e</td><td rowspan="2">Cauliflower mosaic virus:
35S promoter
Tomato bushy stunt virus:
RNA silencing suppressor P19
Bean Yellow Dwarf Virus:
Short intergenic region SIR
Long intergenic region LIR
C1/C2 Replication protein and replication protein A
Tobacco mosaic virus Ω:
5′untranslated region</td><td rowspan="2">The RBD from SARS-CoV-2
The bFGF</td><td rowspan="2">[35]</td></tr>
<tr><td>Chlorella vulgaris</td></tr>
</table>

LTB: Bacterial toxin B subunit of the heat-labile *E. coli* enterotoxin; RAGE: Receptor of Advanced Glycation End products; ZK1: amino acids LDKQSDTQYVCKRTLVDR; ZK2: amino acids FSDLYYLTM; ZK3: amino acids LKGVSYSLCTAAFTFTKI; HER-2 Human Epidermal Growth Factor Receptor-2; MUC1: Mucin-like glycoprotein 1; WT1: Wilms' Tumor Antigen; MAM-A: Mammaglobin-A; RBD: Receptor Binding Domain; SARS-CoV-2: Severe Acute Respiratory Syndrome Coronavirus 2; bFGF: Fibroblast Growth Factor.

6. Design of a Viral-Based Vector for Microalgae Use

In the virosphere, many species are capable of infecting microalgae. In addition to triggering high mortality rates, such species can reprogram host metabolism, including photosynthesis and important cycling processes, such as central carbon metabolism, phosphorus, nitrogen, and sulfur [178].

To date, a total 63 virus that infect eukaryotic microalgae have been isolated and cultured in the laboratory, whereby 50.79% contain dsDNA as genomic material, 15.8% ssDNA, 1.58% dsRNA, and 22.2% ssRNA, whereas 7.93% have not yet been classified [163,179].

Recently, a list of 10 isolated and characterized viruses was published by Sandaa et al. (2022) [180]. These viruses can infect marine haptophytes species. Here, a rational design of a microalgal-specific viral vector to achieve higher protein yields, using viral elements that naturally infect microalgae, could be a promising strategy.

In a study published by Kadono et al. (2015) [181], a set of five potential promoter regions located upstream of the replication-associated protein (VP3) or structural protein (VP2), coding genes for three marine diatom-infecting viruses (DIVs), were evaluated and compared in the Pennales diatom *Phaeodactylum tricornutum* as a heterologous host (Table 4). The gene-encoding fucoxanthin chlorophyll a/c-binding protein (fcp) was used as an endogenous promoter and eGFP as a protein reporter. In addition, the extrinsic promoter, such as *Cauliflower mosaic virus* 35S (CaMV35S), cytomegalovirus (CMV), and nopaline synthase gene (nos) promoter, were also used. The results show the novel promoter ClP1 mediated significantly higher transcription and translation rates according to mRNA transcripts and flow cytometry analysis, respectively. Further, the abundance of eGFP mRNA transcripts in the stationary phase were higher than those found in the log phase under both low and standard nutrient culture conditions.

Table 4. Molecular elements from viruses infecting microalgae tested for the expression of recombinant proteins.

<table>
<tr><th>Viral Genomic Element</th><th>Name</th><th>Viral Source</th><th>Size (bp)</th><th>Type of Expression</th><th>Transformation Method</th><th>Protein Expressed</th><th>Heterologous Host</th><th>Reference</th></tr>
<tr><td rowspan="6">Promoters</td><td rowspan="2">C1P1</td><td rowspan="3">ClorDNAV</td><td rowspan="2">502</td><td rowspan="2">Stable</td><td>Biobalistic</td><td>eGFP</td><td>Pennales diatom Phaeodactylum tricornutum</td><td rowspan="6">[181]</td></tr>
<tr><td>Electroporation</td><td>Sh ble</td><td>Chlamydomonas reinhardtii</td></tr>
<tr><td>ClP2</td><td>474</td><td rowspan="4">Stable</td><td rowspan="4">Biobalistic</td><td rowspan="4">eGFP</td><td rowspan="4">Pennales diatom Phaeodactylum tricornutum</td></tr>
<tr><td>CdP1</td><td>CdebDNAV</td><td>477</td></tr>
<tr><td>TnP1</td><td rowspan="2">TnitDNAV</td><td>424</td></tr>
<tr><td>TnP2</td><td>424</td></tr>
</table>

ClorDNAV: *Chaetoceros lorenzianus*-infecting DNA virus; CdebDNAV: *Chaetoceros debilis*-infecting DNA virus; TnitDNAV: *Thalassionema nitzschioides*-infecting DNA virus; eGFP: enhanced green fluorescence protein; *Sh ble*: bleomycin-resistant gene.

In addition to DIVs, other viruses can help explore their genetic elements and design a novel viral-based vector, particularly those with ssDNA or dsRNA genomes. Among them, viral species infecting the most commonly studied microalgal, such as the genus *Chlorella*, may offer a useful genetic toolbox. For example, the *Paramecium bursaria Chlorella virus 1* (PBCV-1), a large dsDNA virus (>300 kb) infecting the green microalgae *Chlorella variabilis* NC64A, is now a model system for studying DNA virus/algal interactions, which has also been tested for biomass saccharification with subsequent bioethanol production and proteins involved in cell wall degradation [182–185]. Another virus fully sequenced that infects the *Chlorella* genus with potential biotechnology application are those that exclusively multiply in Syngen 2–3 or SAG 3.83 cells, which could lead to specific protein expression in microalgae strains. The prototype viruses are only Syngen viruses—NE5 (OSy-NE5) and *Acanthocystis turfacea* chlorella virus (ATCV-1) [186,187].

7. Challenges and Perspectives

In recent years, the current pandemic has pushed progress of several biomedical technologies, e.g., RNA vaccines and adenovirus-based vaccines. Based on these advances, what are the key insights from the field of algae-based biopharmaceuticals? Biopharmaceuticals using algae are considered a promising alternative for improving global health. Algae offer low production costs and some species are already used at industrial levels in the food industry and thus are considered safe for use as delivery vehicles, especially oral formulations. However, although the proof of concept for using algae to produce and even deliver biopharmaceutical has been reported by several groups, a number of challenges remain to be addressed in this field, including improving recombinant protein yield productivity.

Another critical path that deserves research attention in developing algae-made biopharmaceuticals is related to regulation. Defining the main guidelines for specific regulations applied to this type of biological agent is a major priority task. Performing clinical trials requires translating prototypes generated in academic labs to facilities with good laboratory practices that can approve and perform clinical trials. Moreover, implementing GMP-compliant processes in cooperation with pharmaceutical companies is urgently needed.

The current pandemic has increased support from several countries to invest in biomedical research and strengthen the developmental path for drugs and biologics. For example, several developing and emerging countries are increasing funding for research on innovative platforms for biopharmaceuticals production, including Thailand and México. We consider that the innovative green platforms required to produce biopharmaceuticals are a promising niche that could be accelerated by such initiatives. However, this should be a mid-term goal considering that conventional production systems with well-established regulatory frameworks will be the priority for such countries to provide rapid solutions for immediate needs. As biopharmaceuticals are inherently more complex than conventional chemical drugs, they demand a more complicated manufacturing process with varying quality and demands for extensive processes and product understanding. In addition, downstream processing represents another bottleneck. For algae, eliminating large amounts of lipids present in total extracts should be studied and the impact of differential glycosylation compared with mammalian glycosylation is another aspect that deserves attention.

Although the good manufacturing practice (GMP) standards of various regulatory authorities and international organizations are very similar and appropriate in addressing the manufacturing challenges, introducing innovative platforms always presents challenges. This challenge is exacerbated in developing or emerging countries that require affordable biopharmaceuticals. For instance, a recent study by Rahalkar et al. (2021) revealed that, in several emerging countries, the lack of standardized biosimilar development criteria and regulatory convergence across agencies led to challenges in multi-country biosimilar development, limiting our ability to introduce new, cheaper biosimilars into the market [188]. Unfortunately, for biopharmaceuticals produced in algae, this remains an ongoing challenge.

Although using viral vectors improves efficiency in expression systems, using *Agrobacterium* presents the need for complex purification steps to eliminate bacterial endotoxins. Therefore, expanding stable transformation systems to express viral replicons under an inducible approach is a possible solution to this limitation. Avoiding antibiotic-resistant markers is another challenge when designing vectors. Alternative markers, such as nutrient-selective markers, are accruing more interest. Another possibility is developing oral formulations subjected to less strict regulations. It is clear that this field is still in its infancy; thus, exploring new constructs optimized for model species, especially *C. reinhardtii*, are required. Special emphasis on developing vectors based on new algae viruses is crucial.

8. Concluding Remarks

Although using viral-based expression systems in algae is still new, this technology has immense potential to revolutionize the algae-based biopharmaceuticals field by offering higher yields and shorter production times compared with chloroplast and nuclear stable transformation methods. The following decade will be critical, as technology will benefit from refreshed interest when supporting biomedical research in response to the COVID-19 pandemic. Research and development goals should be focused not only on generating prototypes in academic labs but also on critical regulatory issues to ensure the success of new products that enter the market and ultimately benefit human health, especially in developing and emerging countries. On February 2022, Medicago, a Canadian company, and GlaxoSmithKline (GSK) announced approval by the Health agency in Canada of COVIFENZ®, a COVID-19 vaccine produced in plants. This is a milestone, as it is the first vaccine produced using a green platform approved for human use. Will algae-based products reach the same success? The following decade will be crucial in addressing this goal.

Author Contributions: Conceptualization, S.V. and O.C.B.-M.; writing—original draft preparation, O.C.B.-M., G.M.; S.R.-M.; writing—review and editing, S.V.; O.C.B.-M.; S.R.-M.; supervision, S.V. All authors have read and agreed to the published version of the manuscript.

Funding: This research was funded by Chulalongkorn University (Grant No. ReinUni_65_03_33_18).

Institutional Review Board Statement: Not applicable.

Informed Consent Statement: Not applicable.

Data Availability Statement: Not applicable.

Acknowledgments: The author Omayra C. Bolaños-Martínez is an alumnus of the Second Century Fund (C2F) recipient, Chulalongkorn University.

Conflicts of Interest: The authors declare no conflict of interest.

References

1. De Morais, M.G.; Vaz, B.D.S.; de Morais, E.G.; Costa, J.A.V. Biologically active metabolites synthesized by microalgae. *Biomed. Res. Int.* **2015**, *2015*, 835761. [CrossRef] [PubMed]
2. Grama, S.B.; Liu, Z.; Li, J. Emerging trends in genetic engineering of microalgae for commercial applications. *Mar. Drugs* **2022**, *20*, 285. [CrossRef] [PubMed]
3. Camacho, F.; Macedo, A.; Malcata, F. Potential industrial applications and commercialization of microalgae in the functional food and feed industries: A short review. *Mar. Drugs* **2019**, *17*, 312. [CrossRef] [PubMed]
4. Rasmussen, A.S.B.; Hammou, A.; Poulsen, T.F.; Laursen, M.C.; Hansen, S.F. Definition, categorization, and environ-mental risk assessment of biopharmaceuticals. *Sci. Total Environ.* **2021**, *789*, 147884. [CrossRef]
5. Mizukami, A.; Caron, A.L.; Picanço-Castro, V.; Swiech, K. Platforms for recombinant therapeutic glycoprotein production. *Methods Mol. Biol.* **2018**, *1674*, 1–14. [CrossRef]
6. Baeshen, M.N.; Al-Hejin, A.M.; Bora, R.S.; Ahmed, M.M.; Ramadan, H.A.; Saini, K.S.; Baeshen, N.A.; Redwan, E.M. Production of biopharmaceuticals in *E. coli*: Current scenario and future perspectives. *J. Microbiol. Biotechnol.* **2015**, *25*, 953–962. [CrossRef]
7. Madhavan, A.; Arun, K.B.; Sindhu, R.; Krishnamoorthy, J.; Reshmy, R.; Sirohi, R.; Pugazhendi, A.; Awasthi, M.K.; Szakacs, G.; Binod, P. Customized yeast cell factories for biopharmaceuticals: From cell engineering to process scale up. *Microb. Cell Fact.* **2021**, *20*, 124. [CrossRef]
8. Hanisch, F.G. Recombinant norovirus capsid protein VP1 (GII.4) expressed in H5 insect cells exhibits post-translational modifications with potential impact on lectin activity and vaccine design. *Glycobiology* **2022**, *32*, 496–505. [CrossRef]
9. Rosales-Mendoza, S.; Solís-Andrade, K.I.; Márquez-Escobar, V.A.; González-Ortega, O.; Bañuelos-Hernandez, B. Current advances in the algae-made biopharmaceuticals field. *Expert Opin. Biol. Ther.* **2020**, *20*, 751–766. [CrossRef]
10. Khavari, F.; Saidijam, M.; Taheri, M.; Nouri, F. Microalgae: Therapeutic potentials and applications. *Mol. Biol. Rep.* **2021**, *48*, 4757–4765. [CrossRef]
11. Jareonsin, S.; Pumas, C. Advantages of heterotrophic microalgae as a host for phytochemicals production. *Front. Bioeng. Biotechnol.* **2021**, *9*, 628597. [CrossRef] [PubMed]
12. Gutiérrez, S.; Lauersen, K.J. Gene delivery technologies with applications in microalgal genetic engineering. *Biology* **2021**, *10*, 265. [CrossRef] [PubMed]
13. Gong, Y.; Hu, H.; Gao, Y.; Xu, X.; Gao, H. Microalgae as platforms for production of recombinant proteins and valuable compounds: Progress and prospects. *J. Ind. Microbiol. Biotechnol.* **2011**, *38*, 1879–1890. [CrossRef] [PubMed]
14. Bañuelos-Hernández, B.; Beltrán-López, J.I.; Rosales-Mendoza, S. Production of biopharmaceuticals in microalgae. In *Handbook of Marine Microalgae*; Academic Press: Cambridge, MA, USA, 2015; pp. 281–298.
15. Doron, L.; Segal, N.; Shapira, M. Transgene expression in microalgae-from tools to applications. *Front. Plant Sci.* **2016**, *7*, 505. [CrossRef]
16. Sproles, A.E.; Fields, F.J.; Smalley, T.N.; Le, C.H.; Badary, A.; Mayfield, S.P. Recent advancements in the genetic engineering of microalgae. *Algal Res.* **2021**, *53*, 102158. [CrossRef]

17. Kselíková, V.; Singh, A.; Bialevich, V.; Čížková, M.; Bišová, K. Improving microalgae for biotechnology-From genetics to synthetic biology-Moving forward but not there yet. *Biotechnol. Adv.* **2022**, *58*, 107885. [CrossRef]
18. Kindle, K.L. High-frequency nuclear transformation of *Chlamydomonas reinhardtii*. *Proc. Natl. Acad. Sci. USA* **1990**, *87*, 1228–1232. [CrossRef]
19. Kindle, K.L.; Schnell, R.A.; Fernandez, E.; Lefebvre, P.A. Stable nuclear transformation of *Chlamydomonas* using the *Chlamydomonas* gene for nitrate reductase. *J. Cell Biol.* **1989**, *109*, 2589–2601. [CrossRef]
20. Shimogawara, K.; Fujiwara, S.; Grossman, A.; Usuda, H. High-efficiency transformation of *Chlamydomonas reinhardtii* by electroporation. *Genetics* **1998**, *4*, 1821–1828. [CrossRef]
21. Hwang, H.H.; Yu, M.; Lai, E.M. *Agrobacterium*-mediated plant transformation: Biology and applications. In *Arabidopsis Book*; American Society of Plant Biologists: Rockville, MD, USA, 2017; Volume 15, p. e0186.
22. Kathiresan, S.; Chandrashekar, A.; Ravishankar, G.A.; Sarada, R. *Agrobacterium*-mediated transformation in the green alga *Haematococcus pluvialis* (Chlorophyceae, Volvocales). *J. Phycol.* **2009**, *45*, 642–649. [CrossRef]
23. Kumar, S.V.; Misquitta, R.W.; Reddy, V.S.; Rao, B.J.; Rajam, M.V. Genetic transformation of the green alga-*Chlamydomonas reinhardtii* by *Agrobacterium tumefaciens*. *Plant Sci.* **2004**, *166*, 731–738. [CrossRef]
24. Kumar, S.V.; Rajam, M.V. Induction of *Agrobacterium tumefaciens* vir genes by the green alga, *Chlamydomonas reinhardtii*. *Curr. Sci.* **2007**, *92*, 1727–1729.
25. Wang, P.; Wang, G.; Teng, Y.; Li, X.; Ji, J.; Xu, X.; Li, Y. Effects of cefotaxime and kanamycin on thallus proliferation and differentiation in *Porphyra yezoensis* and their inhibition on *Agrobacterium tumefaciens*. *Mar. Biol. Res.* **2010**, *6*, 100–105. [CrossRef]
26. Sodeinde, O.A.; Kindle, K.L. Homologous recombination in the nuclear genome of *Chlamydomonas reinhardtii*. *Proc. Natl. Acad. Sci. USA* **1993**, *90*, 9199–9203. [CrossRef] [PubMed]
27. Sreenikethanam, A.; Raj, S.; Banu, R.J.; Gugulothu, P.; Bajhaiya, A.K. Genetic engineering of microalgae for secondary metabolite production: Recent developments, challenges, and future prospects. *Front. Bioeng. Biotechnol.* **2022**, *10*, 836056. [CrossRef]
28. Kindle, K.L.; Richards, K.L.; Stern, D.B. Engineering the chloroplast genome: Techniques and capabilities for chloroplast transformation in *Chlamydomonas reinhardtii*. *Proc. Natl. Acad. Sci. USA* **1991**, *88*, 1721–1725. [CrossRef]
29. Economou, C.; Wannathong, T.; Szaub, J.; Purton, S. A simple low-cost method for chloroplast transformation of the green alga *Chlamydomonas reinhardtii*. In *Methods in Molecular Biology*; Springer International Publishing: Cham, Switzerland, 2014; Volume 1132, pp. 401–411. [CrossRef]
30. Rochaix, J.D.; Surzycki, R.; Ramundo, S. Tools for regulated gene expression in the chloroplast of *Chlamydomonas*. In *Methods in Molecular Biology*; Springer International Publishing: Cham, Switzerland, 2014; Volume 1132, pp. 413–424. [CrossRef]
31. Jarvis, E.E.; Brown, L.M. Transient expression of firefly luciferase in protoplasts of the green alga *Chlorella ellipsoidea*. *Curr. Genet.* **1991**, *19*, 317–321. [CrossRef]
32. Wannathong, T.; Waterhouse, J.C.; Young, R.E.; Economou, C.K.; Purton, S. New tools for chloroplast genetic engineering allow the synthesis of human growth hormone in the green alga *Chlamydomonas reinhardtii*. *Appl. Microbiol. Biotechnol.* **2016**, *100*, 5467–5477. [CrossRef]
33. Rosenberg, J.N.; Oyler, G.A.; Wilkinson, L.; Betenbaugh, M.J. A green light for engineered algae: Redirecting metabolism to fuel a biotechnology revolution. *Curr. Opin. Biotechnol.* **2008**, *19*, 430–436. [CrossRef]
34. Barrera, D.J.; Mayfield, S.P. High-value recombinant protein production in microalgae. In *Handbook of Microalgal Culture Applied Phycology and Biotechnology*, 2nd ed.; Richmond, A., Hu, Q., Eds.; John Wiley & Sons: London, UK, 2013; pp. 532–544. [CrossRef]
35. Malla, A.; Rosales-Mendoza, S.; Phoolcharoen, W.; Vimolmangkang, S. Efficient transient expression of recombinant proteins using DNA viral vectors in freshwater microalgal species. *Front. Plant Sci.* **2021**, *12*, 513. [CrossRef]
36. Feng, S.; Xue, L.; Liu, H.; Lu, P. Improvement of efficiency of genetic transformation for *Dunaliella salina* by glass beads method. *Mol. Biol. Rep.* **2009**, *36*, 1433–1439. [CrossRef] [PubMed]
37. Cui, Y.; Wang, J.; Jiang, P.; Bian, S.; Qin, S. Transformation of Platymonas (Tetraselmis) subcordiformis (Prasinophyceae, Chlorophyta) by agitation with glass beads. *World J. Microbiol. Biotechnol.* **2010**, *26*, 1653–1657. [CrossRef]
38. Wang, L.; Yang, L.; Wen, X.; Chen, Z.; Liang, Q.; Li, J.; Wang, W. Rapid and high efficiency transformation of *Chlamydomonas reinhardtii* by square-wave electroporation. *Biosci. Rep.* **2019**, *39*, BSR20181210. [CrossRef] [PubMed]
39. Im, D.J.; Jeong, S.N.; Yoo, B.S.; Kim, B.; Kim, D.P.; Jeong, W.J.; Kang, I.S. Digital microfluidic approach for efficient electroporation with high productivity: Transgene expression of microalgae without cell wall removal. *Anal. Chem.* **2015**, *87*, 6592–6599. [CrossRef] [PubMed]
40. Chen, Y.; Hu, H. High efficiency transformation by electroporation of the freshwater alga *Nannochloropsis limnetica*. *World J. Microbiol. Biotechnol.* **2019**, *35*, 1–10. [CrossRef]
41. Geng, D.; Wang, Y.; Wang, P.; Li, W.; Sun, Y. Stable expression of hepatitis B surface antigen gene in *Dunaliella salina* (Chlorophyta). *J. Appl. Phycol.* **2003**, *15*, 451–456. [CrossRef]
42. Guo, S.L.; Zhao, X.Q.; Tang, Y.; Wan, C.; Alam, M.A.; Ho, S.H.; Bai, F.W.; Chang, J.S. Establishment of an efficient genetic transformation system in *Scenedesmus obliquus*. *J. Biotechnol.* **2013**, *163*, 61–68. [CrossRef]
43. Jaeger, D.; Hübner, W.; Huser, T.; Mussgnug, J.H.; Kruse, O. Nuclear transformation and functional gene expression in the oleaginous microalga *Monoraphidium neglectum*. *J. Biotechnol.* **2017**, *249*, 10–15. [CrossRef]
44. Run, C.; Fang, L.; Fan, J.; Fan, C.; Luo, Y.; Hu, Z.; Li, Y. Stable nuclear transformation of the industrial alga *Chlorella pyrenoidosa*. *Algal Res.* **2016**, *17*, 196–201. [CrossRef]

45. Chen, H.L.; Li, S.S.; Huang, R.; Tsai, H.J. Conditional production of a functional fish growth hormone in the transgenic line of *nannochloropsis oculata* (Eustigmatophyceae) 1. *J. Phycol.* **2008**, *44*, 768–776. [CrossRef]
46. Anila, N.; Chandrashekar, A.; Ravishankar, G.A.; Sarada, R. Establishment of *Agrobacterium tumefaciens*-mediated genetic transformation in *Dunaliella bardawil*. *Eur. J. Phycol.* **2011**, *46*, 36–44. [CrossRef]
47. Srinivasan, R.; Gothandam, K.M. Synergistic action of D-Glucose and acetosyringone on *Agrobacterium* strains for efficient *Dunaliella* transformation. *PLoS ONE* **2016**, *11*, e0158322. [CrossRef] [PubMed]
48. Schiedlmeier, B.; Schmitt, R.; Müller, W.; Kirk, M.M.; Gruber, H.; Mages, W.; Kirk, D.L. Nuclear transformation of *Volvox carteri*. *Proc. Natl. Acad. Sci. USA* **1994**, *91*, 5080–5084. [CrossRef] [PubMed]
49. Boynton, J.E.; Gillham, N.W.; Harris, E.H.; Hosler, J.P.; Johnson, A.M.; Jones, A.R.; Randolph-Anderson, B.L.; Robertson, D.; Klein, T.M.; Shark, K.B.; et al. Chloroplast transformation in *Chlamydomonas* with high velocity microprojectiles. *Science* **1988**, *240*, 1534–1538. [CrossRef]
50. Mini, P.; Demurtas, O.C.; Valentini, S.; Pallara, P.; Aprea, G.; Ferrante, P.; Giuliano, G. *Agrobacterium*-mediated and electroporation mediated transformation of *Chlamydomonas reinhardtii*: A comparative study. *BMC Biotechnol.* **2018**, *18*, 11. [CrossRef]
51. Gregory, J.A.; Topol, A.B.; Doerner, D.Z.; Mayfield, S. Alga-produced cholera toxin-Pfs25 fusion proteins as oral vaccines. *Appl. Environ. Microbiol.* **2013**, *79*, 3917–3925. [CrossRef]
52. Hempel, F.; Maier, U.G. An engineered diatom acting like a plasma cell secreting human IgG antibodies with high efficiency. *Microb. Cell Fact.* **2012**, *11*, 126. [CrossRef]
53. Dunahay, T.G.; Jarvis, E.E.; Roessler, P.G. Genetic transformation of the diatoms *Cyclotella cryptica* and *Navicula saprophila*. *J. Phycol.* **1995**, *31*, 1004–1012. [CrossRef]
54. Fischer, H.; Robl, I.; Sumper, M.; Kröger, N. Targeting and covalent modification of cell wall and membrane proteins heterologously expressed in the diatom *Cylindrotheca fusiformis* (Bacillariophyceae). *J. Phycol.* **1999**, *35*, 113–120. [CrossRef]
55. Dunahay, T.G. Transformation of *Chlamydomonas reinhardtii* with silicon carbide whiskers. *Biotechniques* **1993**, *15*, 452–455.
56. Wang, K.; Drayton, P.; Frame, B.; Dunwell, J.; Thompson, J. Whisker mediated plant transformation an alternative technology. *In Vitro Cell Dev. Biol.* **1995**, *31*, 101–104. [CrossRef]
57. Chai, X.J.; Chen, H.X.; Xu, W.Q.; Xu, Y.W. Expression of soybean Kunitz trypsin inhibitor gene SKTI in *Dunaliella salina*. *J. Appl. Phycol.* **2013**, *25*, 139–144. [CrossRef]
58. Pratheesh, P.T.; Vineetha, M.; Kurup, G.M. An efficient protocol for the *Agrobacterium*-mediated genetic transformation of microalga *Chlamydomonas reinhardtii*. *Mol. Biotechnol.* **2014**, *56*, 507–515. [CrossRef] [PubMed]
59. Steinbrenner, J.; Sandmann, G. Transformation of the green alga *Haematococcus pluvialis* with a phytoene desaturase for accelerated astaxanthin biosynthesis. *Appl. Environ. Microbiol.* **2006**, *72*, 7477–7484. [CrossRef] [PubMed]
60. Cha, T.S.; Yee, W.; Aziz, A. Assessment of factors affecting *Agrobacterium*-mediated genetic transformation of the unicellular green alga, *Chlorella vulgaris*. *World J. Microbiol. Biotechnol.* **2012**, *28*, 1771–1779. [CrossRef]
61. Rathod, J.P.; Prakash, G.; Pandit, R.; Lali, A.M. *Agrobacterium*-mediated transformation of promising oil-bearing marine algae *Parachlorella kessleri*. *Photosynth. Res.* **2013**, *118*, 141–146. [CrossRef]
62. Simon, D.P.; Anila, N.; Gayathri, K.; Sarada, R. Heterologous expression of β-carotene hydroxylase in *Dunaliella salina* by *Agrobacterium*-mediated genetic transformation. *Algal Res.* **2016**, *18*, 257–265. [CrossRef]
63. Khatiwada, B.; Kautto, L.; Sunna, A.; Sun, A.; Nevalainen, H. Nuclear transformation of the versatile microalga *Euglena gracilis*. *Algal Res.* **2019**, *37*, 178–185. [CrossRef]
64. Dautor, Y.; Úbeda-Mínguez, P.; Chileh, T.; García-Maroto, F.; Alonso, D.L. Development of genetic transformation methodologies for an industrially-promising microalga: *Scenedesmus almeriensis*. *Biotechnol. Lett.* **2014**, *36*, 2551–2558. [CrossRef]
65. Bashir, K.M.I.; Kim, M.-S.; Stahl, U.; Cho, M.-G. *Agrobacterium* mediated genetic transformation of *Dictyosphaerium pulchellum* for the expression of erythropoietin. *J. Appl. Phycol.* **2018**, *30*, 3503–3518. [CrossRef]
66. Rathod, J.P.; Gade, R.M.; Rathod, D.R.; Dudhare, M. A review on molecular tools of microalgal genetic transformation and their application for overexpression of different genes. *Int. J. Curr. Microbiol. Appl. Sci.* **2017**, *6*, 3191–3207. [CrossRef]
67. Brown, L.E.; Sprecher, S.L.; Keller, L.R. Introduction of exogenous DNA into *Chlamydomonas reinhardtii* by electroporation. *Mol. Cell Biol.* **1991**, *11*, 2328–2332. [CrossRef] [PubMed]
68. Weaver, J.C. Plant Cell Electroporation and Electrofusion Protocols. In *Methods in Molecular Biology*; Nickoloff, J.A., Ed.; Springer: Totowa, NJ, USA, 1995; Volume 55. [CrossRef]
69. Niu, Y.F.; Zhang, M.H.; Xie, W.H.; Li, J.N.; Gao, Y.F.; Yang, W.D.; Liu, J.S.; Li, H.Y. A new inducible expression system in a transformed green alga, *Chlorella vulgaris*. *Genet. Mol. Res.* **2011**, *10*, 3427–3434. [CrossRef] [PubMed]
70. Liu, J.; Sun, Z.; Gerken, H.; Huang, J.; Jiang, Y.; Chen, F. Genetic engineering of the green alga *Chlorella zofingiensis*: A modified norflurazon-resistant phytoene desaturase gene as a dominant selectable marker. *Appl. Microbiol. Biotechnol.* **2014**, *98*, 5069–5079. [CrossRef] [PubMed]
71. Liu, L.; Wang, Y.; Zhang, Y.; Chen, X.; Zhang, P.; Ma, S. Development of a new method for genetic transformation of the green alga *Chlorella ellipsoidea*. *Mol. Biotechnol.* **2013**, *54*, 211–219. [CrossRef] [PubMed]
72. Yamano, T.; Fukuzawa, H. Transformation of the model microalga *Chlamydomonas reinhardtii* without cell-wall removal. In *Electroporation Protocols*; Li, S., Cutrera, J., Heller, R., Teissie, J., Eds.; Humana: New York, NY, USA, 2020. [CrossRef]
73. Day, A.; Debuchy, R.; van Dillewijn, J.; Purton, S.; Rochaix, J.-D. Studies on the maintenance and expression of cloned DNA fragments in the nuclear genome of the green alga *Chlamydomonas reinhardtii*. *Physiol. Plant.* **1990**, *78*, 254–260. [CrossRef]

74. Remacle, C.; Cardol, P.; Coosemans, N.; Gaisne, M.; Bonnefoy, N. High-efficiency biolistic transformation of *Chlamydomonas* mitochondria can be used to insert mutations in complex I genes. *Proc. Natl. Acad. Sci. USA* **2006**, *103*, 4771–4776. [CrossRef]
75. Jones, C.S.; Luong, T.; Hannon, M.; Tran, M.; Gregory, J.A.; Shen, Z.; Briggs, S.P.; Mayfield, S.P. Heterologous expression of the C terminal antigenic domain of the malaria vaccine candidate Pfs48/45 in the green algae *Chlamydmonas reinhardtii. Appl. Microbiol. Biotechnol.* **2013**, *97*, 1987–1995. [CrossRef]
76. Dreesen, I.A.; Charpin-El Hamri, G.; Fussenegger, M. Heat-stable oral alga-based vaccine protects mice from *Staphylococcus aureus* infection. *J. Biotechnol.* **2010**, *145*, 273–280. [CrossRef]
77. Tan, C.; Qin, S.; Zhang, Q.; Jiang, P.; Zhao, F. Establishment of a micro-particle bombardment transformation system for *Dunaliella salina*. *J. Microbiol.* **2005**, *43*, 361–365.
78. Bayne, A.C.V.; Boltz, D.; Owen, C.; Betz, Y.; Maia, G.; Azadi, P.; Archer-Hartmann, S.; Zirkle, R.; Lippmeier, J.C. Vaccination against influenza with recombinant hemagglutinin expressed by Schizochytrium sp. confers protective immunity. *PLoS ONE* **2013**, *8*, e61790. [CrossRef] [PubMed]
79. Bañuelos-Hernández, B.; Monreal-Escalante, E.; González-Ortega, O.; Angulo, C.; Rosales-Mendoza, S. Algevir: An expression system for microalgae based on viral vectors. *Front. Microbiol.* **2017**, *8*, 1100. [CrossRef] [PubMed]
80. Baek, K.; Kim, D.H.; Jeong, J.; Sim, S.J.; Melis, A.; Kim, J.S.; Jin, E.; Bae, S. DNA-free two-gene knockout in *Chlamydomonas reinhardtii* via CRISPR-Cas9 ribonucleoproteins. *Sci. Rep.* **2016**, *6*, 30620. [CrossRef] [PubMed]
81. Shin, S.E.; Lim, J.M.; Koh, H.G.; Kim, E.K.; Kang, N.K.; Jeon, S.; Kwon, S.; Shin, W.S.; Lee, B.; Hwangbo, K.; et al. CRISPR/Cas9-induced knockout and knock-inmutations in *Chlamydomonas reinhardtii*. *Sci. Rep.* **2016**, *6*, 27810. [CrossRef]
82. Serif, M.; Dubois, G.; Finoux, A.-L.; Teste, M.-A.; Jallet, D.; Daboussi, F. One-step generation of multiple gene knock-outs in the diatom *Phaeodactylum tricornutum* by DNA-free genome editing. *Nat. Commun.* **2018**, *9*, 3924. [CrossRef] [PubMed]
83. Chang, K.S.; Kim, J.; Park, H.; Hong, S.-J.; Lee, C.-G.; Jin, E. Enhanced lipid productivity in AGP knockout marine microalga Tetraselmis sp. using a DNA-free CRISPR-Cas9 RNP method. *Bioresour. Technol.* **2020**, *303*, 122932. [CrossRef]
84. Hawkins, R.L.; Nakamura, M. Expression of human growth hormone by the eukaryotic alga, *Chlorella*. *Curr. Microbiol.* **1999**, *38*, 335–341. [CrossRef]
85. Endo, H.; Yoshida, M.; Uji, T.; Saga, N.; Inoue, K.; Nagasawa, H. Stable nuclear transformation system for the coccolithophorid alga *Pleurochrysis carterae*. *Sci. Rep.* **2016**, *6*, 1–10. [CrossRef]
86. Karas, B.J.; Diner, R.E.; Lefebvre, S.C.; McQuaid, J.; Phillips, A.P.; Noddings, C.M.; Brunson, J.K.; Valas, R.E.; Deerinck, T.J.; Jablanovic, J.; et al. Designer diatom episomes delivered by bacterial conjugation. *Nat. Commun.* **2015**, *6*, 6925. [CrossRef]
87. Muñoz, C.F.; Sturme, M.H.J.; D'Adamo, S.; Weusthuis, R.A.; Wijffels, R.H. Stable transformation of the green algae Acutodesmus obliquus and *Neochloris oleoabundans* based on *E coli* conjugation. *Algal Res.* **2019**, *39*, 101453. [CrossRef]
88. Zhou, X.; Zhang, X.; Boualavong, J.; Durney, A.R.; Wang, T.; Kirschner, S.; Wentz, M.; Mukaibo, H. Electrokinetically controlled fluid injection into unicellular microalgae. *Electrophoresis* **2017**, *38*, 2587–2591. [CrossRef] [PubMed]
89. Zhou, H.C.; Long, J.R.; Yaghi, O.M. Introduction to metal-organic frameworks. *Chem. Rev.* **2012**, *112*, 673–674. [CrossRef]
90. Silva, S.; Almeida, A.J.; Vale, N. Combination of cell-penetrating peptides with nanoparticles for therapeutic application: A review. *Biomolecules* **2019**, *9*, 22. [CrossRef] [PubMed]
91. Lanigan, T.M.; Kopera, H.C.; Saunders, T.L. Principles of Genetic Engineering. *Genes* **2020**, *11*, 291. [CrossRef] [PubMed]
92. Radakovits, R.; Jinkerson, R.E.; Fuerstenberg, S.I.; Tae, H.; Settlage, R.E.; Boore, J.L.; Posewitz, M.C. Draft genome sequence and genetic transformation of the oleaginous alga *Nannochloropis gaditana*. *Nat. Commun.* **2012**, *3*, 686. [CrossRef]
93. León-Bañares, R.; González-Ballester, D.; Galván, A.; Fernández, E. Transgenic microalgae as green cell-factories. *Trends Biotechnol.* **2004**, *22*, 45–52. [CrossRef] [PubMed]
94. Potvin, G.; Zhang, Z. Strategies for high-level recombinant protein expression in transgenic microalgae: A review. *Biotechnol. Adv.* **2010**, *28*, 910–918. [CrossRef] [PubMed]
95. Lichtenthaler, H.K.; Schwender, J.; Disch, A.; Rohmer, M. Biosynthesis of isoprenoids in higher plant chloroplasts proceeds via a mevalonate-independent pathway. *FEBS Lett.* **1997**, *400*, 271–274. [CrossRef]
96. Bowsher, C.G.; Tobin, A.K. Compartmentation of metabolism within mitochondria and plastids. *J. Exp. Bot.* **2001**, *52*, 513–527. [CrossRef]
97. Gallaher, S.D.; Fitz-Gibbon, S.T.; Strenkert, D.; Purvine, S.O.; Pellegrini, M.; Merchant, S.S. High-throughput sequencing of the chloroplast and mitochondrion of *Chlamydomonas reinhardtii* to generate improved de novo assemblies, analyze expression patterns and transcript speciation, and evaluate diversity among laboratory strains and wild isolates. *Plant. J.* **2018**, *93*, 545–565. [CrossRef]
98. Kwon, Y.M.; Kim, K.W.; Choi, T.-Y.; Kim, S.Y.; Kim, J.Y.H. Manipulation of the microalgal chloroplast by genetic engineering for biotechnological utilization as a green biofactory. *World J. Microbiol. Biotechnol.* **2018**, *34*, 183. [CrossRef] [PubMed]
99. Sun, M.; Qian, K.; Su, N.; Chang, H.; Liu, J.; Shen, G. Foot-and-mouth disease virus VP1 protein fused with cholera toxin B subunit expressed in *Chlamydmonas reinhardtii* chloroplast. *Biotechnol. Lett.* **2003**, *25*, 1087–1092. [CrossRef] [PubMed]
100. Bertalan, I.; Munder, M.C.; Weiß, C.; Kopf, J.; Fischer, D.; Johanningmeier, U. A rapid, modular and marker-free chloroplast expression system for the green alga *Chlamydomonas reinhardtii*. *J. Biotechnol.* **2015**, *195*, 60–66. [CrossRef] [PubMed]
101. Xie, W.-H.; Zhu, C.-C.; Zhang, N.-S.; Li, D.-W.; Yang, W.-D.; Liu, J.-S.; Sathishkumar, R.; Li, H.Y. Construction of novel chloroplast expression vector and development of an efficient transformation system for the diatom *Phaeodactylum tricornutum*. *Mar. Biotechnol.* **2014**, *16*, 538–546. [CrossRef] [PubMed]

102. Fiehn, O. Metabolomics—The link between genotypes and phenotypes. *Plant Mol. Biol.* **2002**, *48*, 155–171. [CrossRef] [PubMed]
103. Senousy, H.H.; Abd Ellatif, S.; Ali, S. Assessment of the antioxidant and anticancer potential of different isolated strains of cyanobacteria and microalgae from soil and agriculture drain water. *Environ. Sci. Pollut. Res. Int.* **2020**, *27*, 18463–18474. [CrossRef] [PubMed]
104. Shanab, S.M.; Mostafa, S.S.; Shalaby, E.A.; Mahmoud, G.I. Aqueous extracts of microalgae exhibit antioxidant and anticancer activities. *Asian Pac. J. Trop. Biomed.* **2012**, *2*, 608–615. [CrossRef]
105. Suh, S.S.; Kim, S.M.; Kim, J.E.; Hong, J.M.; Lee, S.G.; Youn, U.J.; Han, S.J.; Kim, I.C.; Kim, S. Anticancer activities of ethanol extract from the Antarctic freshwater microalga, *Botryidiopsidaceae* sp. *BMC Complement. Altern. Med.* **2017**, *17*, 509. [CrossRef]
106. Zaharieva, M.M.; Zheleva-Dimitrova, D.; Rusinova-Videva, S.; Ilieva, Y.; Brachkova, A.; Balabanova, V.; Gevrenova, R.; Kim, T.C.; Kaleva, M.; Georgieva, A.; et al. Antimicrobial and antioxidant potential of *Scenedesmus obliquus* microalgae in the context of integral biorefinery concept. *Molecules* **2022**, *27*, 519. [CrossRef]
107. Hwang, J.; Yadav, D.; Lee, P.C.; Jin, J.O. Immunomodulatory effects of polysaccharides from marine algae for treating cancer, infectious disease, and inflammation. *Phytother. Res.* **2022**, *36*, 761–777. [CrossRef]
108. Mehariya, S.; Goswami, R.K.; Karthikeysan, O.P.; Verma, P. Microalgae for high-value products: A way towards green nutraceutical and pharmaceutical compounds. *Chemosphere* **2021**, *280*, 130553. [CrossRef] [PubMed]
109. Sarkar, S.; Manna, M.S.; Bhowmick, T.K.; Gayen, K. Priority-based multiple products from microalgae: Review on techniques and strategies. *Crit. Rev. Biotechnol.* **2020**, *40*, 590–607. [CrossRef] [PubMed]
110. Castro, J.S.; Calijuri, M.L.; Ferreira, J.; Assemany, P.P.; Ribeiro, V.J. Microalgae based biofertilizer: A life cycle approach. *Sci. Total Environ.* **2020**, *724*, 138138. [CrossRef]
111. Lakatos, G.E.; Ranglová, K.; Manoel, J.C.; Grivalský, T.; Kopecký, J.; Masojídek, J. Bioethanol production from microalgae polysaccharides. *Folia Microbiol.* **2019**, *64*, 627–644. [CrossRef] [PubMed]
112. Plöhn, M.; Spain, O.; Sirin, S.; Silva, M.; Escudero-Oñate, C.; Ferrando-Climent, L.; Allahverdiyeva, Y.; Funk, C. Wastewater treatment by microalgae. *Physiol. Plant.* **2021**, *173*, 568–578. [CrossRef] [PubMed]
113. Hom-Diaz, A.; Llorca, M.; Rodríguez-Mozaz, S.; Vicent, T.; Barceló, D.; Blánquez, P. Microalgae cultivation on wastewater digestate: β-estradiol and 17α-ethynylestradiol degradation and transformation products identification. *J. Environ. Manag.* **2015**, *15*, 106–113. [CrossRef] [PubMed]
114. Heimann, K.; Huerlimann, R. Microalgal classification: Major classes and genera of commercial microalgal species. In *Handbook of Marine Microalgae*; Kim, S.-K., Ed.; Academic Press: Busan, South Korea, 2015; pp. 25–41.
115. Márquez-Escobar, V.A.; Bañuelos-Hernández, B.; Rosales-Mendoza, S. Expression of a Zika virus antigen in microalgae: Towards mucosal vaccine development. *J. Biotechnol.* **2018**, *282*, 86–91. [CrossRef]
116. Castellanos-Huerta, I.; Bañuelos-Hernández, B.; Téllez, G.; Rosales-Mendoza, S.; Brieba, L.G.; Esquivel-Ramos, E.; Beltrán-López, J.I.; Velazquez, G.; Fernandez-Siurob, I. Recombinant Hemagglutinin of Avian Influenza Virus H5 expressed in the chloroplast of *Chlamydomonas reinhardtii* and evaluation of its immunogenicity in chickens. *Avian. Dis.* **2016**, *60*, 784–791. [CrossRef]
117. Demurtas, O.C.; Massa, S.; Ferrante, P.; Venuti, A.; Franconi, R.; Giuliano, G. A *Chlamydomonas*-derived Human Papillomavirus 16 E7 vaccine induces specific tumor protection. *PLoS ONE* **2013**, *8*, e61473. [CrossRef]
118. Barahimipour, R.; Neupert, J.; Bock, R. Efficient expression of nuclear transgenes in the green alga *Chlamydomonas*: Synthesis of an HIV antigen and development of a new selectable marker. *Plant Mol. Biol.* **2016**, *90*, 403–418. [CrossRef]
119. Davis, A.; Crum, L.T.; Corbeil, L.B.; Hildebrand, M. Expression of Histophilus somni IbpA DR2 protective antigen in the diatom *Thalassiosira pseudonana*. *Appl. Microbiol. Biotechnol.* **2017**, *101*, 5313–5324. [CrossRef] [PubMed]
120. Dauvillée, D.; Delhaye, S.; Gruyer, S.; Slomianny, C.; Moretz, S.E.; d'Hulst, C.; Long, C.A.; Ball, S.G.; Tomavo, S. Engineering the chloroplast targeted malarial vaccine antigens in *Chlamydomonas* starch granules. *PLoS ONE* **2010**, *5*, e15424. [CrossRef] [PubMed]
121. Shamriz, S.; Ofoghi, H. Engineering the chloroplast of *Chlamydomonas reinhardtii* to express the recombinant PfCelTOS-Il2 antigen-adjuvant fusion protein. *J. Biotechnol.* **2018**, *266*, 111–117. [CrossRef] [PubMed]
122. Hempel, F.; Maurer, M.; Brockmann, B.; Mayer, C.; Biedenkopf, N.; Kelterbaum, A.; Becker, S.; Maier, U.G. From hybridomas to a robust microalgal-based production platform: Molecular design of a diatom secreting monoclonal antibodies directed against the Marburg virus nucleoprotein. *Microb. Cell Fact.* **2017**, *16*, 131. [CrossRef] [PubMed]
123. Kim, D.H.; Kim, Y.T.; Cho, J.J.; Bae, J.H.; Hur, S.B.; Hwang, I.; Choi, T.J. Stable integration and functional expression of flounder growth hormone gene in transformed microalga, *Chlorella ellipsoidea*. *Mar. Biotechnol.* **2002**, *4*, 63–73. [CrossRef]
124. Tran, M.; Van, C.; Barrera, D.J.; Pettersson, P.L.; Peinado, C.D.; Bui, J.; Mayfield, S.P. Production of unique immunotoxin cancer therapeutics in algal chloroplasts. *Proc. Natl. Acad. Sci. USA* **2013**, *110*, E15–E22. [CrossRef]
125. Wang, X.; Brandsma, M.; Tremblay, R.; Maxwell, D.; Jevnikar, A.M.; Huner, N.; Ma, S. A novel expression platform for the production of diabetes-associated autoantigen human glutamic acid decarboxylase (hGAD65). *BMC Biotechnol.* **2008**, *8*, 87. [CrossRef]
126. Kwon, K.C.; Lamb, A.; Fox, D.; Jegathese, S.J.P. An evaluation of microalgae as a recombinant protein oral delivery platform for fish using green fluorescent protein (GFP). *Fish Shellfish Immunol.* **2019**, *87*, 414–420. [CrossRef]
127. Castellanos-Huerta, I.; Gómez-Verduzco, G.; Tellez-Isaias, G.; Ayora-Talavera, G.; Bañuelos-Hernández, B.; Petrone-García, V.M.; Velázquez-Juárez, G.; Fernández-Siurob, I. Transformation of *Dunaliella salina* by *Agrobacterium tumefaciens* for the expression of the Hemagglutinin of Avian Influenza Virus H5. *Microorganisms* **2022**, *10*, 361. [CrossRef]

128. Feng, S.; Feng, W.; Zhao, L.; Gu, H.; Li, Q.; Shi, K.; Guo, S.; Zhang, N. Preparation of transgenic *Dunaliella salina* for immunization against white spot syndrome virus in crayfish. *Arch. Virol.* **2014**, *159*, 519–525. [CrossRef]
129. Yang, Z.; Chen, F.; Li, D.; Zhang, Z.; Liu, Y.; Zheng, D.; Wang, Y.; Shen, G. Expression of human soluble TRAIL in *Chlamydomonas reinhardtii* chloroplast. *Chin. Sci. Bull.* **2006**, *51*, 1703–1709. [CrossRef]
130. Soria-Guerra, R.E.; Ramírez-Alonso, J.I.; Ibáñez-Salazar, A.; Govea-Alonso, D.O.; Paz-Maldonado, L.M.; Bañuelos-Hernández, B.; Korban, S.S.; Rosales-Mendoza, S. Expression of an HBcAg-based antigen carrying angiotensin II in *Chlamydomonas reinhardtii* as a candidate hypertension vaccine. *Plant Cell Tissue Organ Cult.* **2014**, *116*, 133–139. [CrossRef]
131. El-Ayouty, Y.; El-Manawy, I.; Nasih, S.; Hamdy, E.; Kebeish, R. Engineering Chlamydomonas reinhardtii for expression of functionally active human interferon-α? *Mol. Biotechnol.* **2019**, *61*, 134–144. [CrossRef] [PubMed]
132. Kiataramgul, A.; Maneenin, S.; Purton, S.; Areechon, N.; Hirono, I.; Brocklehurst, T.W.; Unajak, S. An oral delivery system for controlling white spot syndrome virus infection in shrimp using transgenic microalgae. *Aquaculture* **2020**, *521*, 735022. [CrossRef]
133. Charoonnart, P.; Worakajit, N.; Zedler, J.; Meetam, M.; Robinson, C.; Saksmerprome, V. Generation of microalga *Chlamydomonas reinhardtii* expressing shrimp antiviral dsRNA without supplementation of antibiotics. *Sci. Rep.* **2019**, *9*, 3164. [CrossRef] [PubMed]
134. Gregory, J.A.; Li, F.; Tomosada, L.M.; Cox, C.J.; Topol, A.B.; Vinetz, J.M.; Mayfield, S.P. Algae-produced Pfs25 elicits antibodies that inhibit malaria transmission. *PLoS ONE* **2012**, *7*, e37179. [CrossRef]
135. Jarquín-Cordero, M.; Chávez, M.N.; Centeno-Cerdas, C.; Bohne, A.V.; Hopfner, U.; Machens, H.G.; Egaña, J.T.; Nickelsen, J. Towards a biotechnological platform for the production of human pro-angiogenic growth factors in the green alga *Chlamydomonas reinhardtii*. *Appl. Microbiol. Biotechnol.* **2020**, *104*, 725–739. [CrossRef]
136. Baier, T.; Kros, D.; Feiner, R.C.; Lauersen, K.J.; Mu?ller, K.M.; Kruse, O. Engineered fusion proteins for efficient protein secretion and purification of a human growth factor from the green microalga *Chlamydomonas reinhardtii*. *ACS Synth. Biol.* **2018**, *7*, 2547–2557. [CrossRef]
137. Stoffels, L.; Taunt, H.N.; Charalambous, B.; Purton, S. Synthesis of bacteriophage lytic proteins against *Streptococcus pneumoniae* in the chloroplast of *Chlamydomonas reinhardtii*. *Plant Biotechnol. J.* **2017**, *15*, 1130–1140. [CrossRef]
138. Li, A.; Huang, R.; Wang, C.; Hu, Q.; Li, H.; Li, X. Expression of anti-lipopolysaccharide factor isoform 3 in *Chlamydomonas reinhardtii* showing high antimicrobial activity. *Mar. Drugs* **2021**, *19*, 239. [CrossRef]
139. Lima, S.; Webb, C.L.; Deery, E.; Robinson, C.; Zedler, J.A. Human intrinsic factor expression for bioavailable vitamin B12 enrichment in microalgae. *Biology* **2018**, *7*, 19. [CrossRef] [PubMed]
140. Hempel, F.; Lau, J.; Klingl, A.; Maier, U.G. Algae as protein factories: Expression of a human antibody and the respective antigen in the diatom *Phaeodactylum tricornutum*. *PLoS ONE* **2011**, *6*, e28424. [CrossRef] [PubMed]
141. Mayfield, S.P.; Franklin, S.E. Expression of human antibodies in eukaryotic micro-algae. *Vaccine* **2005**, *23*, 1828–1832. [CrossRef] [PubMed]
142. Mayfield, S.P.; Franklin, S.E.; Lerner, R.A. Expression and assembly of a fully active antibody in algae. *Proc. Natl. Acad. Sci. USA* **2003**, *100*, 438–442. [CrossRef]
143. Manuell, A.L.; Beligni, M.V.; Elder, J.H.; Siefker, D.T.; Tran, M.; Weber, A.; McDonald, T.L.; Mayfield, S.P. Robust expression of a bioactive mammalian protein in *Chlamydomonas* chloroplast. *Plant Biotechnol. J.* **2007**, *5*, 402–412. [CrossRef]
144. Su, Z.L.; Qian, K.X.; Tan, C.P.; Meng, C.X.; Qin, S. Recombination and heterologous expression of allophycocyanin gene in the chloroplast of *Chlamydomonas reinhardtii*. *Acta Biochim. Biophy. Sin.* **2005**, *37*, 709–712. [CrossRef]
145. Zhang, Y.K.; Shen, G.F.; Ru, B.G. Survival of human metallothionein–2 transplastomic *Chlamydomonas reinhardtii* to ultraviolet B exposure. *Acta Biochim. Biophy. Sin.* **2006**, *38*, 187–193. [CrossRef]
146. Beltrán-López, J.I.; Romero-Maldonado, A.; Monreal-Escalante, E.; Bañuelos-Hernández, B.; Paz-Maldonado, L.M.; Rosales-Mendoza, S. *Chlamydomonas reinhardtii* chloroplasts express an orally immunogenic protein targeting the p210 epitope implicated in atherosclerosis immunotherapies. *Plant Cell Rep.* **2016**, *35*, 1133–1141. [CrossRef]
147. Gregory, J.A.; Shepley-McTaggart, A.; Umpierrez, M.; Hurlburt, B.K.; Maleki, S.J.; Sampson, H.A.; Mayfield, S.P.; Berin, M.C. Immunotherapy using algal-produced Ara h 1 core domain suppresses peanut allergy in mice. *Plant Biotechnol. J.* **2016**, *14*, 1541–1550. [CrossRef]
148. Hirschl, S.; Ralser, C.; Asam, C.; Gangitano, A.; Huber, S.; Ebner, C.; Bohle, B.; Wolf, M.; Briza, P.; Ferreira, F.; et al. Expression and characterization of functional recombinant Bet v 1.0101 in the chloroplast of *Chlamydomonas reinhardtii*. *Int. Arch. Allergy Immunol.* **2017**, *173*, 44–50. [CrossRef]
149. Rasala, B.A.; Muto, M.; Lee, P.A.; Jager, M.; Cardoso, R.M.; Behnke, C.A.; Kirk, P.; Hokanson, C.A.; Crea, R.; Mendez, M.; et al. Production of therapeutic proteins in algae, analysis of expression of seven human proteins in the chloroplast of *Chlamydomonas reinhardtii*. *Plant Biotechnol. J.* **2010**, *8*, 719–733. [CrossRef]
150. Faè, M.; Accossato, S.; Cella, R.; Fontana, F.; Goldschmidt-Clermont, M.; Leelavathi, S.; Reddy, V.S.; Longoni, P. Comparison of transplastomic *Chlamydomonas reinhardtii* and *Nicotiana tabacum* expression system for the production of a bacterial endoglucanase. *Appl. Microbiol. Biotechnol.* **2017**, *101*, 4085–4092. [CrossRef]
151. Kumar, A.; Falcao, V.R.; Sayre, R.T. Evaluating nuclear transgene expression systems in *Chlamydomonas reinhardtii*. *Algal Res.* **2013**, *2*, 321–332. [CrossRef]

152. Dong, B.; Cheng, R.Q.; Liu, Q.Y.; Wang, J.; Fan, Z.C. Multimer of the antimicrobial peptide Mytichitin-A expressed in *Chlamydomonas reinhardtii* exerts a broader antibacterial spectrum and increased potency. *J. Biosci. Bioeng.* **2018**, *125*, 175–179. [CrossRef] [PubMed]
153. Xue, B.; Dong, C.M.; Hu, H.H.; Dong, B.; Fan, Z.C. *Chlamydomonas reinhardtii*-expressed multimer of ToAMP4 inhibits the growth of bacteria of both Gram-positive and Gram-negative. *Process Biochem.* **2020**, *91*, 311–318. [CrossRef]
154. Pang, X.; Tong, Y.; Xue, W.; Yang, Y.F.; Chen, X.; Liu, J.; Chen, D. Expression and characterization of recombinant human lactoferrin in edible alga *Chlamydomonas reinhardtii. Biosci. Biotechnol. Biochem.* **2019**, *83*, 851–859. [CrossRef] [PubMed]
155. Hernández-Ramírez, J.; Wong-Arce, A.; González-Ortega, O.; Rosales-Mendoza, S. Expression in algae of a chimeric protein carrying several epitopes from tumor associated antigens. *Int. J. Biol. Macromol.* **2020**, *147*, 46–52. [CrossRef] [PubMed]
156. Ortega-Berlanga, B.; Bañuelos-Hernández, B.; Rosales-Mendoza, S. Efficient expression of an Alzheimer's disease vaccine candidate in the microalga Schizochytrium sp. using the Algevir system. *Mol. Biotechnol.* **2018**, *60*, 362–368. [CrossRef]
157. He, Y.; Peng, H.; Liu, J.; Chen, F.; Zhou, Y.; Ma, X.; Chen, H.; Wang, K. Chlorella sp. transgenic with Scy-hepc enhancing the survival of *Sparus macrocephalus* and hybrid grouper challenged with *Aeromonas hydrophila*. *Fish Shellfish Immunol.* **2018**, *73*, 22–29. [CrossRef]
158. Georgianna, D.R.; Hannon, M.J.; Marcuschi, M.; Wu, S.; Botsch, K.; Lewis, A.J.; Hyun, J.; Mendez, M.; Mayfield, S.P. Production of recombinant enzymes in the marine alga *Dunaliella tertiolecta*. *Algal Res.* **2013**, *2*, 2–9. [CrossRef]
159. Wang, K.; Cui, Y.; Wang, Y.; Gao, Z.; Liu, T.; Meng, C.; Qin, S. Chloroplast genetic engineering of a unicellular green alga *Haematococcus pluvialis* with expression of an antimicrobial peptide. *Mar. Biotechnol.* **2020**, *22*, 572–580. [CrossRef] [PubMed]
160. Mardanova, E.S.; Blokhina, E.A.; Tsybalova, L.M.; Peyret, H.; Lomonossoff, G.P.; Ravin, N.V. Efficient transient expression of recombinant proteins in plants by the novel pEff vector based on the genome of Potato Virus X. *Front. Plant Sci.* **2017**, *28*, 247. [CrossRef] [PubMed]
161. Vidyasagar, P.; Sridevi, V.N.; Rajan, S.; Praveen, A.; Srikanth, A.; Abhinay, G.; Siva Kumar, V.; Verma, R.R.; Rajendra, L. Generation and characterization of neutralizing monoclonal antibodies against baculo-expressed HPV 16 VLPs. *Eur. J. Microbiol. Immunol.* **2014**, *4*, 56–64. [CrossRef]
162. Van Doremalen, N.; Lambe, T.; Spencer, A.; Belij-Rammerstorfer, S.; Purushotham, J.N.; Port, J.R.; Avanzato, V.A.; Bushmaker, T.; Flaxman, A.; Ulaszewska, M.; et al. ChAdOx1 nCoV-19 vaccine prevents SARS-CoV-2 pneumonia in rhesus macaques. *Nature* **2020**, *586*, 578–582. [CrossRef]
163. D'Adamo, S.; Kormelink, R.; Martens, D.; Barbosa, M.J.; Wijffels, R.H. Prospects for viruses infecting eukaryotic microalgae in biotechnology. *Biotechnol. Adv.* **2022**, *54*, 107790. [CrossRef]
164. Ryabov, E.V.; Robinson, D.J.; Taliansky, M.E. A plant virus-encoded protein facilitates long-distance movement of heterologous viral RNA. *Proc. Natl. Acad. Sci. USA* **1999**, *96*, 1212–1217. [CrossRef]
165. McCormick, A.A.; Reinl, S.J.; Cameron, T.I.; Vojdani, F.; Fronefield, M.; Levy, R.; Tuse, D. Individualized human scFv vaccines produced in plants: Humoral anti-idiotype responses in vaccinated mice confirm relevance to the tumor Ig. *J. Immunol. Methods* **2003**, *278*, 5–104. [CrossRef]
166. Kumagai, M.H.; Donson, J.; della-Cioppa, G.; Grill, L.K. Rapid, high-level expression of glycosylated rice alpha-amylase in transfected plants by an RNA viral vector. *Gene* **2000**, *245*, 169–174. [CrossRef]
167. Gleba, Y.; Klimyuk, V.; Marillonnet, S. Magnifection-a new platform for expressing recombinant vaccines in plants. *Vaccine* **2005**, *23*, 2042–2048. [CrossRef]
168. Salazar-González, J.A.; Rosales-Mendoza, S.; Bañuelos-Hernández, B. Viral vector-based expression strategies. In *Genetically Engineered Plants as a Source of Vaccines against Wide Spread Diseases*; Rosales-Mendoza, S., Ed.; Springer: New York, NY, USA, 2014; pp. 43–60. [CrossRef]
169. Naseri, Z.; Ghaffar, K.; Seyed Javad, D.; Hamideh, O. Virus-based vectors: A new approach for the production of recombinant proteins. *J. Appl. Biotechnol. Rep.* **2019**, *6*, 6–14. [CrossRef]
170. Hanley-Bowdoin, L.; Bejarano, E.R.; Robertson, D.; Mansoor, S. Geminiviruses: Masters at redirecting and reprogramming plant processes. *Nat. Rev. Microbiol.* **2013**, *11*, 777–788. [CrossRef]
171. Rojas, M.R.; Macedo, M.A.; Maliano, M.R.; Soto-Aguilar, M.; Souza, J.O.; Briddon, R.W.; Kenyon, L.; Rivera Bustamante, R.F.; Zerbini, F.M.; Adkins, S.; et al. World management of geminiviruses. *Annu. Rev. Phytopathol.* **2018**, *56*, 637–677. [CrossRef] [PubMed]
172. Huang, Z.; Phoolcharoen, W.; Lai, H.; Piensook, K.; Cardineau, G.; Zeitlin, L.; Whaley, K.J.; Arntzen, C.J.; Mason, H.S.; Chen, Q. High-level rapid production of full-size monoclonal antibodies in plants by a single-vector DNA replicon system. *Biotechnol. Bioeng.* **2010**, *106*, 9–17. [CrossRef] [PubMed]
173. Phakham, T.; Bulaon, C.J.I.; Khorattanakulchai, N.; Shanmugaraj, B.; Buranapraditkun, S.; Boonkrai, C.; Sooksai, S.; Hirankarn, N.; Abe, Y.; Strasser, R.; et al. Functional characterization of pembrolizumab produced in *Nicotiana benthamiana* using a rapid transient expression system. *Front. Plant Sci.* **2021**, *12*, 736299. [CrossRef] [PubMed]
174. Moon, K.B.; Jeon, J.H.; Choi, H.; Park, J.S.; Park, S.J.; Lee, H.J.; Park, J.M.; Cho, H.S.; Moon, J.S.; Oh, H.; et al. Construction of SARS-CoV-2 virus-like particles in plant. *Sci. Rep.* **2022**, *12*, 1005. [CrossRef] [PubMed]
175. Berndt, A.J.; Smalley, T.N.; Ren, B.; Simkovsky, R.; Badary, A.; Sproles, A.E.; Fields, F.J.; Torres-Tiji, Y.; Heredia, V.; Mayfield, S.P. Recombinant production of a functional SARS-CoV-2 spike receptor binding domain in the green algae *Chlamydomonas reinhardtii*. *PLoS ONE* **2021**, *16*, e0257089. [CrossRef]

176. Gimpel, J.A.; Hyun, J.S.; Schoepp, N.G.; Mayfield, S.P. Production of recombinant proteins in microalgae at pilot greenhouse scale. *Biotechnol. Bioeng.* **2015**, *112*, 339–345. [CrossRef]
177. Zhao, Z.; Anselmo, A.C.; Mitragotri, S. Viral vector-based gene therapies in the clinic. *Bioeng. Transl. Med.* **2021**, *7*, e10258. [CrossRef]
178. Short, S.M.; Staniewski, M.A.; Chaban, Y.V.; Long, A.M.; Wang, D. Diversity of viruses infecting eukaryotic algae. *Curr. Issues Mol. Biol.* **2020**, *39*, 29–62. [CrossRef]
179. Coy, S.R.; Gann, E.R.; Pound, H.L.; Short, S.M.; Wilhelm, S.W. Viruses of eukaryotic algae: Diversity, methods for detection, and future directions. *Viruses* **2018**, *10*, 487. [CrossRef]
180. Sandaa, R.A.; Saltvedt, M.R.; Dahle, H.; Wang, H.; Våge, S.; Blanc-Mathieu, R.; Steen, I.H.; Grimsley, N.; Edvardsen, B.; Ogata, H.; et al. Adaptive evolution of viruses infecting marine microalgae (haptophytes), from acute infections to stable coexistence. *Biol. Rev. Camb. Philos. Soc.* **2022**, *97*, 179–194. [CrossRef] [PubMed]
181. Kadono, T.; Miyagawa-Yamaguchi, A.; Kira, N.; Tomaru, Y.; Okami, T.; Yoshimatsu, T.; Hou, L.; Ohama, T.; Fukunaga, K.; Okauchi, M.; et al. Characterization of marine diatom-infecting virus promoters in the model diatom *Phaeodactylum tricornutum*. *Sci. Rep.* **2015**, *5*, 18708. [CrossRef] [PubMed]
182. Blanc, G.; Duncan, G.; Agarkova, I.; Borodovsky, M.; Gurnon, J.; Kuo, A.; Lindquist, E.; Lucas, S.; Pangilinan, J.; Polle, J.; et al. The Chlorella variabilis NC64A genome reveals adaptation to photosymbiosis, coevolution with viruses, and cryptic sex. *Plant Cell.* **2010**, *22*, 2943–2955. [CrossRef] [PubMed]
183. Jeanniard, A.; Dunigan, D.D.; Gurnon, J.R.; Agarkova, I.V.; Kang, M.; Vitek, J.; Duncan, G.; McClung, O.W.; Larsen, M.; Claverie, J.M.; et al. Towards defining the chloroviruses: A genomic journey through a genus of large DNA viruses. *BMC Genom.* **2013**, *14*, 158. [CrossRef] [PubMed]
184. Cheng, Y.S.; Zheng, Y.; Labavitch, J.M.; VanderGheynst, J.S. Virus infection of Chlorella variabilis and enzymatic saccharification of algal biomass for bioethanol production. *Bioresour. Technol.* **2013**, *137*, 326–331. [CrossRef]
185. Agarkova, I.V.; Lane, L.C.; Dunigan, D.D.; Quispe, C.F.; Duncan, G.A.; Milrot, E.; Minsky, A.; Esmael, A.; Ghosh, J.S.; Van Etten, J.L. Identification of a Chlorovirus PBCV-1 protein involved in degrading the host cell wall during virus infection. *Viruses* **2021**, *13*, 782. [CrossRef]
186. Quispe, C.F.; Esmael, A.; Sonderman, O.; McQuinn, M.; Agarkova, I.; Battah, M.; Duncan, G.A.; Dunigan, D.D.; Smith, T.; De Castro, C.; et al. Characterization of a new chlorovirus type with permissive and non-permissive features on phylogenetically related algal strains. *Virology* **2017**, *500*, 103–113. [CrossRef]
187. Fitzgerald, L.A.; Graves, M.V.; Li, X.; Hartigan, J.; Pfitzner, A.J.; Hoffart, E.; Van Etten, J.L. Sequence and annotation of the 288-kb ATCV-1 virus that infects an endosymbiotic chlorella strain of the heliozoon *Acanthocystis turfacea*. *Virology* **2007**, *362*, 350–361. [CrossRef]
188. Rahalkar, H.; Sheppard, A.; Lopez-Morales, C.A.; Lobo, L.; Salek, S. Challenges faced by the biopharmaceutical industry in the development and marketing authorization of biosimilar medicines in BRICS-TM countries: An exploratory study. *Pharmaceut. Med.* **2021**, *35*, 235–251. [CrossRef]

 MDPI

Article

Variations in the Composition, Antioxidant and Antimicrobial Activities of *Cystoseira compressa* during Seasonal Growth

Martina Čagalj [1], Danijela Skroza [2], María del Carmen Razola-Díaz [3], Vito Verardo [3], Daniela Bassi [4], Roberta Frleta [5], Ivana Generalić Mekinić [2], Giulia Tabanelli [6] and Vida Šimat [1,*]

[1] University Department of Marine Studies, University of Split, R. Boškovića 37, HR-21000 Split, Croatia; martina.cagalj@unist.hr
[2] Department of Food Technology and Biotechnology, Faculty of Chemistry and Technology, University of Split, R. Boškovića 35, HR-21000 Split, Croatia; danci@ktf-split.hr (D.S.); gene@ktf-split.hr (I.G.M.)
[3] Department of Nutrition and Food Science, Campus of Cartuja, University of Granada, 18071 Granada, Spain; carmenrazola@ugr.es (M.d.C.R.-D.); vitoverardo@ugr.es (V.V.)
[4] Department for Sustainable Food Process (DISTAS), Università Cattolica del Sacro Cuore, 26100 Cremona, Italy; daniela.bassi@unicatt.it
[5] Center of Excellence for Science and Technology-Integration of Mediterranean Region (STIM), Faculty of Science, University of Split, HR-21000 Split, Croatia; roberta@stim.unist.hr
[6] Department of Agricultural and Food Sciences, University of Bologna, 40127 Bologna, Italy; giulia.tabanelli2@unibo.it
* Correspondence: vida@unist.hr; Tel.: +385-21-510-192

Citation: Čagalj, M.; Skroza, D.; Razola-Díaz, M.d.C.; Verardo, V.; Bassi, D.; Frleta, R.; Generalić Mekinić, I.; Tabanelli, G.; Šimat, V. Variations in the Composition, Antioxidant and Antimicrobial Activities of *Cystoseira compressa* during Seasonal Growth. *Mar. Drugs* **2022**, *20*, 64. https://doi.org/10.3390/md20010064

Academic Editors: Marco García-Vaquero and Brijesh K. Tiwari

Received: 17 December 2021
Accepted: 7 January 2022
Published: 11 January 2022

Abstract: The underexplored biodiversity of seaweeds has recently drawn great attention from researchers to find the bioactive compounds that might contribute to the growth of the blue economy. In this study, we aimed to explore the effect of seasonal growth (from May to September) on the in vitro antioxidant (FRAP, DPPH, and ORAC) and antimicrobial effects (MIC and MBC) of *Cystoseira compressa* collected in the Central Adriatic Sea. Algal compounds were analyzed by UPLC-PDA-ESI-QTOF, and TPC and TTC were determined. Fatty acids, among which oleic acid, palmitoleic acid, and palmitic acid were the dominant compounds in samples. The highest TPC, TTC and FRAP were obtained for June extract, 83.4 ± 4.0 mg GAE/g, 8.8 ± 0.8 mg CE/g and 2.7 ± 0.1 mM TE, respectively. The highest ORAC value of 72.1 ± 1.2 μM TE was obtained for the August samples, and all samples showed extremely high free radical scavenging activity and DPPH inhibition (>80%). The MIC and MBC results showed the best antibacterial activity for the June, July and August samples, when sea temperature was the highest, against *Listeria monocytogenes*, *Staphylococcus aureus*, and *Salmonella enteritidis*. The results show *C. compressa* as a potential species for the industrial production of nutraceuticals or functional food ingredients.

Keywords: *Cystoseira compressa*; microwave-assisted extraction; green extraction; biological activity; seaweed; seasonal variations; nutraceuticals; fatty acids

1. Introduction

Among seaweeds, the brown macroalgae (Phaeophyceae) have been identified as an outstanding source of phenolic compounds, from simple phenolic acids to more complex polymers such as tannins (mainly phlorotannins). Algal phlorotannins, a group of phenolic compounds restricted to the polymers of phloroglucinol, present a heterogeneous and high molecular weight group of compounds (from 126 Da to 650 kDa) which are verified in terrestrial plants [1,2]. The phlorotannins play an important role in the cellular and ecological growth and tissue healing of alga but also show strong antioxidant, antimicrobial, cytotoxic, and antitumor properties [3–6].

Brown fucoid algae of the genus *Cystoseira sensu lato* (Sargassaceae) consist of 40 species of large marine canopy-forming macroalgae found along the Atlantic–Mediterranean coasts [7,8]. So far, a total of 214 compounds have been isolated from sixteen *Cystoseira* species, and the chemical constituents of *Cystoseira* spp. were found to contain fatty acids and derivatives, terpenoids, steroids, carbohydrates, phlorotannins, phenolic compounds, pigments and vitamins [7]. The chemical composition of the alga depends on numerous ecological factors such as temperature, salinity, UV irradiation, collecting season, depth, geographic location, thallus development, etc. However, their individual and synergistic effect on the brown alga chemical profile and biological activity is still relatively unknown. Recent studies showed that seasonality and thallus vegetative parts significantly affect the nutritional and chemical profile of alga [9]. It is considered that higher nutritional and phenolic content, higher polyunsaturated fatty acid (PUFA) content, higher vitamin and mineral content, as well as the antiproliferative properties of brown algae from brown fucoid algae were obtained during hot and dry summer seasons and higher sea temperatures [10–13]. On the other hand, no seasonal effect was recorded for the pigment profile and fucoxanthin content, nor total phenolic content and antimicrobial activity of the genera *Padina*, *Colpomenia*, *Saccharina* or *Dictyota* [14,15]. So far, there are no reports on seasonal variations in chemical profile nor the biological activity of *Cystoseira* spp.

Cystoseira spp. composition suggests their high nutritional value with potential applications in the nutraceutical industry. A range of 29–46% of PUFA, a low *n-6* PUFA/*n-3* PUFA ratio as well as favorable unsaturation, atherogenicity, and thrombogenicity indices were observed in several *Cystoseira* species [16]. Compounds from *Cystoseira* species are important sources of nutraceuticals and may be considered as functional foods, such as extracts of *C. tamariscifolia* and *C. nodicaulis* that were able to protect a human dopaminergic cell line from hydrogen peroxide-induced cytotoxicity and inhibit cholinesterases, while those from *C. crinita* showed significant cytotoxic activity against human breast adenocarcinoma (MCF-7 cells), inducing apoptosis and autophagy [17,18]. Besides non-volatiles, the essential oil constituents of *C. compressa* and their seasonal changes have been identified and among them for a large number of compounds a broad range of biological activities have been already proved [19]. So far, over 50 biological properties have been attributed to compounds found in genus *Cystoseira*, and the most reported are antioxidant, anti-inflammatory, cytotoxic, anticancer, cholinesterase inhibition, antidiabetic, and antiherpetic activities [7,20–24]. Phlorotannins are regarded as responsible for high antioxidant activity (e.g., free radical scavenging ability) [1,25–27]. Besides, there is little information on the antimicrobial activity of *Cystoseira* spp. extracts against major foodborne Gram-positive and Gram-negative bacteria [28].

The aim of this study was to investigate the chemical composition of *C. compressa*, one of the most widely distributed algae in the Adriatic Sea, to determine changes in its antioxidant and antimicrobial activity over the seasonal growth (May–September) when the algae are in the growing and reproductive phases, and the development of dense thallus occurs.

2. Results and Discussion

2.1. Total Phenolic Content, Total Tannin Content and Antioxidant Activity

Seaweed extracts were screened for total phenolic content (TPC), total tannin content (TTC) and antioxidant activity measured by ferric reducing/antioxidant power (FRAP), 2,2-diphenyl-1-picrylhydrazyl radical scavenging ability (DPPH) and the oxygen radical absorbance capacity (ORAC).

The results of TPC and TTC for *C. compressa* are shown in Figure 1. The results for TPC varied from 48.2 ± 0.5 to 83.4 ± 4.0 mg GAE/g. The highest TPC was found in June samples. On the other hand, the TTC values ranged from 2.0 ± 0.3 to 8.8 ± 0.8 mg CE/g with the highest value found also in June, followed by the extract from May. The FRAP values, shown in Figure 2, ranged from 1.0 ± 0.0 to 2.7 ± 0.1 mM TE. Similar to TPC and TTC, the highest FRAP result was obtained for June, showing the reducing activity of >2.5 mM TE. TPC and FRAP results were in high correlation (0.956; $p < 0.01$). The ORAC results are shown in Figure 2. The seaweed extracts were 200-fold diluted for ORAC assay. Among the investigated samples, the highest ORAC value of 72.1 ± 1.2 μM TE was found in the August extract, with extracts from May having the second best. June and July extracts had the lowest ORAC values, more than 3-fold lower in comparison to the August extract. The DPPH radical inhibitions (in percentages) are shown in Figure 2. The extract from May had the highest inhibition (90.2%) while the August extract had the lowest inhibition (77.3%). The activity of other extracts was similar, around 85%. In the growing season, the sea temperature was the lowest in May (18.3 °C) and it rose every month till August when it peaked at 26.9 °C. Finally, a decrease in the temperature by 2.2 °C was observed in September (Table 1).

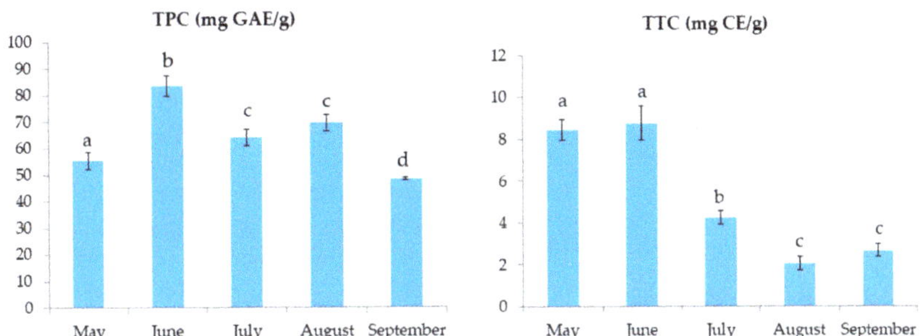

Figure 1. Total phenolic content (TPC) and total tannin content (TTC) of *C. compressa* extracts from May to September. $^{a–d}$ different letters denote statistically significant difference (n = 4).

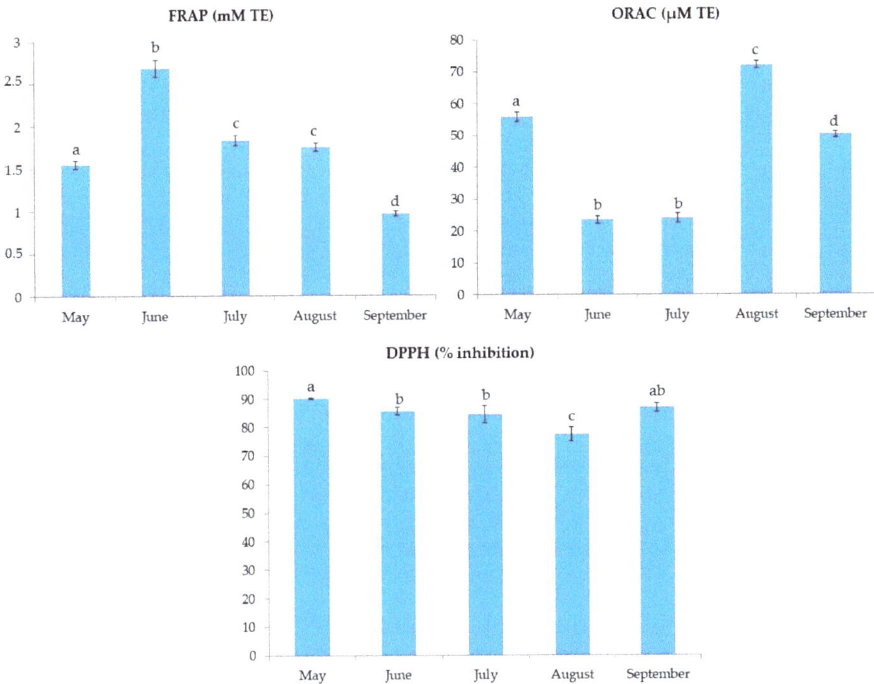

Figure 2. FRAP, ORAC and DPPH inhibition results for *C. compressa* extracts from May to September. $^{a–d}$ different letters denote statistically significant difference (n = 4).

Table 1. Sea temperature and salinity recorded during the harvest of the algal samples.

	May	June	July	August	September
Temperature (°C)	18.3	21.8	22.4	26.9	24.7
Salinity (PSU)	37.4	38.1	38.3	38.3	38.3

The TPC of alga varies with seasonal changes of sea temperature, salinity, light intensity, geographical location and depth, as well as other biological factors such as age, size, the life cycle of the seaweed, presence of herbivores [1]. In this study, the geographical location and depth were eliminated as a factor as samples were collected from the same area and depth each month. The TPC, TTC and antioxidant activity results showed no correlation to the sea temperature and salinity. If the growth of alga is considered, in June when the TPC, TTC and FRAP were the highest, *C. compressa* had a fully developed, densely ramified thalli with aerocysts. In May the thalli are not yet fully developed, while in July–September it is less dense, aerocysts appear in fewer numbers [29].

The TPC of *Cystoseira* species was previously investigated and researchers reported a strong effect of harvesting location and seasonal changes, especially temperature. Mancuso et al. [12] investigated TPC in *C. compressa* from eight locations along the Italian coast and confirmed the change of TPC with geographical location. The TPC ranged between different locations from 0.1 to 0.5% of algal dry weight (DW). The authors observed the increase in TPC with the rise of sea temperature (measured at different locations). Accordingly, the highest TPC of 0.53% DW was recorded at 28 °C. In contrast, Mannino et al. [30] investigated the effect of sea temperature seasonal variation on the TPC of *C. amentacea*. They harvested algae once in every season (winter, spring, summer and autumn) and measured the sea temperature. The authors observed the highest TPC in winter (0.8% DW) when the sea temperature was the lowest. In summer and autumn, when the sea temperatures were above 20 °C, the TPCs were the lowest, 0.4 and 0.37% DW, respectively. In their study, the TPC values showed a negative correlation with sea temperature. *Cystoseira compressa* extracts, from Urla (Turkey) [31], were screened for TPC, total flavonoid content (TFC), antioxidant and antimicrobial activity. The highest TPC of 1.5 mg GAE/g and TFC of 0.8 mg QE/g were found for hexane extract while the antioxidant activity of the hexane extract measured by DPPH radical inhibition was only 21.2%, more than four-fold lower than results in our study for hydroalcoholic extracts. In comparison, the methanolic extracts (similar polarity like ethanol) showed the TPC and TFC of 0.2 mg GAE/g and 0.3 mg QE/g, respectively and two-fold lower DPPH inhibition.

Abu-Khudir et al. [18] evaluated the antioxidant, antimicrobial, and anticancer activities of cold methanolic extract, hot methanolic extract, cold aqueous extract, and hot aqueous extract from *C. crinita* and *Sargassum linearifolium*. The highest TPC was found for the cold methanolic extract of *C. crinita*, 15.0 ± 0.58 mg GAE/g of dried extract, which is more than two-fold lower than the amount detected in the September extract from our study which contained the lowest TPC. The authors also found a high content of fatty acids (44%) and their esters in *C. crinita* cold methanolic extract. Both seaweeds showed similar DPPH and ABTS radical scavenging activity with *C. crinita* cold methanolic extract having IC_{50} of 125.6 µg/mL and 254.8 µg/mL, respectively. De La Fuente et al. [32] extracted *C. amentacea* var. *stricta* with dimethyl sulfoxide (DMSO) and 50% ethanol for determining TPC, TFC and antioxidant activity of extracts by DPPH radical scavenging, FRAP, OH scavenging, and nitric oxide (NO) scavenging methods. The TPC and TFC of DMSO extracts were 65.9 µg GAE/mg and 15.8 µg QE/mg, 3.2- and 5.1-fold higher than ethanolic extracts. Similar to our results, both investigated extracts had DPPH radical scavenging activity higher than 90%. Furthermore, the DMSO extracts showed a reducing activity of almost 90% while ethanolic extract showed a higher OH radical scavenging activity. Both extracts showed very low cytotoxicity, enabling their possible use as nutraceuticals.

Oucif et al. [20] screened six seaweed species (including *C. compressa* and *C. stricta*) for TPC, DPPH radical scavenging activity and reducing power. The highest TPCs were found for *C. compressa* methanolic and ethanolic extracts, 10.24 ± 0.09 and 15.70 ± 0.72 mg GAE/g DW, respectively. *Cystoseira compressa* ethanol extract had over 90% inhibition activity for DPPH radical and the highest reducing power, which can be compared with our results. Mhadhebi et al. [24] determined TPC, DPPH and FRAP in *C. crinita, C. sedoides* and *C. compressa* extracts. Among the three alga, *C. compressa* extract had the highest TPC of 61.0 mg GAE/g, which is comparable to our results, the lowest DPPH IC_{50} of 12.0 µg/mL, and the highest FRAP value, 2.6 mg GAE/g.

2.2. Antimicrobial Activity

The results of minimal inhibitory concentration (MIC) and minimal bactericidal concentration (MBC) of the *C. compressa* extracts against common foodborne pathogens are shown in Table 2. Gram-positive bacteria were more susceptible to seaweed extracts than Gram-negative bacteria. The lowest MIC results were found against *L. monocytogenes*, for June, July, and August samples with the lowest MBC in June, and against *S. aureus* in July and August with the same MBC. There was no difference in MIC and MBC values for *E. coli* among the investigated months. June, July, and August extracts had the lowest MIC values for *S. enteritidis*. The results showed higher antimicrobial activity from June to August when the sea temperature was the highest, against all bacteria.

Table 2. Results of the minimal inhibitory concentration (MIC, mg/mL) and minimal bactericidal concentration (MBC, mg/mL) of the seaweed extracts against foodborne pathogens (n = 3).

	May		June		July		August		September	
	MIC	MBC	MIC	MBC	MIC	MBC	MIC	MBC	MIC	MBC
Escherichia coli	10	10	10	10	10	10	10	10	10	>10
Salmonella enteritidis	10	10	5	10	5	10	5	10	10	>10
Enterococcus faecalis	10	10	10	10	10	10	5	10	10	10
Listeria monocytogenes	10	>10	2.5	2.5	2.5	5	2.5	5	5	>10
Staphylococcus aureus	10	>10	5	5	2.5	2.5	2.5	2.5	10	10
Bacillus cereus	>10	n.d.	10	>10	10	>10	10	>10	>10	n.d.

n.d.—not determined

Alghazeer et al. [33] performed microwave-assisted extraction (MAE) on *P. pavonica* and *C. compressa*. Flavonoid-rich extracts (110.92 ± 11.38 mg rutin equivalents /g for *C. compressa*) were tested for antibacterial activity against multidrug-resistant (MDR) isolates of *S. aureus* subsp. *aureus, Bacillus pumilus, B. cereus, Salmonella enterica* subsp. *enteric*, and enterohemorrhagic *E. coli* using the well diffusion method, MIC and MBC. *Cystoseira compressa* extract showed stronger antibacterial activity than *P. pavonica* with inhibition zones against 14 tested isolates. The largest inhibition zones were 20.5 mm for *S. aureus* and *B. cereus*, 31 mm for *S. enterica* and 17 mm for *E. coli*. Furthermore, *C. compressa* extract had the lowest MIC (31.25 µg/mL) and MBC (62.5 µg/mL) values against *S. aureus* and *S. enterica*. Against *B. cereus*, it had an MIC value of 62.5 µg/mL and an MBC value of 125 µg/mL. The highest MIC (125 µg/mL) and MBC (500 µg/mL) values were found against *E. coli*. Maggio et al. [28] evaluated the antibacterial activity of eight brown seaweeds, six belonging to the genus *Cystoseira* (including *C. compressa*) and two belonging to the Dictyotaceae family, against *E. coli, Kocuria rhizophila, S. aureus* and a toxigenic and MDR *S. aureus* using the disk diffusion method. None of the seaweed extracts inhibited the growth of *E. coli*. *Cystoseira compressa* and *Carpodesmia amentacea* extracts showed antibacterial activity against *K. rhizophila, S. aureus* and MDR *S. aureus*. Abdeldjebbar et al. [34] tested the antibacterial effect of *C. compressa* and *P. pavonica* acetonic extracts against *E. coli* and *S. aureus*. The antibacterial activity was measured by disk diffusion method and MIC determination. *Cystoseira compressa* extract had 14 mm inhibition diameter for *E. coli*, showing better antibacterial activity than *P. pavonica* (12 mm). However, MIC values were not detected for *C. compressa* against both bacteria. *Padina pavonica* extract had an MIC of 50 µL for tested strains. Both extracts had a 10 mm inhibition diameter for *S. aureus*. The authors also tested

Table 3. *Cont.*

N°	RT (min)	Observed (*m/z*)	Theorical (*m/z*)	Error (ppm)	Score (%)	Molecular Formula	In Source Fragments	Tentative Compound	May (%)	June (%)	July (%)	August (%)	September (%)
37	17.71	301.2158	301.2168	−3.3	99.56	$C_{20}H_{30}O_2$	283.2283; 275.1972	Eicosapentanoic acid isomer a (C20:5n-3)	0.65	0.65	1.09	0.81	0.75
38	17.73	301.2156	301.2168	−4.0	98.12	$C_{20}H_{30}O_2$	283.2287; 275.1957	Eicosapentanoic acid isomer b (C20:5n-3)	0.62	0.63	1.06	0.80	0.74
39	17.77	227.2001	227.2011	−4.4	93.6	$C_{14}H_{28}O_2$	-	Tetradecanoic acid (C14:0)	5.04	5.19	5.34	5.17	5.22
40	17.81	271.2266	271.2273	−2.6	97.75	$C_{16}H_{32}O_3$	253.0954; 225.2211	Hydroxy-palmitic acid	0.47	0.41	0.45	0.56	0.66
41	17.85	253.2156	253.2168	−4.7	96.47	$C_{16}H_{30}O_2$	-	Palmitoleic acid isomer a (C16:1n-7)	12.65	11.93	11.45	11.57	11.74
42	17.94	241.2170	241.2168	0.8	100	$C_{15}H_{30}O_2$	223.2081	Pentadecanoic acid (C15:0)	3.68	3.89	3.87	3.72	3.67
43	17.97	279.2314	279.2324	−3.6	98.25	$C_{18}H_{32}O_2$	267.2340; 275.2037	Octadeca-10,12-dienoic acid (C18:2n-6)	1.07	1.14	1.29	1.31	1.27
44	18.01	267.2318	267.2324	−2.2	99.96	$C_{17}H_{32}O_2$	249.0437; 223.0291	9-Heptadecenoic acid (C17:1n-8)	3.73	3.99	3.67	3.78	3.60
45	18.08	255.2321	255.2324	−1.2	99.9	$C_{16}H_{32}O_2$	227.2015	Hexadecanoic acid (palmitic acid)(C16:0)	10.46	10.46	10.24	9.92	9.83
46	18.12	281.2486	281.2481	1.8	96.88	$C_{18}H_{34}O_2$	-	Oleic acid (C18:1n-9)	15.87	15.06	15.39	15.33	15.12
47	18.22	269.2476	269.2481	−5.6	99.96	$C_{17}H_{34}O_2$	255.2325	Heptadecanoic acid (C17:0)	5.30	5.23	5.18	5.08	4.95
48	18.33	283.2618	283.2637	−1.9	99.21	$C_{18}H_{36}O_2$	-	Octadecanoic acid (stearic acid) C18:0	5.44	4.92	5.01	5.11	5.24
49	18.54	311.2944	311.2950	−2.0	90.87	$C_{20}H_{40}O_2$	255.2307; 225.0060	Arachidic acid	0.68	0.62	0.67	0.66	0.68

The most dominant compound tentatively identified was oleic acid (C18:1n-9) with a content more than 15% in all tested samples, highest in May. The other two dominant compounds were palmitoleic acid (C16:1n-7) and palmitic acid (C16:0) also showing highest content in the May extract. Except for highly represented fatty acids, ω-3 eicosapentaenoic acid (EPA) was also found, with the highest content in July.

Low molecular weight phenolic compounds were not identified. This does not confirm that the phenolics are not present, but the main phenolics in algae are probably present as tannins (phlorotannins) that cannot be determined by HPLC-ESI-TOF-MS because they cannot be ionized due to their high molecular weight. Maggio et al. [28] reported citric acid, isocitric acid, vanillic acid methyl ester, vanillic acid sulfate, gallic acid, dihydroxybenzoic acid, 2-hydroxy-6-oxo-6-(2-hydroxyphenoxy)-hexa-2,4-dienoate, phloracetophenone, bromo-phloroglucinol, vanillylmandelic acid and exifone in *C. compressa*. The compounds were identified without quantification. Previously, vanilic acid, hydroxybenzoic acid, gallocatechin, carnosic acid, phloroglucinol, and hydroxytyrosol 4-*O*-glucoside were identified as main phenolics in fucoidan algae *Sargassum* sp. [36].

Jerković et al. [37] investigated fucoidal brown alga *Fucus virsoides* and found 42.28% oleic acid, 15.00% arachidonic acid and 10.51% myristic acid in its fatty acid composition. The authors used high performance liquid chromatography–high-resolution mass spectrometry (HPLC-ESI-HRMS) to determine the composition of less polar non-volatile compounds. The major compounds tentatively identified belonged to five groups, steroids, terpenoids, fatty acid glycerides, carotenoids, and chlorophyll derivatives. Fatty acid glycerides were dominant, which is comparable to our study.

Ristivojević et al. [38] identified the bioactive compounds responsible for the radical scavenging and antimicrobial activities of *Undaria pinnatifida* and *Saccharina japonica* methanolic extracts using the high-performance thin layer chromatography (HPTLC)-bioautography assay and ultra-high-performance liquid chromatography (UHPLC)-LTQ-MS/MS combined. They reported eicosapentaenoic, stearidonic and arachidonic acids as major compounds accountable for these activities. Their findings are in accordance with previous reports on PUFAs having antimicrobial activity against bacteria, viruses and fungi [39,40].

PUFAs, such as EPA, docosahexaenoic acid (DHA) and linolenic acid (LNA), showed in vitro antibacterial activity against *Helicobacter pylori*, *S. aureus*, Methicillin-resistant *S. aureus* (MRSA), *Vibrio vulnificus*, and *Streptococcus mutans*, inhibiting bacterial growth or altering their cell morphology [40]. To deactivate microbial cells, PUFAs directly affected the cell membranes, enhanced free radical generation, and increased the formation of cytotoxic lipid peroxides and their bioactive metabolites increasing the leukocytes' and macrophages' phagocytic action [39]. EPA and DHA extracts showed antimicrobial activity against foodborne pathogenic bacteria, *L. monocytogenes*, *B. subtilis*, Enterobacter aerogenes, *E. coli*, *S. aureus*, *S. enteritidis*, *S. typhimurium*, and *P. aeruginosa* [41]. The authors reported the lowest MIC value of 250 μg/mL for DHA extract against *P. aeruginosa*. A low MIC value of 350 μg/mL was found for EPA extract against *L. monocytogenes*, *B. subtilis* and *P. aeruginosa*, and for DHA extract against *L. monocytogenes* and *B. subtilis*. Besides, Cvitković et al. [42] investigated the extraction of lipid fractions from *C. compressa*, *C. barbata*, *F. virsoides*, and *Codium bursa*. In agreement with our results, the dominant fatty acids in all seaweeds were palmitic, oleic and linolenic fatty acids. *Cystoseira compressa* and *C. barbata* had the highest amounts of omega-3 EPA and DHA. *Cystoseira compressa* had 20.35% oleic acid, 17.66% arachidonic acid, 14.86% linoleic acid, 11.92% palmitic acid and 8.72% linolenic acid. Bacteria *S. aureus* can be inhibited by most free fatty acids: Lacey and Lord [43] seeded this bacterium on human skin and then applied LNA to the skin which resulted in the rapid death of the seeded bacteria. EPA (C20:5n-3) was found to successfully inhibit the growth of *S. aureus* and *B. cereus* with a 64 mg/L MIC value [44]. Oleic acid was confirmed in vitro and in vivo to effectively eliminate MRSA by disrupting its cell wall [45].

3. Materials and Methods

3.1. Sample Collection

Cystoseira compressa samples were collected off the south coast of the island Čiovo in the Adriatic Sea from May to September 2020 (43.493389° N, 16.272505° E). Sampling was done throughout a lagoon at 25 points in a depth range of 20 to 80 cm. The sea temperature and salinity were measured during sampling using a YSI Pro2030 probe (Yellow Springs, OH, USA). A sample of this species is deposited in the herbarium at the University Department of Marine Studies in Split.

3.2. Pre-Treatment and Extraction

Prior to the extraction, harvested algal samples were washed with tap water to remove epiphytes. Samples were then freeze-dried (FreeZone 2.5, Labconco, Kansas City, MO, USA) and ground. Based on the previous research [46] seaweeds were extracted using MAE in the advanced microwave extraction system (ETHOS X, Milestone Srl, Sorisole, Italy). Seaweeds were mixed with 50% ethanol, using 1:10 (*w*:*v*) algae to solvent ratio and extracted for 15 min at 200 W and 60 °C. The extracts were further centrifuged at 5000 rpm for 8 min at room temperature and the supernatant was filtered. The ethanolic solvent was evaporated at 50 °C and the rest of the extracts freeze dried.

3.3. Determination of Total Phenolics, Total Tannins and Antioxidant Activity

The crude algal extracts were dissolved in 50% ethanol prior to analyses in the concentration of 20 mg/mL. Folin–Ciocalteu method [47] was used for determining the TPC. Briefly, 25 μL of the extract was mixed with 1.5 mL distilled water and 125 μL Folin–Ciocalteu reagent. The solution was stirred and 375 μL 20% sodium carbonate solution and 475 μL distilled water was added after one minute. Samples were left in the dark at room temperature for 2 h. The absorbance was read using a spectrophotometer (SPECORD 200 Plus, Edition 2010, Analytik Jena AG, Jena, Germany) at 765 nm. Results were expressed as gallic acid equivalents in mg/g of freeze-dried extract (mg GAE/g).

The TTC was measured according to Zhong et al. [36] with some modifications. Briefly, 25 μL of the sample, 150 μL 4% (*w*/*v*) ethanolic vanillin solution, and 25 μL 32% sulfuric acid (diluted with ethanol) were added to the 96-well plate and mixed. The plate was incubated for 15 min at room temperature and absorbance was read at 500 nm using the microplate reader (Synergy HTX Multi-Mode Reader, BioTek Instruments, Inc., Winooski, VT, USA). The TTC results were expressed as mg catechin equivalents per g of dried extract (mg CE/g).

The reducing activity was measured as FRAP (ferric reducing/antioxidant power) [48]. Briefly, 300 μL of FRAP reagent solution was pipetted into the microplate wells, and absorbance at 592 nm was recorded. Then, 10 µL of the sample was added to the FRAP reagent and the change in absorbance after 4 min was measured. The change in absorbance, calculated as the difference between the final value of the absorbance of the reaction mixture after a certain reaction time (4 min) and the absorbance of FRAP reagent before sample addition, was compared with the values obtained for the standard solutions of Trolox. Results were expressed as micromoles of Trolox equivalents per liter of extract (µM TE).

The 2,2-diphenyl-1-picrylhydrazyl (DPPH) radical scavenging ability of extracts was also measured in 96-well microplates [49]. DPPH radical solution with the initial absorbance of 1.2 (290 µL) was pipetted into microplate wells, and absorbance was measured at 517 nm. Then, 10 µL of the sample was added to the wells and the decrease in the absorbance was measured after 1 h using the plate reader. The antioxidant activity of extracts was expressed as DPPH radical inhibition percentages (% inhibition).

The oxygen radical absorbance capacity (ORAC) method [50,51] was performed to determine the antioxidant capacity of extracts by monitoring the inhibition of the action of free peroxyl radicals formed by the decomposition of 2,2-azobis (2-methylpropionamide)-dihydrochloride (AAPH) against the fluorescent compound fluorescein. Briefly, 150 µL of fluorescein and 25 µL of the sample in 1:200 dilution (or Trolox in the case of standard compound, or puffer in the case of blank) were pipetted into microplate wells and thermostated for 30 min at 37 °C. After 30 min, 25 µL of AAPH was added and measurements were performed at excitation and emission wavelengths of 485 and 520 nm every minute for 80 min. The results were expressed as µM of Trolox Equivalents (µM TE).

3.4. Determination of the Antimicrobial Activity

The foodborne pathogens *Escherichia coli* ATCC 25922, *Salmonella enteritidis* ATCC 13076, *Enterococcus faecalis* ATCC 29212, *Listeria monocytogenes* ATCC 7644, *Staphylococcus aureus* ATCC 25923, and *Bacillus cereus* ATCC 14579 were used in this study.

The microdilution method was used to determine the extracts' MICs against foodborne pathogens. The extracts were dissolved in 4% DMSO (10 mg/mL) and diluted with Mueller–Hinton broth (MHB). Then, 100 µL of the mixture was added to the first well of the 96-well microtiter plate. Two-fold dilutions were done in the next wells (10–0.16 mg/mL). The 50 µL of prepared inoculum (1×10^5 colony forming units (CFU)/mL determined by using the growth curves of bacteria in the *log* phase) was added to each well and plates were mixed on a microtiter plate shaker for 1 min at 600 rpm (Plate Shaker-Thermostat PST-60 HL, Biosan, Riga, Latvia). Positive control (50 µL of inoculum and 50 µL of broth media), negative control (50 µL of broth media and 50 µL of extract), blank (100 µL of broth media) and 4% DMSO were also tested. After 24 h of incubation, 20 µL of the indicator of bacterial metabolic activity, 2-*p*-iodophenyl-3-*p*-nitrophenyl-5-phenyl tetrazolium chloride (INT, in 2 mg/mL concentration) was added to each well. Plates were mixed on a plate shaker and incubated for 1 h in the dark. MIC values were read visually as the lowest concentration of the extract at which there was no detection of bacterial growth seen as the reduction of INT to red formazan [52].

MBC of the seaweed extracts was determined as the lowest concentration at which no microbial growth was detected on agar plates after subcultivation of bacterial suspension pipetted from wells where MIC was determined and from wells with higher extract concentrations [53].

3.5. Compound Analysis by UPLC-PDA-ESI-QTOF

Dried extract (3 mg) of algae was dissolved in 1 mL of MeOH/H_2O 1/1 *v/v*. The analysis of compounds from algae was carried out with the use of an ACQUITY Ultra Performance LC system equipped with a photodiode array detector with a binary solvent manager (Waters Corporation, Milford, MA, USA) series with a mass detector Q/TOF micro mass spectrometer (Waters) equipped with an electrospray ionization (ESI) source operating in negative mode at the following conditions: capillary voltage, 2300 kV; source temperature, 100 °C; cone gas flow, 40 L/h; desolvation temperature, 500 °C; desolvation gas flow, 11,000 L/h; and scan range, *m/z* 50–1500. Separation of individual compounds was carried out using an ACQUITY UPLC BEH Shield RP18 column (1.7 μm, 2.1 mm × 100 mm; Waters Corporation, Milford, MA, USA) at 40 °C. The elution gradient test was carried out using water containing 1% acetic acid (A) and acetonitrile (B), and applied as follows: 0 min, 1% B; 2.3 min, 1% B; 4.4 min, 7% B; 8.1 min, 14% B; 12.2 min, 24% B; 16 min, 40% B; 18.3 min, 100% B, 21 min, 100% B; 22.4 min, 1% B; 25 min, 1% B. The sample volume injected was 2 μL and the flow rate used was 0.6 mL/min. The compounds were monitored at 280 nm. Integration and data elaboration were performed using MassLynx 4.1 software (Waters Corporation, Milford, MA, USA) [54].

3.6. Statistical Analyses

The results of antioxidant analyses were expressed as mean ± standard deviation and antimicrobial results as a mean of 3 replicas. Analysis of variance (one-way ANOVA) was used to assess the difference between TPC, TTC and antioxidant assays, followed by a least significance difference test at 95% confidence level to evaluate differences between sets of mean values [55]. Pearson's correlation coefficient was used to determine the relation between the variables. Analyses were carried out using Statgraphics Centurion-Ver.16.1.11 (StatPoint Technologies, Inc., Warrenton, VA, USA).

4. Conclusions

The results obtained for the brown fucoidal macroalgae *C. compressa* from the Adriatic Sea indicated that it was a good source of compounds. The TPC and TTC content reflected a variation over the growing season, with the highest values in June. The detected FRAP showed high correlation with TPC and TTC content. The DPPH values were >80% inhibition over the whole sampling period, while the highest antioxidant activity with regards to ORAC was in August when the sea temperature was the highest. No evident correlation existed between the temperature and salinity change and TPC, TTC or antioxidant activity. From June to August, higher antimicrobial activity against foodborne pathogens was observed, especially against *L. monocytogenes*, *S. aureus* and *S. enteritidis*. Further investigations are needed to gain insight into the effect of abiotic factors, growth and thallus development of the alga on its biological potential and to discover the compounds responsible for the different biological activities.

Author Contributions: Conceptualization, V.Š. and G.T.; methodology, M.Č., D.S., V.V. and D.B.; validation, D.S. and D.B.; formal analysis M.Č., D.S., R.F., I.G.M., D.B. and M.d.C.R.-D.; resources, V.Š., D.B. and V.V.; data curation, M.Č., D.S., R.F., I.G.M., D.B. and M.d.C.R.-D.; writing—original draft preparation, M.Č. and V.Š.; writing—review and editing, V.Š., G.T., I.G.M. and V.V.; supervision, V.Š. and D.S.; project administration, V.Š. and G.T.; funding acquisition, V.Š., D.B. and V.V. All authors have read and agreed to the published version of the manuscript.

Funding: This research is supported by the PRIMA program under project BioProMedFood (Project ID 1467). The PRIMA programme is supported by the European Union.

Institutional Review Board Statement: Not applicable.

Informed Consent Statement: Not applicable.

Data Availability Statement: Data available on request.

Acknowledgments: The authors would like to thank Ivan Šimat from Centaurus d.o.o., Solin for administrative support and technical help during the investigations and Matilda Šprung for technical help during antimicrobial analyses.

Conflicts of Interest: The authors declare no conflict of interest.

References

1. Generalić Mekinić, I.; Skroza, D.; Šimat, V.; Hamed, I.; Čagalj, M.; Perković, Z.P. Phenolic Content of Brown Algae (Pheophyceae) Species: Extraction, Identification, and Quantification. *Biomolecules* **2019**, *9*, 244. [CrossRef]
2. Mannino, A.M.; Micheli, C. Ecological Function of Phenolic Compounds from Mediterranean Fucoid Algae and Seagrasses: An Overview on the Genus Cystoseira Sensu Lato and Posidonia Oceanica (L.) Delile. *J. Mar. Sci. Eng.* **2020**, *8*, 19. [CrossRef]
3. Stiger-Pouvreau, V.; Jégou, C.; Cérantola, S.; Guérard, F.; Le Lann, K. Phlorotannins in Sargassaceae Species from Brittany (France). In *Advances in Botanical Research*; Bourgougnon, N., Ed.; Academic Press: Cambridge, MA, USA; Elsevier: Amsterdam, The Netherlands, 2014; pp. 379–411. [CrossRef]
4. Messina, C.; Renda, G.; Laudicella, V.; Trepos, R.; Fauchon, M.; Hellio, C.; Santulli, A. From Ecology to Biotechnology, Study of the Defense Strategies of Algae and Halophytes (from Trapani Saltworks, NW Sicily) with a Focus on Antioxidants and Antimicrobial Properties. *Int. J. Mol. Sci.* **2019**, *20*, 881. [CrossRef]
5. Abdelhamid, A.; Jouini, M.; Bel Haj Amor, H.; Mzoughi, Z.; Dridi, M.; Ben Said, R.; Bouraoui, A. Phytochemical Analysis and Evaluation of the Antioxidant, Anti-Inflammatory, and Antinociceptive Potential of Phlorotannin-Rich Fractions from Three Mediterranean Brown Seaweeds. *Mar. Biotechnol.* **2018**, *20*, 60–74. [CrossRef] [PubMed]
6. Šimat, V.; Elabed, N.; Kulawik, P.; Ceylan, Z.; Jamroz, E.; Yazgan, H.; Čagalj, M.; Regenstein, J.M.; Özogul, F. Recent Advances in Marine-Based Nutraceuticals and Their Health Benefits. *Mar. Drugs* **2020**, *18*, 627. [CrossRef] [PubMed]
7. Bruno de Sousa, C.; Gangadhar, K.N.; Macridachis, J.; Pavão, M.; Morais, T.R.; Campino, L.; Varela, J.; Lago, J.H.G. Cystoseira Algae (Fucaceae): Update on Their Chemical Entities and Biological Activities. *Tetrahedron Asymmetry* **2017**, *28*, 1486–1505. [CrossRef]
8. Orellana, S.; Hernández, M.; Sansón, M. Diversity of Cystoseira Sensu Lato (Fucales, Phaeophyceae) in the Eastern Atlantic and Mediterranean Based on Morphological and DNA Evidence, Including Carpodesmia Gen. Emend. and Treptacantha Gen. Emend. *Eur. J. Phycol.* **2019**, *54*, 447–465. [CrossRef]
9. Kumar, S.; Sahoo, D.; Levine, I. Assessment of Nutritional Value in a Brown Seaweed Sargassum Wightii and Their Seasonal Variations. *Algal Res.* **2015**, *9*, 117–125. [CrossRef]
10. Praiboon, J.; Palakas, S.; Noiraksa, T.; Miyashita, K. Seasonal Variation in Nutritional Composition and Anti-Proliferative Activity of Brown Seaweed, Sargassum Oligocystum. *J. Appl. Phycol.* **2018**, *30*, 101–111. [CrossRef]
11. Gosch, B.J.; Paul, N.A.; de Nys, R.; Magnusson, M. Seasonal and Within-Plant Variation in Fatty Acid Content and Composition in the Brown Seaweed Spatoglossum Macrodontum (Dictyotales, Phaeophyceae). *J. Appl. Phycol.* **2015**, *27*, 387–398. [CrossRef]
12. Mancuso, F.P.; Messina, C.M.; Santulli, A.; Laudicella, V.A.; Giommi, C.; Sarà, G.; Airoldi, L. Influence of Ambient Temperature on the Photosynthetic Activity and Phenolic Content of the Intertidal Cystoseira Compressa along the Italian Coastline. *J. Appl. Phycol.* **2019**, *31*, 3069–3076. [CrossRef]
13. Britton, D.; Schmid, M.; Revill, A.T.; Virtue, P.; Nichols, P.D.; Hurd, C.L.; Mundy, C.N. Seasonal and Site-Specific Variation in the Nutritional Quality of Temperate Seaweed Assemblages: Implications for Grazing Invertebrates and the Commercial Exploitation of Seaweeds. *J. Appl. Phycol.* **2021**, *33*, 603–616. [CrossRef]
14. Karkhaneh Yousefi, M.; Seyed Hashtroudi, M.; Mashinchian Moradi, A.; Ghassempour, A.R. Seasonal Variation of Fucoxanthin Content in Four Species of Brown Seaweeds from Qeshm Island, Persian Gulf and Evaluation of Their Antibacterial and Antioxidant Activities. *Iran. J. Fish. Sci.* **2020**, *19*, 2394–2408. [CrossRef]
15. Marinho, G.S.; Sørensen, A.D.M.; Safafar, H.; Pedersen, A.H.; Holdt, S.L. Antioxidant Content and Activity of the Seaweed Saccharina Latissima: A Seasonal Perspective. *J. Appl. Phycol.* **2019**, *31*, 1343–1354. [CrossRef]
16. Vizetto-Duarte, C.; Pereira, H.; De Sousa, C.B.; Rauter, A.P.; Albericio, F.; Custódio, L.; Barreira, L.; Varela, J. Fatty Acid Profile of Different Species of Algae of the Cystoseira Genus: A Nutraceutical Perspective. *Nat. Prod. Res.* **2015**, *29*, 1264–1270. [CrossRef]
17. Custódio, L.; Silvestre, L.; Rocha, M.I.; Rodrigues, M.J.; Vizetto-Duarte, C.; Pereira, H.; Barreira, L.; Varela, J. Methanol Extracts from Cystoseira Tamariscifolia and Cystoseira Nodicaulis Are Able to Inhibit Cholinesterases and Protect a Human Dopaminergic Cell Line from Hydrogen Peroxide-Induced Cytotoxicity. *Pharm. Biol.* **2016**, *54*, 1687–1696. [CrossRef]
18. Abu-Khudir, R.; Ismail, G.A.; Diab, T. Antimicrobial, Antioxidant, and Anti-Tumor Activities of Sargassum Linearifolium and Cystoseira Crinita from Egyptian Mediterranean Coast. *Nutr. Cancer* **2021**, *73*, 829–844. [CrossRef] [PubMed]
19. Generalić Mekinić, I.; Čagalj, M.; Tabanelli, G.; Montanari, C.; Barbieri, F.; Skroza, D.; Šimat, V. Seasonal Changes in Essential Oil Constituents of Cystoseira Compressa: First Report. *Molecules* **2021**, *26*, 6649. [CrossRef]
20. Oucif, H.; Adjout, R.; Sebahi, R.; Boukortt, F.O.; Ali-Mehidi, S.; El, S.-M.; Abi-Ayad, A. Comparison of In Vitro Antioxidant Activity of Some Selected Seaweeds from Algerian West Coast. *Afr. J. Biotechnol.* **2017**, *16*, 1474–1480. [CrossRef]
21. Dulger, G.; Dulger, B. Antibacterial Activity of Two Brown Algae (Cystoseira Compressa and Padina Pavonica) against Methicillin-Resistant Staphylococcus Aureus. *Br. Microbiol. Res. J.* **2014**, *4*, 918–923. [CrossRef]
22. Mhadhebi, L.; Dellai, A.; Clary-Laroche, A.; Said, R.B.; Robert, J.; Bouraoui, A. Anti-Inflammatory and Antiproliferative Activities of C. Compressa. *Drug Dev. Res.* **2012**, *73*, 82–89. [CrossRef]
23. Kosanić, M.; Ranković, B.; Stanojković, T. Biological Potential of Marine Macroalgae of the Genus Cystoseira. *Acta Biol. Hung.* **2015**, *66*, 374–384. [CrossRef]
24. Mhadhebi, L.; Mhadhebi, A.; Robert, J.; Bouraoui, A. Antioxidant, Anti-Inflammatory and Antiproliferative Effects of Aqueous Extracts of Three Mediterranean Brown Seaweeds of the Genus Cystoseira. *Iran. J. Pharm. Res.* **2014**, *13*, 207–220. [PubMed]

25. Hermund, D.B. Antioxidant Properties of Seaweed-Derived Substances. In *Bioactive Seaweeds for Food Applications*; Elsevier Inc.: Amsterdam, The Netherlands, 2018; pp. 201–221. [CrossRef]
26. Jacobsen, C.; Sørensen, A.D.M.; Holdt, S.L.; Akoh, C.C.; Hermund, D.B. Source, Extraction, Characterization, and Applications of Novel Antioxidants from Seaweed. *Annu. Rev. Food Sci. Technol.* **2019**, *10*, 541–568. [CrossRef] [PubMed]
27. Polat, S.; Trif, M.; Rusu, A.; Šimat, V.; Čagalj, M.; Alak, G.; Meral, R.; Özogul, Y.; Polat, A.; Özogul, F. Recent Advances in Industrial Applications of Seaweeds. *Crit. Rev. Food Sci. Nutr.* **2021**, 1–30. [CrossRef]
28. Maggio, A.; Alduina, R.; Oddo, E.; Piccionello, A.P.; Mannino, A.M. Antibacterial Activity and HPLC Analysis of Extracts from Mediterranean Brown Algae. *Plant Biosyst.* **2020**, 1–17. [CrossRef]
29. Falace, A.; Zanelli, E.; Bressan, G. Morphological and Reproductive Phenology of *Cystoseira Compressa* (Esper) Gerloff & Nizamuddin (Fucales, Fucophyceae) in the Gulf of Trieste (North Adriatic). *Ann. Ser. Hist. Nat.* **2005**, *5*, 5–12.
30. Mannino, A.M.; Vaglica, V.; Cammarata, M.; Oddo, E. Effects of Temperature on Total Phenolic Compounds in Cystoseira Amentacea (C. Agardh) Bory (Fucales, Phaeophyceae) from Southern Mediterranean Sea. *Plant Biosyst.—Int. J. Deal. Asp. Plant Biol.* **2016**, *150*, 152–160. [CrossRef]
31. Güner, A.; Köksal, Ç.; Erel, Ş.B.; Kayalar, H.; Nalbantsoy, A.; Sukatar, A.; Karabay Yavaşoğlu, N.Ü. Antimicrobial and Antioxidant Activities with Acute Toxicity, Cytotoxicity and Mutagenicity of Cystoseira Compressa (Esper) Gerloff & Nizamuddin from the Coast of Urla (Izmir, Turkey). *Cytotechnology* **2015**, *67*, 135–143. [CrossRef]
32. De La Fuente, G.; Fontana, M.; Asnaghi, V.; Chiantore, M.; Mirata, S.; Salis, A.; Damonte, G.; Scarfi, S. The Remarkable Antioxidant and Anti-Inflammatory Potential of the Extracts of the Brown Alga Cystoseira Amentacea Var. Stricta. *Mar. Drugs* **2021**, *19*, 2. [CrossRef] [PubMed]
33. Alghazeer, R.; Elmansori, A.; Sidati, M.; Gammoudi, F.; Azwai, S.; Naas, H.; Garbaj, A.; Eldaghayes, I. In Vitro Antibacterial Activity of Flavonoid Extracts of Two Selected Libyan Algae against Multi-Drug Resistant Bacteria Isolated from Food Products. *J. Biosci. Med.* **2017**, *5*, 26–48. [CrossRef]
34. Abdeldjebbar, F.Z.; Bennabi, F.; Ayache, A.; Berrayah, M.; Tassadiat, S. Synergistic Effect of Padina Pavonica and Cystoseira Compressa in Antibacterial Activity and Retention of Heavy Metals. *Ukr. J. Ecol.* **2021**, *11*, 36–40. [CrossRef]
35. Bouafif, C.; Messaoud, C.; Boussaid, M.; Langar, H. Fatty Acid Profile of Cystoseira C. Agardh (Phaeophyceae, Fucales) Species from the Tunisian Coast: Taxonomic and Nutritional Assessments. *Ciencias Mar.* **2018**, *44*, 169–183. [CrossRef]
36. Zhong, B.; Robinson, N.A.; Warner, R.D.; Barrow, C.J.; Dunshea, F.R.; Suleria, H.A.R. LC-ESI-QTOF-MS/MS Characterization of Seaweed Phenolics and Their Antioxidant Potential. *Mar. Drugs* **2020**, *18*, 331. [CrossRef] [PubMed]
37. Jerković, I.; Cikoš, A.-M.; Babić, S.; Čižmek, L.; Bojanić, K.; Aladić, K.; Ul'yanovskii, N.V.; Kosyakov, D.S.; Lebedev, A.T.; Čož-Rakovac, R.; et al. Bioprospecting of Less-Polar Constituents from Endemic Brown Macroalga Fucus Virsoides J. Agardh from the Adriatic Sea and Targeted Antioxidant Effects In Vitro and In Vivo (Zebrafish Model). *Mar. Drugs* **2021**, *19*, 235. [CrossRef]
38. Ristivojević, P.; Jovanović, V.; Opsenica, D.M.; Park, J.; Rollinger, J.M.; Velicković, T.Ć. Rapid Analytical Approach for Bioprofiling Compounds with Radical Scavenging and Antimicrobial Activities from Seaweeds. *Food Chem.* **2021**, *334*, 127562. [CrossRef]
39. Das, U.N. Arachidonic Acid and Other Unsaturated Fatty Acids and Some of Their Metabolites Function as Endogenous Antimicrobial Molecules: A Review. *J. Adv. Res.* **2018**, *11*, 57–66. [CrossRef]
40. Chanda, W.; Joseph, T.P.; Guo, X.F.; Wang, W.D.; Liu, M.; Vuai, M.S.; Padhiar, A.A.; Zhong, M.T. Effectiveness of Omega-3 Polyunsaturated Fatty Acids against Microbial Pathogens. *J. Zhejiang Univ. Sci. B* **2018**, *19*, 253–262. [CrossRef]
41. Shin, S.Y.; Bajpai, V.K.; Kim, H.R.; Kang, S.C. Antibacterial Activity of Bioconverted Eicosapentaenoic (EPA) and Docosahexaenoic Acid (DHA) against Foodborne Pathogenic Bacteria. *Int. J. Food Microbiol.* **2007**, *113*, 233–236. [CrossRef] [PubMed]
42. Cvitković, D.; Dragović-Uzelac, V.; Dobrinčić, A.; Čož-Rakovac, R.; Balbino, S. The Effect of Solvent and Extraction Method on the Recovery of Lipid Fraction from Adriatic Sea Macroalgae. *Algal Res.* **2021**, *56*, 102291. [CrossRef]
43. Lacey, R.W.; Lord, V.L. Sensitivity of Staphylococci to Fatty Acids: Novel Inactivation of Linolenic Acid by Serum. *J. Med. Microbiol.* **1981**, *14*, 41–49. [CrossRef] [PubMed]
44. Le, P.; Desbois, A. Antibacterial Effect of Eicosapentaenoic Acid against Bacillus Cereus and Staphylococcus Aureus: Killing Kinetics, Selection for Resistance, and Potential Cellular Target. *Mar. Drugs* **2017**, *15*, 334. [CrossRef]
45. Chen, C.-H.; Wang, Y.; Nakatsuji, T.; Liu, Y.-T.; Zouboulis, C.C.; Gallo, R.L.; Zhang, L.; Hsieh, M.-F.; Huang, C.-M. An Innate Bactericidal Oleic Acid Effective against Skin Infection of Methicillin-Resistant Staphylococcus Aureus: A Therapy Concordant with Evolutionary Medicine. *J. Microbiol. Biotechnol.* **2011**, *21*, 391–399. [CrossRef] [PubMed]
46. Čagalj, M.; Skroza, D.; Tabanelli, G.; Özogul, F.; Šimat, V. Maximizing the Antioxidant Capacity of Padina Pavonica by Choosing the Right Drying and Extraction Methods. *Processes* **2021**, *9*, 587. [CrossRef]
47. Amerine, M.A.; Ough, C.S. *Methods for Analysis of Musts and Wines*; Wiley: Hoboken, NJ, USA, 1980.
48. Benzie, I.F.F.; Strain, J.J. The Ferric Reducing Ability of Plasma (FRAP) as a Measure of "Antioxidant Power": The FRAP Assay. *Anal. Biochem.* **1996**, *239*, 70–76. [CrossRef] [PubMed]
49. Milat, A.M.; Boban, M.; Teissedre, P.L.; Šešelja-Perišin, A.; Jurić, D.; Skroza, D.; Generalić-Mekinić, I.; Ljubenkov, I.; Volarević, J.; Rasines-Perea, Z.; et al. Effects of Oxidation and Browning of Macerated White Wine on Its Antioxidant and Direct Vasodilatory Activity. *J. Funct. Foods* **2019**, *59*, 138–147. [CrossRef]
50. Prior, R.L.; Hoang, H.; Gu, L.; Wu, X.; Bacchiocca, M.; Howard, L.; Hampsch-Woodill, M.; Huang, D.; Ou, B.; Jacob, R. Assays for Hydrophilic and Lipophilic Antioxidant Capacity (Oxygen Radical Absorbance Capacity (ORACFL)) of Plasma and Other Biological and Food Samples. *J. Agric. Food Chem.* **2003**, *51*, 3273–3279. [CrossRef]

51. Burčul, F.; Generalić Mekinić, I.; Radan, M.; Rollin, P.; Blažević, I. Isothiocyanates: Cholinesterase Inhibiting, Antioxidant, and Anti-Inflammatory Activity. *J. Enzym. Inhib. Med. Chem.* **2018**, *33*, 577–582. [CrossRef] [PubMed]
52. Skroza, D.; Šimat, V.; Smole Možina, S.; Katalinić, V.; Boban, N.; Generalić Mekinić, I. Interactions of Resveratrol with Other Phenolics and Activity against Food-Borne Pathogens. *Food Sci. Nutr.* **2019**, *7*, 2312–2318. [CrossRef] [PubMed]
53. Elez Garofulić, I.; Malin, V.; Repajić, M.; Zorić, Z.; Pedisić, S.; Sterniša, M.; Smole Možina, S.; Dragović-Uzelac, V. Phenolic Profile, Antioxidant Capacity and Antimicrobial Activity of Nettle Leaves Extracts Obtained by Advanced Extraction Techniques. *Molecules* **2021**, *26*, 6153. [CrossRef]
54. Verni, M.; Pontonio, E.; Krona, A.; Jacob, S.; Pinto, D.; Rinaldi, F.; Verardo, V.; Díaz-de-Cerio, E.; Coda, R.; Rizzello, C.G. Bioprocessing of Brewers' Spent Grain Enhances Its Antioxidant Activity: Characterization of Phenolic Compounds and Bioactive Peptides. *Front. Microbiol.* **2020**, *11*, 1831. [CrossRef] [PubMed]
55. Šimat, V.; Vlahović, J.; Soldo, B.; Generalić Mekinić, I.; Čagalj, M.; Hamed, I.; Skroza, D. Production and Characterization of Crude Oils from Seafood Processing by-Products. *Food Biosci.* **2020**, *33*, 100484. [CrossRef]

MDPI

Review

Fucosterol of Marine Macroalgae: Bioactivity, Safety and Toxicity on Organism

Maria Dyah Nur Meinita [1,2,3,*], Dicky Harwanto [1,4], Gabriel Tirtawijaya [1], Bertoka Fajar Surya Perwira Negara [1,5], Jae-Hak Sohn [1,5], Jin-Soo Kim [6,*] and Jae-Suk Choi [1,5,*]

1 Seafood Research Center, Industry Academy Cooperation Foundation (IACF), Silla University, 606, Advanced Seafood Processing Complex, Wonyang-ro, Amnam-dong, Seo-gu, Busan 49277, Korea; dickyharwanto@lecturer.undip.ac.id (D.H.); ftrnd6@silla.ac.kr (G.T.); ftrnd12@silla.ac.kr (B.F.S.P.N.); jhsohn@silla.ac.kr (J.-H.S.)
2 Faculty of Fisheries and Marine Science, Jenderal Soedirman University, Purwokerto 53123, Indonesia
3 Center for Maritime Bioscience Studies, Jenderal Soedirman University, Purwokerto 53123, Indonesia
4 Faculty of Fisheries and Marine Science, Diponegoro University, Semarang 50275, Indonesia
5 Department of Food Biotechnology, College of Medical and Life Sciences, Silla University, 140, Baegyang-daero 700 beon-gil, Sasang-gu, Busan 46958, Korea
6 Department of Seafood and Aquaculture Science, Gyeongsang National University, 38 Cheong-daegukchi-gil, Tongyeong-si 53064, Korea
* Correspondence: maria.meinita@unsoed.ac.id (M.D.N.M.); jinsukim@gnu.ac.kr (J.-S.K.); jsc1008@silla.ac.kr (J.-S.C.); Tel.: +62-281-642-360 (M.D.N.M.); +82-557-729-146 (J.-S.K.); +82-512-487-789 (J.-S.C)

Abstract: Fucosterol (24-ethylidene cholesterol) is a bioactive compound belonging to the sterol group that can be isolated from marine algae. Fucosterol of marine algae exhibits various biological activities including anti-osteoarthritic, anticancer, anti-inflammatory, anti-photoaging, immunomodulatory, hepatoprotective, anti-neurological, antioxidant, algicidal, anti-obesity, and antimicrobial. Numerous studies on fucosterol, mainly focusing on the quantification and characterization of the chemical structure, bioactivities, and health benefits of fucosterol, have been published. However, there is no comprehensive review on safety and toxicity levels of fucosterol of marine algae. This review aims to discuss the bioactivities, safety, and toxicity of fucosterol comprehensively, which is important for the application and development of fucosterol as a bioactive compound in nutraceutical and pharmaceutical industries. We used four online databases to search for literature on fucosterol published between 2002 and 2020. We identified, screened, selected, and analyzed the literature using the Preferred Reporting Items for Systematic Reviews and Meta-Analyses method and identified 43 studies for review. Despite the potential applications of fucosterol, we identified the need to fill certain related research gaps. Fucosterol exhibited low toxicity in animal cell lines, human cell lines, and animals. However, studies on the safety and toxicity of fucosterol at the clinical stage, which are required before fucosterol is developed for the industry, are lacking.

Keywords: fucosterol; seaweed; algae; toxicity; in vivo; in vitro

Citation: Meinita, M.D.N.; Harwanto, D.; Tirtawijaya, G.; Negara, B.F.S.P.; Sohn, J.-H.; Kim, J.-S.; Choi, J.-S. Fucosterol of Marine Macroalgae: Bioactivity, Safety and Toxicity on Organism. *Mar. Drugs* **2021**, *19*, 545. https://doi.org/10.3390/md19100545

Academic Editors: Marco García-Vaquero and Brijesh K. Tiwari

Received: 10 September 2021
Accepted: 24 September 2021
Published: 27 September 2021

1. Introduction

Fucosterol is abundant and one of the dominant sterols in marine macroalgae [1]. The purest form of fucosterol and its potency were first identified and observed in the brown macroalga *Fucus vesiculosus* by Heilbron et al. [2]. Fucosterol is a stigmasterol bond isomer expressed by the empirical formula $C_{29}H_{48}O$. The fucosterol content in macroalgae ranges from 4 to 95% of the total phytosterol content [3]. Brown macroalgae contain higher levels of fucosterol than green and red macroalgae. The fucosterol content in brown macroalga *Ecklonia radiata* ranged between 312.0 μg/g dry weight in leaves and 378.1 μg/g dry weight in stipes (98.6 and 98.9% of total sterols, respectively) [4]. In the brown macroalgae *Himanthalia elongata*, *Undaria pinnatifida*, and *Laminaria ochroleuca*, fucosterol was observed predominantly in 83–97% of the total sterol content [5]. Fucosterol was also reported to be dominant in *Stephanocystis hakodatensis* (formerly *Cytoseira hakodatensis*) and *Sargassum fusiforme*, which contained 65.9% and 67% of fucosterol, respectively [6].

Brown macroalgae are widely used as food and herbal medicine in Southeast Asia and several European countries. In Europe, brown macroalgae have been used to treat goiter and obesity [7]. In East Asia, brown macroalgae from the genera *Laminaria*, *Undaria*, and *Sargassum* (formerly *Hizikia*) are widely consumed daily and used as herb medicine [8–12]. Hence, the bioactive properties of brown macroalgae have drawn the attention of researchers. Previous studies have investigated fucosterol properties and their potential bioactivities. The antioxidant effect of fucosterol was reported by Lee et al. [13]. Furthermore, Jung et al. [14] investigated the anti-inflammatory properties of fucosterol in LPS-stimulated conditions. Fucosterol has the potential to inhibit particulate-induced inflammation and oxidative stress in the alveolar cell line A549. Through regulation of the FoxO signaling pathway, fucosterol exhibits anti-obesity characteristics by suppressing adipogenesis in 3T3-L1 preadipocytes [15]. In addition, fucosterol protects human neuroblastoma cell line SH-SY5Y cells from amyloid-induced neurotoxicity [16] and affects human lung cancer cells by inducing apoptosis and cell cycle arrest and targeting the Raf/signaling mitogen-activated protein kinase/extracellular-signal-regulated kinase (MEK/ERK) pathway [17]. Based on these studies, fucosterol can potentially be developed for use in nutraceutical and pharmaceutical fields. However, before further developing fucosterol properties, information on the safety and toxicity of fucosterol is required to comprehend the optimum and sustainable benefits of fucosterol as a functional agent. This study reviews the current scientific literature regarding the bioactivity, safety, and toxicity of fucosterol extracted from marine macroalgae. In addition to the bioactivity of fucosterol, we investigated the safety and toxicity of fucosterol in various organisms, including bacteria/fungi, animal cell lines, human cell lines, and animals. Through this review article, we express our hope that the applications of fucosterol from marine algae can be further developed in the nutraceutical and pharmaceutical industries.

2. Results and Discussion

Studies on the bioactivity, safety, and toxicity levels of fucosterol from marine macroalgae conducted from 2002 to 2020 were reviewed, and an increasing trend was observed (Figure 1). This indicates that fucosterol has drawn additional research attention in recent years.

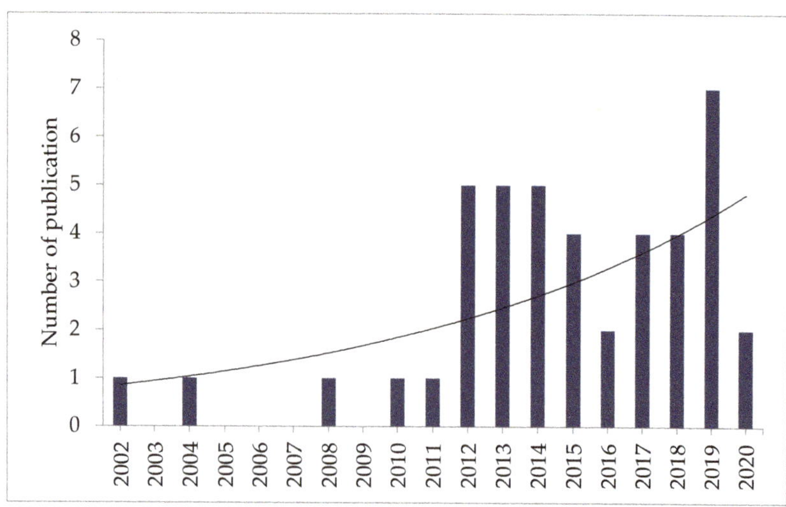

Figure 1. Number of publications on the safety and toxicity of fucosterol published in each year.

Articles discussing the bioactivity, safety, and toxicity of fucosterol were categorized with respect to the various organisms or cells whose treatment they describe (Figure 2). Published studies on the safety and toxicity of fucosterol from marine algae have focused on its effects on bacteria and fungi (21%), animal cell lines (14%), human cell lines (38%), and animals (26%). The safety and toxicity of fucosterol have been studied both in vitro and in vivo. No clinical study of fucosterol has been conducted to date. Therefore, the investigation of the safety and toxicity of fucosterol at the clinical stage might be challenging.

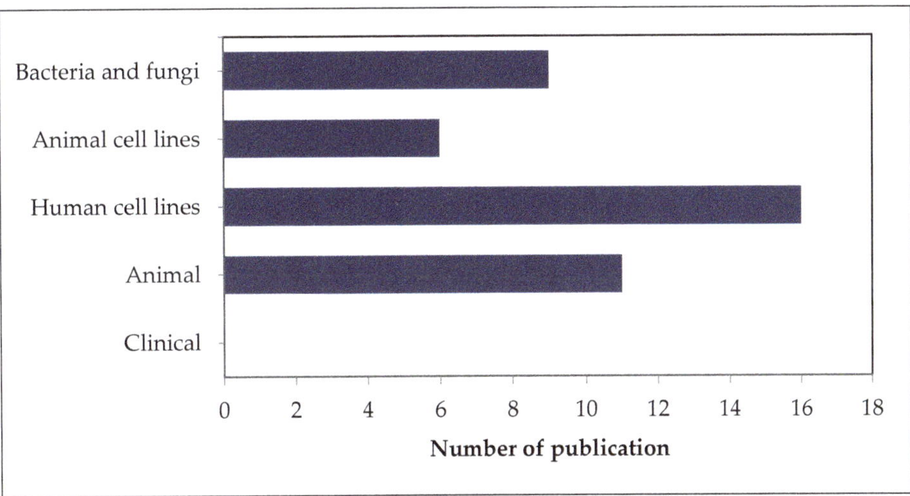

Figure 2. Numbers of publications on safety and toxicity of fucosterol categorized according to the treated organisms or cells.

The numbers of publications on the safety and toxicity of fucosterol from various sources of macroalgae are shown in Figure 3. Sixteen marine algae species belonging to Dictyotaceae, Sargassaceae, Alariaceae, Lessoniaceae, and Fucaceae families have been studied in relation to their fucosterol bioactivity, safety, and toxicity on bacteria, fungi, animal cells, human cells, and animals. Among the 16 marine algae, *Sargassum fusiforme* (formerly *Hizikia fusiformis*) has become the greatest marine macroalgal source for fucosterol that has been studied. Over the last 10 years, studies on the bioactivities and nutritional and pharmacological properties of *S. fusiforme* have increased steadily. Most of the studies on the bioactivity of *S. fusiforme* have focused on its antioxidant (15.09%), anticancer and antitumor (15.09%), anti-inflammatory (11.32%), photoprotective (11.32%), and neuroprotective (11.32%) properties [18]. Figure 3 shows that the study of fucosterol in *S. fusiforme* has focused on its antioxidant, anti-osteoarthritic, anti-inflammatory, anti-photoaging, antidiabetic, hepatoprotective, and algicidal effects. *Ecklonia cava* subsp. *stolonifera* (formerly *Ecklonia stolonifera*) is the second most frequently reported macroalgal species that has been studied for its fucosterol content. *Ecklonia* species have been known as potential source of bioactive compounds [19]. Studies on the safety and toxicity levels of fucosterol obtained from *Ecklonia stolonifera* were mostly conducted on animal cell lines, human cell lines, and animals. Fucosterol from *Ecklonia stolonifera* has been studied for its antidiabetic, anti-obesity, anti-neurological, and hepatoprotective effects. The safety and toxicity levels of fucosterol obtained from *Ecklonia cava* subsp. *stolonifera* were mostly studied in animals.

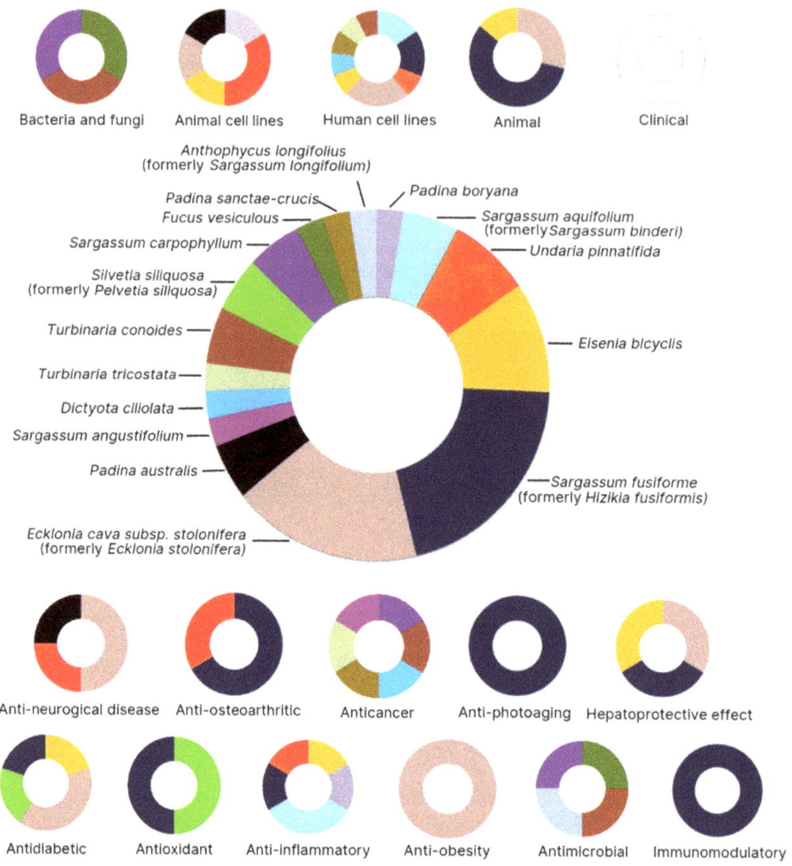

Figure 3. Numbers of publications on safety and toxicity of fucosterol classified according to sources of macroalgae.

2.1. Characteristics and Structure of Fucosterol

Generally, fucosterol could be obtained by extracting dry powder of macroalgae using MeOH, EtOH, or *n*-hexane solvents [13,20–24]. The extracts were then partitioned via solvent fractionation [13,25]. After solvent dissolution under reduced pressure, the organic extracts were fractionated using silica gel column chromatography with a mixture of solvents of increasing polarity [14,22,23,26]. The fraction was eluted using a solvent to remove fatty acids, which were then analyzed further [26].

Currently, fucosterol analysis and identification are carried out using physical properties and spectroscopic methods, including ^{1}H-NMR and ^{13}C-NMR, as well as via comparison with published data, and thin-layer chromatography (TLC) analysis [20]. According to the molecular formula of fucosterol, one hydroxy group must be attached to C-28 in an *R* or *S* configuration. An olefin proton signal and two sets of two olefin signals indicate the presence of a tri-substituted double in the fucosterol side chain [22]. Infrared (IR) absorption peaks of fucosterol at 3400 and 1600 cm^{-1} were attributed to the hydroxyl and olefin groups, respectively. Fucosterol derivatives, namely 24*R*,28*R*- and 24*S*,28*R*-epoxy-24-ethylcholesterol, and 24*R*-saryngosterol have also been used in octadecyl silica gel (ODS) column chromatography [21]. The structure of fucosterol is shown in Figure 4.

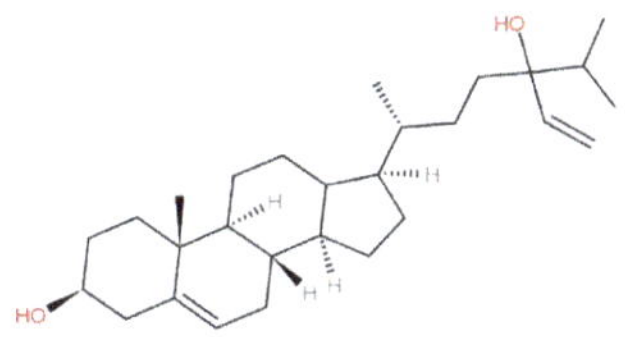

Figure 4. Chemical structure of fucosterol.

2.2. *Bioactivity of Fucosterol*

Fucosterol of marine macroalgae exhibited antidiabetic, anti-obesity, anti-osteoarthritic, immunomodulatory, anticancer, anti-inflammatory, anti-photoaging, hepatoprotective, anti-neurological, antioxidant, and antimicrobial activities (Figure 5).

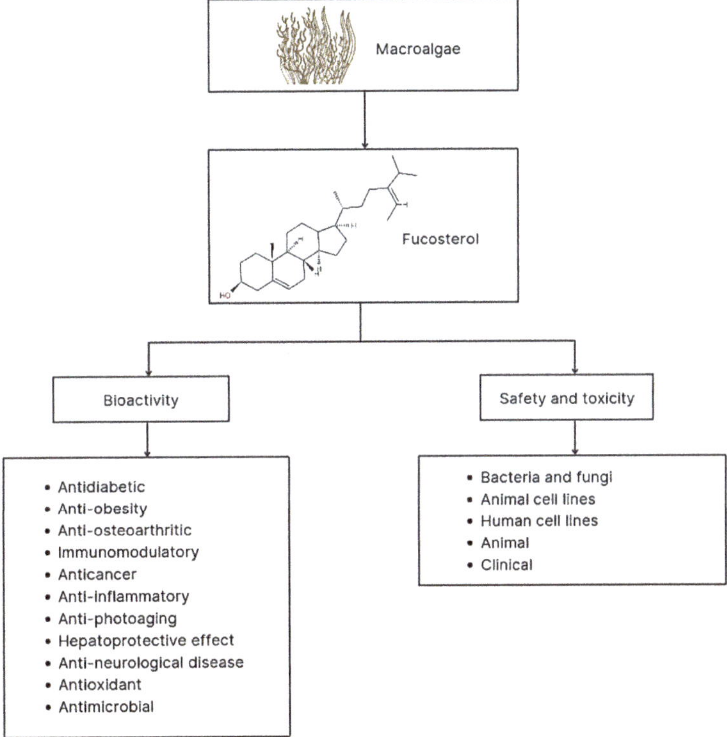

Figure 5. Bioactivities, safety, and toxicity levels of fucosterol derived from macroalgae.

2.2.1. Antidiabetic Activity

Diabetes is a chronic disease that occurs when the pancreas does not produce enough insulin or when the body cannot use insulin effectively. One of the antidiabetic effects of fucosterol obtained from *Eisenia bicyclis* and *Ecklonia cava* subsp. *stolonifera* is characterized by inhibition of enzymes, such as rat lens aldose reductase (RLAR), human recombinant aldose reductase (HRAR), α -glucosidase, and PTP1B. The docking simulations clearly demonstrated negative binding energy for fucosterol (-8.2 kcal mol^{-1} for RLAR and -8.5 kcal mol^{-1} for HRAR), implying a higher affinity and stronger binding competence for the active site of the enzyme [20]. These results were confirmed by Jung et al. [27], who stated that *Ecklonia stolonifera*-derived fucosterol reduced insulin resistance by decreasing PTP1B expression and activating insulin signaling pathways. Fucosterol from *Sargassum fusiforme* also showed strong PTP1B inhibition at low concentrations [28]. Streptozotocin-induced diabetic rats treated with fucosterol from *Pelvetia siliquosa* via an orally administered dose of 30 mg/kg showed a decrease in serum glucose levels and inhibition of its accumulation. However, orally administered doses of 300 mg/kg are required for epinephrine-induced diabetic rats to inhibit blood glucose levels and degrade glycogen [29]. These findings illustrate that fucosterol extracts from *Eisenia bicyclis*, *Ecklonia cava* subsp. *stolonifera*, *Sargassum fusiforme*, and *Silvetia siliquosa* have antidiabetic potential that can be developed in the future.

2.2.2. Anti-Obesity Activity

In addition to being antidiabetic, fucosterol also exhibits anti-adipogenic or anti-obesity properties. Adipocytes have important cardiovascular-related roles; therefore, understanding their development and regulation is important for treating obesity and related diseases. Obesity is associated with hypercholesterolemia, diabetes, and other chronic diseases. Hypercholesterolemia is associated with a higher incidence of liver damage, especially non-alcoholic fatty liver disease (NAFLD) [30]. As reported by Jung et al. [25], following treatment with fucosterol extracted from *Ecklonia cava* subsp. *stolonifera*, lipid accumulation in 3T3-L1 pre-adipocytes decreased because the expression levels of adipocyte markers, proteins peroxisome proliferator-activated receptor (PPAR), and CCAAT/enhancer-binding protein (C/EBP), decreased [25]. Similar results were reported by Lee et al. [15], who found that fucosterol of *Ecklonia cava* subsp. *stolonifera* inhibits adipogenesis of 3T3-L1 preadipocytes via modulation of the FoxO signaling pathway.

2.2.3. Anti-Osteoarthritis Activity

The anti-osteoarthritis activity of fucosterol has been demonstrated both in vitro and in vivo [21,31,32]. Anti-osteoarthritis is a biological activity characterized by the ability of a compound to reduce or prevent bone disease. *Sargassum fusiforme*-derived fucosterol has been shown to increase proliferative activity in osteosarcoma MG63 cells for the treatment of bone-absorbing metabolic bone diseases, including osteoporosis and periodontitis [21]. Similar results were observed in ovariectomized rat osteoporosis and fucosterol of *Sargassum fusiforme* triggered bone regeneration and activation of bone formation [31], and Bang et al. [32] tested fucosterol in vitro and reported that fucosterol from *Undaria pinnatifida* inhibited osteoclast differentiation. Through this biological activity, fucosterol may play an important role in preventing osteoporosis and may be useful as a supplement.

2.2.4. Immunomodulatory Activity

In immunomodulation, regulating many immune cells through signaling molecules to enhance the immune system, is necessary. Based on previous studies, several marine metabolites have been reported to have regulatory effects on the immune system [33]. Fucosterol is one of the compounds that is regarded as a candidate immunomodulator, as reported by Park et al. [34] in vitro and in vivo; fucosterol from *Sargassum fusiforme* can increase the secretion of tumor necrosis factor alpha (TNF-α), NO production, and phagocytosis activity. Generally, fucosterol has the potential to regulate immune function and may offer positive therapeutic effects in immune system diseases.

2.2.5. Anticancer Properties

Certain studies about the anticancer properties of fuscosterol extracted from marine macroalgae, such as *Sargassum carpophyllum*, *Turbinaria conoides*, *Dictyota ciliolata*, and *Padina sanctae-crucis*, have been published [21,35–39]. Based on the study by Jiang et al. [39], commercial fucosterols exhibit anticancer activity by inhibiting the PI3K/Akt/mTOR signaling pathway in cervical cancer cell lines. In addition, fucosterol from *Sargassum fusiforme* has been observed to slow the progression of human ovarian cancer [35] and inhibit the proliferation of osteosarcoma-derived cell MG63 [21]. Tang et al. [22] isolated steroids from *Sargassum carpophyllum* using activity-guided fractionation to determine the effect of fucosterol and other active substances on cancer cell lines. This activity is indicated by IC_{50}, which is the concentration that results in 50% inhibition of cell growth [37]. They found that fucosterol has an IC_{50} value of 7.8 μg/mL against HL-60 cancer cells. A compound can exhibit one of three types of cytotoxicity: (1) potential cytotoxicity, if $IC_{50} < 100$ μg/mL, (2) moderate cytotoxicity, if 100 g/mL < IC_{50} < 1000 μg/mL, and it can be (3) non-toxic, if $IC_{50} > 1000$ μg/mL. Agents from a group of potentially cytotoxic compounds can be utilized as anticancer drugs, while moderately cytotoxic compounds can be used for chemoprevention to stop cancer cell growth [40]. According to the National Cancer Institute (NCI), a compound is classified as having anticancer properties if its IC_{50} is <20 μg/mL. Published research studies prove that fucosterol as a metabolite compound in macroalgae has anticancer properties; however, further clinical studies are required.

2.2.6. Anti-Inflammatory Activity

Inflammation is a biological response to noxious stimuli. It is a protective response involving immune cells such as macrophages, blood vessels, molecular mediators such as NO, pro-inflammatory cytokines (TNF-α, interleukin-1β (IL-1β), and interleukin-6 (IL-6)), and prostaglandins [41]. Previous studies have identified anti-inflammatory compounds in macroalgae. Fucosterol derived from the methanolic extract of the brown alga *Eisenia bicyclis* and *Undaria pinnatifida* exhibits anti-inflammatory properties by suppressing the production of cyclooxygenase-2 (COX-2) and inducible nitric oxide synthase (iNOS) in LPS-stimulated RAW 264.7 macrophages [14,42]. Fucosterol also inhibited t-BHP-induced ROS production and suppressed iNOS and COX-2 expression. Furthermore, *E. bicyclis* has strong anti-inflammatory properties with the potential to inhibit NO and ROS production, as well as the NF-κB pathway [14]. Fucosterol from *Padina boryana* was reported to have anti-inflammatory properties via the regulation of NF-κB/MAPK and involvement of Nrf2/HO-1 pathways in PM-induced inflammation/oxidative stress in RAW 264.7 macrophage cells [43]. Furthermore, fucosterol extracted from *Sargassum aquifolium* (formerly *Sargassum binderi*) decreased ROS levels in PM-induced HaCaT cells and human lung epithelial cells [15,44]. Another study also showed inhibition of hypoxia-inducible factor through the PI3K/Akt pathway in keratinocytes (HaCaT) cells induced by cobalt chloride ($CoCl_2$). Based on the research conducted, fucosterol from brown macroalgae has the potential to be developed further in the pharmacological field.

2.2.7. Anti-Photoaging Effect

Photoaging is a result of chronic ultraviolet irradiation, which is one of the most harmful environmental factors affecting the skin. Recently, studies on marine compounds as safe anti-photoaging alternatives have been published by Hwang et al. [23]. They reported that fucosterol obtained from the brown alga *Sargassum fusiforme* was observed to have anti-photoaging properties. Fucosterol therapy decreased ultraviolet B (UVB)-induced production of matrix metalloproteinase-1 (MMP-1), IL-6, p-c-Jun, and p-c-Fos and increased type I and transforming growth factor-1 (TGF-1) procollagen expression in normal human dermal fibroblast cells. In addition, fucosterol derived from *Sargassum fusiforme* was observed to regulate the expression of MMPs and type-I pro-collagen in UV-irradiated HaCaT cells by modulating, microtubule associated protein kinase (MAPK) [45]. These findings suggest that fucosterol extracted from *Sargassum fusiforme* is a potential candidate for the prevention and treatment of skin aging.

2.2.8. Hepatoprotective Effect

The liver is the largest organ in the abdominal cavity and performs critical physiological functions. Acute liver failure is caused by liver injury, which is mainly caused by viral infections, drugs, food additives, alcohol, or radioactivity [46]. Fucosterol has a potent hepatoprotective effect by increasing GSH levels and decreasing ROS production, thereby preventing liver damage and increasing liver enzyme levels alanine aminotransferase/ aspartate aminotransferase (ALT/AST) [47]. Fucosterol is a dual-LXR agonist that regulates the expression of key genes involved in cholesterol homeostasis in several cell lines without causing triglyceride accumulation in the liver [48,49]. In addition, a study by Mo et al. [50] showed that fucosterol can relieve acute liver injury induced by ConA by inhibiting P38 MAPK/PPARγ/NF-κB signaling, suggesting that fucosterol is a promising potential therapeutic agent for acute liver injury.

2.2.9. Anti-Neurological Disease

Neurological disorders are diseases of the nervous system, such as brain tumors, epilepsy, Parkinson's disease, stroke, and Alzheimer's disease (AD) [51]. In a study conducted by Yoon et al. [24], fucosterol extracted from *Ecklonia cava* subsp. *stolonifera* showed the presence of cholinesterase inhibitors against AchE and BchE. BChE was inhibited by fucosterol and 24-hydroperoxy 24-vinylcholesterol, with IC_{50} values of 421.72 ± 1.43 and 176.46 ± 2.51 μM, respectively. The effect of the compound on amyloid-induced neurotoxicity can be used to determine the potential of the compound as an anti-AD [16,52]. Consequently, *Padina australis*-derived fucosterol reduced intracellular amyloid levels and increased neuroglobin mRNA expression in amyloid-induced SH-SY5Y cells [16]. Fucosterol extracted from *Ecklonia cava* subsp. *stolonifera* co-infusion attenuated cognitive impairment-induced $sA\beta_{1\text{-}42}$ in aging mice via the downregulation of GRP78 expression [52]. Furthermore, the anti-AD properties of fucosterol from macroalgae have also been reported by Jung et al. [53] and Wong et al. [54]. Research on effects of fucosterol against Parkinson's disease has been described by Paudel et al. [55], and it was found to exhibit a mild dopamine D4 antagonist effect by inhibiting the dopamine agonist effect by 32% at 100 μM. Furthermore, fucosterol extracted from *Sargassum fusiforme* has also been reported to inhibit epilepsy and act as an antidepressant. The group treated with 20 mg/kg fucosterol showed a significant increase in the hippocampal brain-derived neurotrophic factor (BDNF) levels ($p < 0.05$). Published studies show that fucosterol from marine algae can be an alternative compound for the treatment of neurological diseases.

2.2.10. Antioxidant Activity

Excessive ROS formation can trigger oxidative stress, which causes cell damage and changes cell functions. Antioxidants are required to maintain a balance and prevent negative effects from excessive ROS formation. Based on the literature reviewed, macroalgae showed antioxidant activity, which can be useful for preventing excessive ROS formation. Fucosterol from *Sargassum fusiforme* exhibited antioxidant properties by downregulating serum transaminase activity in CCl_4-intoxicated rats. Sequentially, sGOT and sGPT activity decreased by 25.57% and 63.16%, respectively. In addition, fucosterol treatment of CCl4-intoxicated rats also increased hepatic cytosolic SOD, catalase, and GSH-px [13]. Oktaviani et al. [56] reported that fucosterol from *Hizikia fusiformis* prolonged the lifespan of *Caenorhabditis elegans* (Nematoda). Based on these studies, fucosterols extracted from macroalgae are potential candidates for antioxidants that can be used in functional foods and medicines.

2.2.11. Antimicrobial Activity

Antimicrobial activities can be defined as the process of inhibiting or destroying the growth of microorganisms, especially pathogenic microorganisms. The antimicrobial properties, including antibacterial and antifungal properties of marine macroalgae, are associated with various groups of bioactive lipids, such as fucosterol. The results of the study by Tyskiewicz et al. [57] showed that fucosterol from *Fucus vesiculosus* at a concentration of 1.0% completely inhibited the germination of macroconidia in *Fusarium culmorum* (Fungi, Ascomycota). Furthermore, when macroconidia were exposed to low doses of fucosterol (0.05–0.2%), their growth was inhibited, and structural degradation occurred. Furthermore, fucosterol from *Sargassum carpophyllum* cultured with *Pyricularia oryzae* (Fungi, Ascomycota) mycelia caused abnormal morphological changes [22]. Previous studies confirmed the antibacterial and antifungal activity of 3,6,17-trihydroxy-stigmasta-4,7,24(28)-triene, fucosterol, and 14,15,18,20-diepoxyturbinarin compounds from *Turbinaria conoides*, with MICs ranging from 2 to 16 μg/mL, against *Staphylococcus aureus*, *S. epidermidis*, *Escherichia coli*, *Pseudomonas aeruginosa*, *Aspergillus niger*, and *Candida albicans* [58]. The antibacterial properties of *Sargassum longifolium* fucosterol were also tested against the human pathogen *Vibrio parahaemolyticus* and the fish pathogens *V. vulnificus*, *V. harveyii*, and *Aeromonas hydrophililla*. Interestingly, among these bacteria, only *P. fluorescens* was not susceptible to the effect of fucosterol [59]. Overall, fucosterol could potentially be a strong and promising antimicrobial agent.

2.3. Safety and Toxicity of Fucosterol in Bacteria and Fungi

Several studies on fucosterol in bacteria and fungi have been published [22,57,58]. From these articles, data extraction was performed, as shown in Table 1.

Table 1. Studies on safety and toxicity of fucosterol extracted from macroalgae, tested in bacteria and fungi.

Bacteria/ Fungi	Extract or Chemical	Sources	Method	Concentration	Toxicity	Ref.
Pyricularia oryzae	Fucosterol extract	*Sargassum carpophyllum*	Screening	Nd	Affected the morphology	[22]
Staphylococcus epidermidis	Fucosterol extract	*Turbinaria conoides*	Broth dilution susceptibility assay	2–256 μg/mL	Inhibited bacteria growth	[58]
Aspergillus niger	Fucosterol extract	*Turbinaria conoides*	Broth dilution susceptibility assay	2–256 μg/mL	Inhibited fungal growth	[58]
Candida albicans	Fucosterol extract	*Turbinaria conoides*	Broth dilution susceptibility assay	2–256 μg/mL	Inhibited fungal growth	[58]
Escherichia coli	Fucosterol extract	*Turbinaria conoides*	Broth dilution susceptibility assay	2–256 μg/mL	Inhibited bacteria growth	[58]
Staphylococcus aureus	Fucosterol extract	*Turbinaria conoides*	Broth dilution susceptibility assay	2–256 μg/mL	Inhibited bacteria growth	[58]
Pseudomonas aeruginosa	Fucosterol extract	*Turbinaria conoides*	Broth dilution susceptibility assay	2–256 μg/mL	Inhibited bacteria growth	[58]
Fusarium culmorum	Commercial fucosterol	Nd	Determined on a liquid RB medium	0.05–1.0%	Inhibited fungal growth and caused total degradation	[57]

Nd: not determined.

According to Tang et al. [58], fucosterol isolated from *Sargassum carpophyllum* showed low toxicity, with IC_{50} = 250 μg/mL, and was able to induce morphological changes in *Pyricularia oryzae*. Furthermore, fucosterol extract from *Turbinaria conoides* was used to test the level of growth inhibition in bacteria (*S. aureus*, *S. epidermidis*, *E. coli*, and *P. Aeruginosa*) and fungi (*C. albicans* and *A. niger*). In the tested bacteria, the MIC values ranged from 8 to 16 μg/mL, which indicated that fucosterol was able to inhibit the growth of the tested bacteria well. In addition, fucosterol showed the highest growth inhibition in *C. albicans*, with MIC = 8 μg/mL. Furthermore, research by Tyskiewicz et al. [57] showed that at a concentration of 1.0% fucosterol was able to optimally inhibit the growth of *F. culmorum* macroconidia. Moreover, macroconidia showed shorter length and structural degradation at lower fucosterol concentrations (0.05–0.2%).

From these studies, we conclude that fucosterol can potentially be developed as a new agent for combating the problem of infection due to bacteria and fungi that are pathogenic because of its excellent biological activity as an inhibitor of bacteria and fungi. Based on our literature review, only the genera *Turbinaria* and *Sargassum* have been studied for the treatment of pathogenic bacteria and fungi. No other genera have been reported with respect to their safety and toxicity in bacteria and fungi. Further research on other bacteria and fungi species is required to comprehensively elucidate the safety and toxicity of fucosterol in bacteria and fungi.

2.4. Safety and Toxicity of Fucosterol in Cell Lines

Several studies have demonstrated safety and toxicity in human cell lines [17,27,28,35–38,47,53,60–62] and animal cell lines [15,42,43,45,54,55]. The safety and toxicity of fucosterol in human and animal cell lines are summarized in Table 2.

Table 2. Studies on safety and toxicity of fucosterol extracted from macroalgae, tested in human and animal cell lines.

Cell Lines	Extract or Chemical	Sources	Method	Concentration	Toxicity	Ref.
RAW 264.7 macrophage cells	Fucosterol extract	*Padina boryana*	MMT assay and incubated for 23 h	12.5, 25, and 50 μg/mL	Increased cell viability	[43]
A549 human lung epithelial cells exposed to CPM	Fucosterol extract	*Sargassum aquifolium* (formerly *Sargassum binderi*)	MMT assay and incubated for 24 h	3.125, 6.25, 12.5, 25, 50, and 100 μg/mL	Low toxicity and increased cell viability	[44]
Chinese hamster ovary (CHO) cells, rat basophil leukemia (RBL) cells, U373 cells, and BA/F3 cells	Fucosterol extract	*Undaria pinnatifida* and *Eisenia bicyclis*	Human monoamine oxidase (*h*MAO) inhibition and functional assay	500 μM	Had no effects on the viability of the cells	[55]
Human dermal fibroblasts (HDF) and HaCaT cells	Fucosterol extract	*Sargassum aquifolium* (formerly *Sargassum binderi*)	MMT assay and incubated for 3 h	3.125, 6.25, 12.5, 25, 50, and 100 μg/mL	Not toxic and increased cell viability	[60]
Human recombinant PTP1B	Fucosterol extract	*Sargassum fusiforme*	Docking simulation	0 to 2 mM	Inhibited PTP1B and α-glucosidase	[28]
RAW 264.7 macrophages	Fucosterol extract	*Undaria pinnatifida*	Western blot analysis	10, 25, or 50 μM	Had no effects on the viability of the cells	[42]
Murine 3T3-L1 preadipocytes	Fucosterol extract	*Ecklonia cava* subsp. *stolonifera* (formerly *Ecklonia stolonifera*)	Western blot analysis	25 and 50 μM	No significant effects up to 50 μM	[15]
β-Site amyloid precursor protein cleaving enzyme 1 (BACE1)	Fucosterol extract	*Undaria pinnatifida* and *Ecklonia cava* subsp. *stolonifera* (formerly *Ecklonia stolonifera*)	Kinetics and molecular docking simulation	0, 5.0, 20, and 100 μM	Inhibited BACE1 and not toxic	[53]
Insulin-resistant HepG2 (human hepatocarcinoma) cells	Fucosterol extract	*Ecklonia cava* subsp. *stolonifera* (formerly *Ecklonia stolonifera*)	MMT assay and incubated for 2 h	12.5, 25, 50, 100, and 200 μM	No significant effect up to 100 μM	[27]
HepG2 cells induced t-BHP and tacrine	Fucosterol extract	*Ecklonia cava* subsp. *stolonifera* (formerly *Ecklonia stolonifera*)and *Eisenia bicyclis*	MMT assay and incubated for 2 h	0, 25, 50 and 100 μM	Low toxicity and increased cell viability	[47]
HaCaT cells induced cobalt chloride ($CoCl_2$)	Fucosterol extract	*Sargassum fusiforme* (formerly *Hizikia fusiformis*)	MMT assay and incubated for 2 h	1, 2, 5, and 10 μM	Low toxicity and increased cell viability	[61]
C8-B4 microglial cells	Fucosterol extract	*Padina australis*	MMT assay and incubated for 4 h	12, 24, 48, 96, and 192 μM	Had no effects on the viability of the cells	[54]
Colon carcinoma (HT-29), colorectal adenocarcinoma (Caco-2), and breast ductal carcinoma (T47D) cell lines	Fucosterol extract	*Sargassum angustifolium*	MMT assay and incubated for 4 h	4.5, 18, 36, and 72 μg/mL	Low toxicity	[38]
Oral carcinoma (KB), epithelial carcinoma of the larynx (Hep-2), MCF-7, cervix adenocarcinoma (SiHa), and a human cell embryonic kidney cell line (HEK-293)	Fucosterol extract	*Dictyota ciliolata Padina sanctae-crucis*, and *Turbinaria tricostata*	MMT assay and incubated for 2 h	Nd	Only inactive on HEK-293	[37]

Table 2. *Cont.*

Cell Lines	Extract or Chemical	Sources	Method	Concentration	Toxicity	Ref.
Dalton's Lymphoma Ascites (DLA) cells	Fucosterol extract	*Turbinaria conoides*	Trypan blue viability assay	100 and 200 μg/mL	Not toxic	[36]
Lung cancer cell and human normal cell	Commercial fucosterol	Nd	MMT assay and incubated for 24 h	1.55, 3.12, 6.25, 12.5, 25, 50, and 100 μg/mL	Decreased cell viability in cancer cell and low toxicity in normal cell	[17]
Human cancer cell lines (HT29 and HCT116) and CCD-18Co fibroblasts	Commercial fucosterol	Nd	MMT assay and incubated for 24 h	5 and 10 μM	Decreased cell viability in HT29 cells, but no effect in HCT116 and CCD-18Co	[62]
Human promyelocytic leukemia cells (HL-60)	Commercial fucosterol	Nd	MMT assay and incubated for 4 h	7.55, 15.1, 30.2, 60.4, and 120.8 μM	Not toxic	[63]
Human ovarian cancer (ES2 and OV90) cells	Commercial fucosterol	Nd	2′,7′-dichlorofluorescin diacetate assay	0, 20, 40, 60, 80, and 100 μM	Not toxic	[35]
HaCaT cells and monkey kidney COS-7 cells	Commercial fucosterol	Nd	MMT assay and incubated for 3 h	0.5, 1, and 5 μM	Had no effects on the viability of the cells	[45]

Nd: not determined.

In a study involving the use of commercial fucosterol for the treatment of RAW 264.7 macrophage cell line stimulated by particulate matter (PM), Jayawardena et al. [43] demonstrated the inhibition of NO production levels by observing inflammatory mediators, such as iNOS, COX-2, and pro-inflammatory cytokines (i.e., IL-6, interleukin-1β (IL-1β), and tumor necrosis factor-α (TNF-α)), including prostaglandin E2 (PGE2)). Furthermore, the effect of fucosterol was amplified by the decreased expression of mitogen-activated protein kinase (MAPK) and NF-κB signaling pathway molecules and ROS regulation. Fernando et al. [44] showed that increasing concentrations (>100 μg/mL) of Chinese fine dust PM (CPM) in A549 cells significantly increased ROS levels and caused cell death. CPM-induced A549 cells treated with fucosterol of *Sargassum aquifolium* at a concentration of 3.125–50 μg/mL caused an increase in cell viability of up to 94.98 ± 1.26%, and the IC_{50} value was estimated to be 21.74 ± 0.67 μg/mL [60]. In addition, *Sargassum aquifolium*-derived fucosterol was also reported to increase cell viability in HaCaT and HDF cells and yielded non-toxic results.

Another study reported that the addition of fucosterol isolated from *Sargassum fusiforme* in HaCaT cells induced by 500 μM $CoCl_2$ increased cell viability up to 56% at a concentration of 10 μM [61]. Furthermore, as reported by Choi et al. [47], there was no effect on HepG2 cells after treatment with the crude extract of fucosterol up to a concentration of 100 μM. However, in HepG2 cells induced by t-BHP and tacrine, fucosterol showed low toxicity and increased cell viability. In addition, LPS-induced RAW 264.7 macrophages treated with fucosterol of *Undaria pinnatifida* showed an unclear toxic effect because the 3-(4, 5-dimethyl thiazole-2-yl)-2, 5-diphenyltetrazolium bromide MTT test results showed that fucosterol did not affect cell viability at a concentration of 10–50 μM [42]. These findings are similar to those of Wong et al. [54] at concentrations ranging from 12 to 192 μM, fucosterol of *Padina australis* did not have a significant cytotoxic effect on C8-B4 cells when compared with control (untreated cells) (approximately 100% cell viability).

Research conducted by Paudel et al. [55] on activity of fucosterol from *Undaria pinnatifida* and *Eisenia bicyclis* reported that there was no visible effect of the crude fucosterol extract on MAO-A and MAO-B (half-maximal inhibitory concentration (IC_{50}) > 500 μM). MAO is a catecholamine-degrading enzyme with a long-standing therapeutic profile; MAO-A and MAO-B are isoenzymes. In addition, the results of the functional assay showed that fucosterol did not show agonist activity at any of the receptors tested. Moreover, the crude extract of fucosterol can potentially inhibit PTP1B and α-glucosidase with an IC_{50} value of 50.58 ± 1.86 μM [28], and BACE1 with an IC_{50} value of 64.12 ± 1.0 μM [53] without causing side effects.

In a report by Lee et al. [15], treatment with fucosterol from *Ecklonia cava* subsp. *stolonifera* of 3T3-L1 preadipocytes had no effect on inhibiting cell proliferation up to a concentration of 50 μM [15]. This finding is complemented by previous investigations that stated that the survival of HepG2 cells was not affected up to a concentration of 100 μM *Ecklonia cava* subsp. *stolonifera*-derived fucosterol for 24 h. However, cell survival was reduced to 48 h at a concentration of 200 μM. Based on these results, they recommended that additional in vitro studies on the anti-diabetic activity of fucosterol be carried out using non-toxic concentrations of 50, 25, and 12.5 μM [27].

The cytotoxicity of fucosterol in various human cancer cells has been widely published [17,62,64,65]. According to Mao et al. [17], commercial fucosterol inhibited the growth of all human lung cancer cells tested with IC_{50} values ranging from 15 to 60 μM. However, interestingly, fucosterol showed low toxicity in all normal cells with IC_{50} > 100 μM, which indicated that fucosterol can selectively inhibit lung cancer cell growth, induce cell cycle arrest, and target the Raf/MEK/ERK signaling pathway. Other studies have shown that commercial fucosterol reduces cell viability and enhances the cytotoxic effect of 5-Fu in HT29 cancer cells without affecting normal colon fibroblasts (CCD-18Co). Studies on the toxicity of fucosterol on HL-60 [63]; ES2 and OV90 [35], HT-29, Caco-2, and T47D [38]; KB, Hep-2, MCF-7, and SiHa [37] yielded low to no toxicity results in DLA cells [36].

In previous studies, the safety and toxicity of fucosterol in human and animal cell lines have been extensively investigated. Most of the studies focused only on the brown seaweeds of genera *Sargassum*, *Undaria*, *Turbinaria*, and *Ecklonia*. Based on the literature reviewed, other genera, such as *Ulva* and *Enteromorpha*, have not been tested in cell lines. Fucosterol from these genera has been reported to have biological activity, but its safety and toxicity levels in cell lines have not been reported. Therefore, the mechanisms of other genera in cell lines should be investigated further.

2.5. Safety and Toxicity of Fucosterol in Animals

Several studies on the safety and toxicity of fucosterol isolated from marine macroalgae in animals have been published. The brown seaweed *Ecklonia cava* subsp. *stolonifera* [56,57] and *Hizikia fusiformis* or *Sargassum fusiformis* [31,34,56,66] were the most frequently discussed. A summary of the safety and toxicity of fucosterol in animals is presented in Table 3.

The research of Mo et al. [50], complements the data about the anti-diabetic and anti-obesity properties of fucosterol that inhibits necrosis and apoptosis in a process mediated by PPARγ activation and inhibition of NF-B, which reduces inflammatory factors. Fucosterol also inhibits apoptosis and autophagy by upregulating Bcl-2 via PPARγ, thereby decreasing functional Bax and Beclin-1. Park et al. [34] reported that administration of 200 mg/kg body weight to mice increased splenocyte proliferation and NO production without cytotoxicity. Based on the research by Oktaviani et al. [56], fucosterol from *Sargassum fusiforme* showed low toxicity because it significantly affected the survival of *C. elegans* (1.54-fold and 1.23-fold increase), at a concentration of 0.05 mg/mL.

Some studies have been conducted to test the effect of fucosterol on the nervous system. Oh et al. [52] reported no toxicological response induced by the administration of fucosterol derived from *Ecklonia stolonifera* when it was injected at 10 mol/h into the dorsal hippocampus for four weeks. After training of aging rats, there was an increase in latency to reach the platform. Furthermore, Zhen et al. [66] showed no neurotoxic effect at the same dose levels administered after 0.5 and 4 h. Conversely, fucosterol from *Sargassum fusiforme*, showed no neurotoxic activity at the doses used in the forced or tail suspension tests (10, 20, 30, and 40 mg/kg). Lee et al. [31] found that all doses of fucosterol from *Sargassum fusiforme* had no toxic effects in OVX mice. The results of the study showed that fucosterol treatment significantly improved the loss of bone density caused by ovariectomy. Furthermore, the research conducted by Choi et al. [47] showed that there were no deaths or gross appearance abnormalities, and no fucosterol-induced abnormal behavioral changes, seizures, or death over 24 h due to *Ecklonia cava* subsp. *stolonifera* and *Eisenia bicyclis*. However, pretreatment with fucosterol at doses of 25, 50, and 100 mg/kg body weight markedly attenuated this cytotoxic effect of tacrine. In addition, the hepatoprotective effect of fucosterol at the highest dose (100 mg/kg body weight) resulted in serum ALT levels similar to those of the control group, suggesting that fucosterol has the potential to reduce tacrine-induced hepatotoxicity. The published results of fucosterol studies indicate that the number of in vivo tests involving algal metabolites is very limited. The experimental model that has been used thus far has focused on mice. Therefore, we propose that future research should focus on determining the full in vivo potency of fucosterol.

Table 3. Studies on safety and toxicity of fucosterol extracted from macroalgae, tested in animals.

Animal	Extract or Chemical	Sources	Method	Concentration	Toxicity	Ref.
E18 aging rats	Fucosterol extract	*Ecklonia cava* subsp. *stolonifera*	Dorsal hippocampus injected by fucosterol for 4 weeks	10 µmol/h	Increased the latency to reach the platform	[52]
C57BL/6 mice	Fucosterol extract	*Sargassum fusiforme*	Oral administration	50, 100, and 200 mg/kg	Increased splenocyte proliferation and increased NO production with no cytotoxicity	[34]
Balb/e mice	Fucosterol extract	*Sargassum fusiforme*	Administered by via gastric intubation route	0.1 mL/20 g of mouse	Not neurotoxic	[66]
Ovariectomized (OVX) rats	Fucosterol extract	*Sargassum fusiforme*	Oral, for 7 weeks beginning 12 weeks post-operation	25, 50, and 100 mg/kg	Had no toxic effects	[31]
Caenorhabditis elegans	Fucosterol extract	*Sargassum fusiforme*	Measured on both NGM agar and broth containing fucosterol	Up to 0.1 mg/mL in 2% dimethyl sulfoxide (DMSO)	Low toxicity	[56]
Institute of Cancer Research (ICR) mice	Fucosterol extract	*Ecklonia cava* subsp. *stolonifera* and *Eisenia bicyclis*	Oral, for 3 consecutive days	200 µL fucosterol (25, 50, and 100 mg/kg)	No mortality	[47]
BALB/c mice weighing	Commercial fucosterol	Nd	Oral, administered daily for 3 days	25, 50, or 100 mg/kg	Inhibited ConA-induced acute liver injury significantly	[50]

3. Materials and Methods

3.1. Literature Search

The preferred reporting items for systematic reviews and meta-analyses (PRISMA) method [67] were used for the collection, identification, screening, selection, and analysis of the studies reviewed. A literature search was performed using four databases: PubMed, Science Direct, Wiley, and Web of Science. The search criteria included scientific articles on fucosterol published between 2002 and 2020. The keywords used in the literature search were "fucosterol" and "bioactivites OR "biological activities" OR "safety" OR "toxicity" OR "characteristics" OR "structure" OR "cell lines" OR "microalgae" OR "macroalgae" OR "plant" OR "bacteria" OR "fungi" OR "invertebrates" OR "animals" OR "human." The total number of articles found was 1251, which, upon further screening by checking the title and keywords and removing similar articles, was decreased to 621.

3.2. Selection Criteria

In the second screening stage, a total of 532 articles from the initial 621 articles were excluded after screening, based on in-depth observations of the abstract content of the articles. The second screening yielded 89 articles that met the criteria. Of the 89 articles, additional in-depth observations of the full text of all the articles were made. A total of 46 articles remained after this stage 3 screening. Upon completion of this final stage, a total of 38 publications and 5 additional articles through manual reference tracing were included in the final data collection and further analyzed. All stages of systematic screening of the articles are shown in Figure 6.

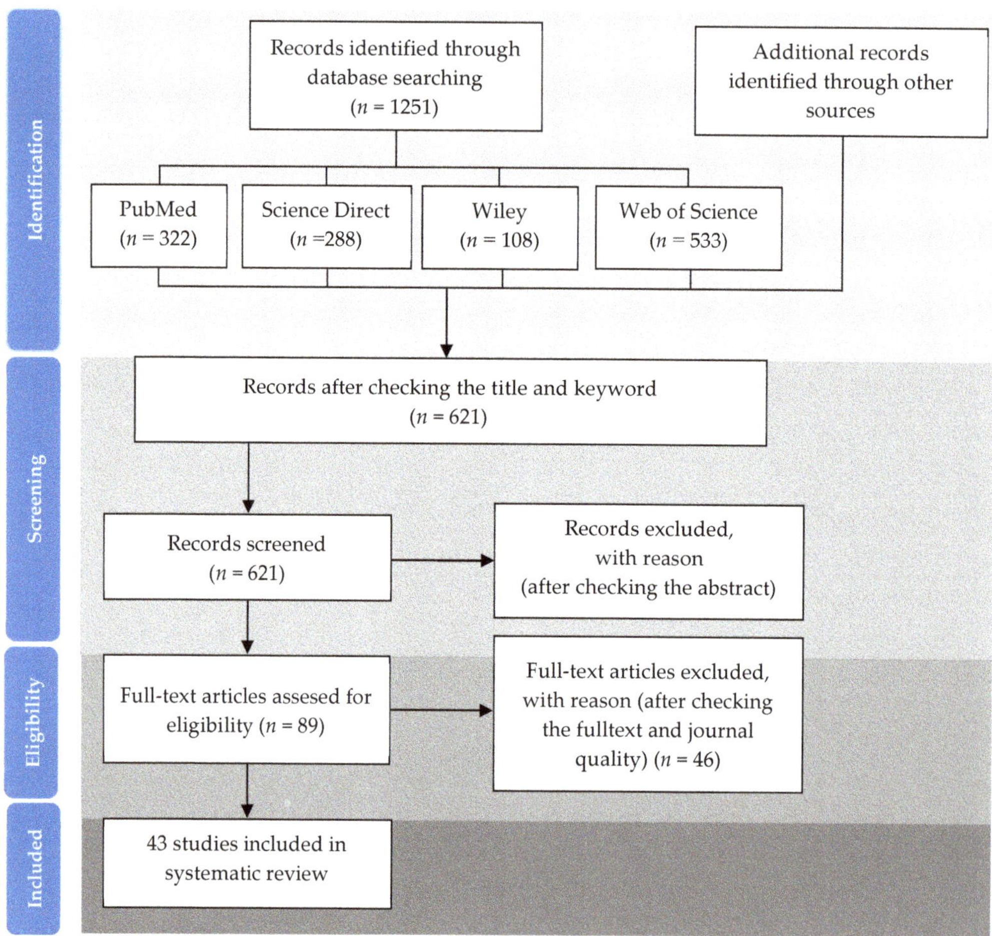

Figure 6. Summarized search method based on PRISMA.

3.3. Data Extraction

During the final selection, data were drawn from 43 studies on fucosterol. The data collection process was based on two main themes: (1) the characteristics of the articles, and (2) the criteria used to determine the bioactivity, safety, and toxicity of fucosterol. Specifically, the following features were extracted: year of publication, name of journal, geographic area where the research was conducted, author affiliation, publisher, biological activity of fucosterol, source of fucosterol, method used, experimental model, concentration of fucosterol, safety and toxicity of fucosterol.

4. Conclusions

Our review revealed that fucosterols derived from macroalgae can potentially be applied in the nutraceutical and pharmacological industries based on its bioactivities, safety, and toxicity. Fucosterol derived from brown macroalgae exhibits potentially beneficial activities; however, certain research gaps should be addressed. Despite the uncovered biological activities, most studies remain at a preliminary stage, and some of them have not included the study of safety and toxicity to organisms; hence, more in-depth studies are required. The studies of bioactivity, safety, and toxicity of fucosterol have been carried out mostly at the in vitro level (52%), and only 26% have been conducted at the in vivo level using a mouse model. An in vivo experiment can provide fundamental data for understanding the mechanism of nutraceutical and pharmacological properties of sterol before it is used in clinical trials. However, we did not find any clinical studies on fucosterol published between 2002 and 2020. Studies on the safety and toxicity levels of fucosterol at the clinical stage are very important for the development of this sterol for the nutraceuticals and pharmaceutical industry. Brown macroalgae (Phaeophyta) are a major source of fucosterol. A total of 2071 species belonging to the class Phaeophyta have been recorded [68]. However, the study of fucosterol in brown macroalgae continues to be limited to only certain species. *Hizikia fusiformis/Sargassum fusiforme* and *Ecklonia cava* subsp. *stolonifera* are the two species with the highest number of publications. Additional studies on other Phaeophyta species are required to elucidate the fucosterol content in specific algal classes. Future comprehensive research on fucosterol, including the study of macroalgae sources, chemical characterization, pharmacokinetic mechanisms, in vitro, in vivo, and clinical experiments will elucidate the role of fucosterol as a potent bioactive compound derived from marine sources.

Author Contributions: Conceptualization, M.D.N.M., J.-S.K. and J.-S.C.; formal analysis, M.D.N.M., D.H., B.F.S.P.N. and G.T.; investigation, M.D.N.M., D.H., J.-S.C., B.F.S.P.N. and G.T.; methodology, M.D.N.M., J.-H.S. and J.-S.C.; software, D.H.; supervision, M.D.N.M., J.-S.K. and J.-S.C.; visualization, M.D.N.M., J.-H.S. and D.H.; writing—original draft, M.D.N.M. and D.H.; writing—review and editing, M.D.N.M. and J.-S.C. All authors have read and agreed to the published version of the manuscript.

Funding: This research was funded by the Ministry of Oceans and Fisheries, Korea, under the Project No. PJT200885 entitled: "Development and commercialization of traditional seafood products based on the Korean coastal marine resources".

Institutional Review Board Statement: Not applicable.

Data Availability Statement: Data supporting reported results are available upon request.

Conflicts of Interest: The authors declare no conflict of interest.

Abbreviations

Abbreviations

5-Fu, 5-fluorouracil; AchE, acetylcholinesterase; ALT, alanine aminotransferase; AR, aldose reductase; RLAR, rat lens aldose reductase; AST, aspartate aminotransferase; BACE1, β-secretase 1; BchE, butyrylcholinesterase; Bcl-2, B-cell lymphoma-2; ConA, concanavalin A; COX-2, cyclooxygenase-2; ERK, extracellular-signal-regulated kinase; EtOAc, ethyl acetate; FoxO, forkhead box protein O; GRP78, glucose regulatory protein 78; GSH, glutathione; GSH-px, glutathione peroxidase; HAB, harmful algal blooms; HDL, high density lipoprotein; HO-1, heme oxygenase-1; HRAR, human recombinant aldose reductase;IC_{50}, half-maximal inhibitory concentration; IL-6, interleukin-6; IL-1β, interleukin-1β; iNOS, inducible nitric oxide synthase; LDL, low density lipoprotein; LPS, lipopolysaccharide; LXR, liver X receptor; MAO, monoamine oxidase; MAPK, microtubule associated protein kinase; MEK, mitogen-activated protein kinase; MIC, minimum inhibitory concentration; MMP-1, matrix metalloproteinase-1; MMPs, matrix metalloproteinases; mRNA, messenger RNA; MTT, (3-(4, 5-dimethyl thiazole-2-yl)-2, 5-diphenyltetrazolium bromide); mTOR, mammalian target of rapamycin; NF-κB, nuclear factor-κB; NGM, nematode growth medium; NO, nitric oxide; Nrf2, nuclear factor erythroid 2-related factor 2; P53, protein 53; p-c-Fos, phospho-c-Fos; p-c-Jun, phospho-c-Jun; PI3K/Akt, phosphatidylinositol 3-kinase/protein kinase B; PM, particulate matter; PPAR, proteins peroxisome proliferator-activated receptor; PPARγ, peroxisome proliferator- activated receptor gamma; PTP1B, protein tyrosine phosphatase 1B; Raf, mitogen-activated protein kinase; ROS, reactive oxygen species; sGOT, serum glutamic oxaloacetic transaminase; sGPT, serum glutamic pyruvic transaminase; SIRT1, sirtuin 1; SOD, superoxide dismutase; t-BHP, tert-butyl hydroperoxide; TGF-1, transforming growth factor-1; TLC, thin layer chromatography; TNF-α, tumor necrosis factor alpha; UVB, ultraviolet B.

References

1. Hannan, M.A.; Sohag, A.A.M.; Dash, R.; Haque, M.N.; Mohibbullah, M.; Oktaviani, D.F.; Hossain, M.T.; Choi, H.J.; Moon, I.S. Phytosterols of marine algae: Insights into the potential health benefits and molecular pharmacology. *Phytomedicine* **2020**, *69*, 153201. [CrossRef]
2. Heilbron, I.; Phipers, R.F.; Wright, H.R. The chemistry of the algae. Part I. The algal sterol fucosterol. *J. Chem. Soc.* **1934**, 1572–1576. [CrossRef]
3. Lopes, G.; Sousa, C.; Bernardo, J.; Andrade, P.B.; Valentão, P.; Ferreres, F.; Mouga, T. Sterol profiles in 18 macroalgae of the portuguese coast. *J. Phycol.* **2011**, *47*, 1210–1218. [CrossRef] [PubMed]
4. Saini, R.K.; Mahomoodally, M.F.; Sadeer, N.B.; Keum, Y.S.; Rengasamy, K.R. Characterization of nutritionally important lipophilic constituents from brown kelp *Ecklonia radiata* (C. Ag.) J. Agardh. *Food Chem.* **2021**, *340*, 127897. [CrossRef] [PubMed]
5. Sánchez-Machado, D.I.; López-Hernández, J.; Paseiro-Losada, P.; López-Cervantes, J. An HPLC method for the quantification of sterols in edible seaweeds. *Biomed. Chromatogr.* **2004**, *18*, 183–190. [CrossRef] [PubMed]
6. Terasaki, M.; Hirose, A.; Narayan, B.; Baba, Y.; Kawagoe, C.; Yasui, H.; Saga, N.; Hosokawa, M.; Miyashita, K. Evaluation of recoverable functional lipid components of several brown seaweeds (phaeophyta) from Japan with special reference to fucoxanthin and fucosterol contents1. *J. Phycol.* **2009**, *45*, 974–980. [CrossRef]
7. Kılınç, B.; Cirik, S.; Turan, G.; Tekogul, H.; Koru, E. Seaweed for Food and Industrial Application. In *Food Industry*; IntechOpen: London, UK, 2013; pp. 735–748.
8. McHugh, D.J. *Seaweeds uses as Human Foods*; Food and Agriculture Organization of United Nations: Rome, Italy, 2003; ISBN 92-5-104958-0.
9. Choi, J.S.; Seo, H.J.; Lee, Y.R.; Kwon, S.J.; Moon, S.H.; Park, S.M.; Sohn, J.H. Characteristics and in vitro anti-diabetic properties of the Korean rice wine, Makgeolli fermented with *Laminaria japonica*. *Prev. Nutr. Food Sci.* **2014**, *19*, 98–107. [CrossRef]
10. Choi, J.S.; Lee, K.; Lee, B.B.; Kim, Y.C.; Cho, K.K.; Choi, I.S. Safety, functionality and stability tests of *Ecklonia cava* extract for dermal application. *Toxicol. Environ. Health Sci.* **2014**, *6*, 61–66. [CrossRef]
11. Negara, B.F.S.P.; Sohn, J.H.; Kim, J.S.; Choi, J.S. Antifungal and larvicidal activities of phlorotannins from brown seaweeds. *Mar. Drugs* **2021**, *19*, 223. [CrossRef]
12. Khan, M.N.A.; Yoon, S.J.; Choi, J.S.; Park, N.G.; Lee, H.H.; Cho, J.Y.; Hong, Y.K. Anti-edema effects of brown seaweed (*Undaria pinnatifida*) extract on phorbol 12-myristate 13-Acetate-Induced mouse ear inflammation. *Am. J. Chin. Med.* **2009**, *37*, 373–381. [CrossRef]
13. Lee, S.; Yeon, S.L.; Sang, H.J.; Sam, S.K.; Kuk, H.S. Anti-oxidant activities of fucosterol from the marine algae *Pelvetia siliquosa*. *Arch. Pharm. Res.* **2003**, *26*, 719–722. [CrossRef]
14. Jung, H.A.; Jin, S.E.; Ahn, B.R.; Lee, C.M.; Choi, J.S. Anti-inflammatory activity of edible brown alga *Eisenia bicyclis* and its constituents fucosterol and phlorotannins in LPS-stimulated RAW264.7 macrophages. *Food Chem. Toxicol.* **2013**, *59*, 199–206. [CrossRef]
15. Lee, J.H.; Jung, H.A.; Kang, M.J.; Choi, J.S.; Kim, G. Do Fucosterol, isolated from *Ecklonia stolonifera*, inhibits adipogenesis through modulation of FoxO1 pathway in 3T3-L1 adipocytes. *J. Pharm. Pharmacol.* **2017**, *69*, 325–333. [CrossRef]
16. Gan, S.Y.; Wong, L.Z.; Wong, J.W.; Tan, E.L. Fucosterol exerts protection against amyloid β-induced neurotoxicity, reduces intracellular levels of amyloid β and enhances the mRNA expression of neuroglobin in amyloid β-induced SH-SY5Y cells. *Int. J. Biol. Macromol.* **2019**, *121*, 207–213. [CrossRef]
17. Mao, Z.; Shen, X.; Dong, P.; Liu, G.; Pan, S.; Sun, X.; Hu, H.; Pan, L.; Huang, J. Fucosterol exerts antiproliferative effects on human lung cancer cells by inducing apoptosis, cell cycle arrest and targeting of Raf/MEK/ERK signalling pathway. *Phytomedicine* **2019**, *61*, 152809. [CrossRef] [PubMed]
18. Meinita, M.D.N.; Harwanto, D.; Sohn, J.H.; Kim, J.S.; Choi, J.S. Hizikia fusiformis: Pharmacological and Nutritional Properties. *Foods* **2021**, *10*, 1660. [CrossRef] [PubMed]

19. Negara, B.F.P.S.; Sohn, J.H.; Kim, J.S.; Choi, J.S. Effects of Phlorotannins on Organisms: Focus on the Safety, Toxicity, and Availability of Phlorotannins. *Foods*. **2021**, *10*, 452. [CrossRef] [PubMed]
20. Jung, H.A.; Islam, M.N.; Lee, C.M.; Oh, S.H.; Lee, S.; Jung, J.H.; Choi, J.S. Kinetics and molecular docking studies of an anti-diabetic complication inhibitor fucosterol from edible brown algae *Eisenia bicyclis* and *Ecklonia stolonifera*. *Chem. Biol. Interact.* **2013**, *206*, 55–62. [CrossRef]
21. Huh, G.W.; Lee, D.Y.; In, S.J.; Lee, D.G.; Park, S.Y.; Yi, T.H.; Kang, H.C.; Seo, W.D.; Baek, N.I. Fucosterols from *Hizikia fusiformis* and their proliferation activities on osteosarcoma-derived cell MG63. *J. Korean Soc. Appl. Biol. Chem.* **2012**, *55*, 551–555. [CrossRef]
22. Tang, H.F.; Yi, Y.H.; Yao, X.S.; Xu, Q.Z.; Zhang, S.Y.; Lin, H.W. Bioactive steroids from the brown alga *Sargassum carpophyllum*. *J. Asian Nat. Prod. Res.* **2002**, *4*, 95–101. [CrossRef]
23. Hwang, E.; Park, S.Y.; Sun, Z.W.; Shin, H.S.; Lee, D.G.; Yi, T.H. The Protective Effects of Fucosterol Against Skin Damage in UVB-Irradiated Human Dermal Fibroblasts. *Mar. Biotechnol.* **2014**, *16*, 361–370. [CrossRef]
24. Yoon, N.Y.; Chung, H.Y.; Kim, H.R.; Choi, J.S. Acetyl- and butyrylcholinesterase inhibitory activities of sterols and phlorotannins from *Ecklonia stolonifera*. *Fish. Sci.* **2008**, *74*, 200–207. [CrossRef]
25. Jung, H.A.; Jung, H.J.; Jeong, H.Y.; Kwon, H.J.; Kim, M.S.; Choi, J.S. Anti-adipogenic activity of the edible brown alga *Ecklonia stolonifera* and its constituent fucosterol in 3T3-L1 adipocytes. *Arch. Pharm. Res.* **2014**, *37*, 713–720. [CrossRef] [PubMed]
26. Bouzidi, N.; Viano, Y.; Ortalo-Magné, A.; Seridi, H.; Alliche, Z.; Daghbouche, Y.; Culioli, G.; El Hattab, M. Sterols from the brown alga *Cystoseira foeniculacea*: Degradation of fucosterol into saringosterol epimers. *Arab. J. Chem.* **2019**, *12*, 1474–1478. [CrossRef]
27. Jung, H.A.; Bhakta, H.K.; Min, B.S.; Choi, J.S. Fucosterol activates the insulin signaling pathway in insulin resistant HepG2 cells via inhibiting PTP1B. *Arch. Pharm. Res.* **2016**, *39*, 1454–1464. [CrossRef] [PubMed]
28. Seong, S.H.; Nguyen, D.H.; Wagle, A.; Woo, M.H.; Jung, H.A.; Choi, J.S. Experimental and Computational Study to Reveal the Potential of Non-Polar Constituents from *Hizikia fusiformis* as Dual Protein Tyrosine Phosphatase 1B and α-Glucosidase Inhibitors. *Mar. Drugs* **2019**, *17*, 302. [CrossRef] [PubMed]
29. Lee, Y.S.; Shin, K.H.; Kim, B.K.; Lee, S. Anti-diabetic activities of fucosterol from *Pelvetia siliquosa*. *Arch. Pharm. Res.* **2004**, *27*, 1120–1122. [CrossRef]
30. Noureddin, M.; Mato, J.M.; Lu, S.C. Nonalcoholic fatty liver disease: Update on pathogenesis, diagnosis, treatment and the role of S-adenosylmethionine. *Exp. Biol. Med.* **2015**, *240*, 809–820. [CrossRef]
31. Lee, D.G.; Park, S.Y.; Chung, W.S.; Park, J.H.; Shin, H.S.; Hwang, E.; Kim, I.H.; Yi, T.H. The bone regenerative effects of fucosterol in in vitro and in vivo models of postmenopausal osteoporosis. *Mol. Nutr. Food Res.* **2014**, *58*, 1249–1257. [CrossRef]
32. Bang, M.H.; Kim, H.H.; Lee, D.Y.; Han, M.W.; Baek, Y.S.; Chung, D.K.; Baek, N.I. Anti-osteoporotic activities of fucosterol from sea mustard (*Undaria pinnatifida*). *Food Sci. Biotechnol.* **2011**, *20*, 343–347. [CrossRef]
33. Mayer, A.M.S.; Rodríguez, A.D.; Taglialatela-Scafati, O.; Fusetani, N. Marine pharmacology in 2009–2011: Marine compounds with antibacterial, antidiabetic, antifungal, anti-inflammatory, antiprotozoal, antituberculosis, and antiviral activities; affecting the immune and nervous systems, and other miscellaneous mechanisms of action. *Mar. Drugs* **2013**, *11*, 2510–2573. [CrossRef]
34. Park, S.; Hwang, E.; Shin, Y.; Lee, D.; Yang, J.; Park, J.; Yi, T. Immunostimulatory Effect of Enzyme-Modified *Hizikia fusiforme* in a Mouse Model In Vitro and Ex Vivo. *Mar. Biotechnol.* **2017**, *19*, 65–75. [CrossRef]
35. Bae, H.; Lee, J.Y.; Song, G.; Lim, W. Fucosterol suppresses the progression of human ovarian cancer by inducing mitochondrial dysfunction and endoplasmic reticulum stress. *Mar. Drugs* **2020**, *18*, 261. [CrossRef]
36. Kala, K.J.; Prashob Peter, K.J.; Chandramohanakumar, N. Cyto-toxic potential of fucosterol isolated from *Turbinaria conoides* against dalton's lymphoma ascites. *Int. J. Pharmacogn. Phytochem. Res.* **2015**, *7*, 1217–1221.
37. Caamal-Fuentes, E.; Moo-Puc, R.; Freile-Pelegrín, Y.; Robledo, D. Cytotoxic and antiproliferative constituents from *Dictyota ciliolata*, *Padina sanctae-crucis* and *Turbinaria tricostata*. *Pharm. Biol.* **2014**, *52*, 1244–1248. [CrossRef]
38. Khanavi, M.; Gheidarloo, R.; Sadati, N.; Shams Ardekani, M.R.; Bagher Nabavi, S.M.; Tavajohi, S.; Ostad, S.N. Cytotoxicity of fucosterol containing fraction of marine algae against breast and colon carcinoma cell line. *Pharmacogn. Mag.* **2012**, *8*, 60–64. [CrossRef]
39. Jiang, H.; Li, J.; Chen, A.; Li, Y.; Xia, M.; Guo, P.; Yao, S.; Chen, S. Fucosterol exhibits selective antitumor anticancer activity against hela human cervical cell line by inducing mitochondrial mediated apoptosis, cell cycle migration inhibition and downregulation of m-TOR/PI3K/Akt signalling pathway. *Oncol. Lett.* **2018**, *15*, 3458–3463. [CrossRef] [PubMed]
40. Prayong, P.; Barusrux, S.; Weerapreeyakul, N. Cytotoxic activity screening of some indigenous Thai plants. *Fitoterapia* **2008**, *79*, 598–601. [CrossRef] [PubMed]
41. Heo, S.J.; Yoon, W.J.; Kim, K.N.; Ahn, G.N.; Kang, S.M.; Kang, D.H.; Oh, C.; Jung, W.K.; Jeon, Y.J. Evaluation of anti-inflammatory effect of fucoxanthin isolated from brown algae in lipopolysaccharide-stimulated RAW 264.7 macrophages. *Food Chem. Toxicol.* **2010**, *48*, 2045–2051. [CrossRef] [PubMed]
42. Yoo, M.S.; Shin, J.S.; Choi, H.E.; Cho, Y.W.; Bang, M.H.; Baek, N.I.; Lee, K.T. Fucosterol isolated from *Undaria pinnatifida* inhibits lipopolysaccharide- induced production of nitric oxide and pro-inflammatory cytokines via the inactivation of nuclear factor-κB and p38 mitogen-activated protein kinase in RAW264.7 macrophages. *Food Chem.* **2012**, *135*, 967–975. [CrossRef] [PubMed]
43. Jayawardena, T.U.; Sanjeewa, K.K.A.; Lee, H.G.; Nagahawatta, D.P.; Yang, H.W.; Kang, M.C.; Jeon, Y.J. Particulate Matter-Induced Inflammation/Oxidative Stress in Macrophages: Fucosterol from Padina boryana as a Potent Protector, Activated via NF-κB/MAPK Pathways and Nrf2/HO-1 Involvement. *Mar. Drugs* **2020**, *18*, 628. [CrossRef]
44. Fernando, I.P.S.; Jayawardena, T.U.; Kim, H.S.; Lee, W.W.; Vaas, A.P.J.P.; De Silva, H.I.C.; Abayaweera, G.S.; Nanayakkara, C.M.; Abeytunga, D.T.U.; Lee, D.S.; et al. Beijing urban particulate matter-induced injury and inflammation in human lung epithelial

cells and the protective effects of fucosterol from *Sargassum binderi* (Sonder ex J. Agardh). *Environ. Res.* **2019**, *172*, 150–158. [CrossRef]

45. Kim, M.S.; Oh, G.H.; Kim, M.J.; Hwang, J.K. Fucosterol inhibits matrix metalloproteinase expression and promotes type-1 procollagen production in UVB-induced HaCaT cells. *Photochem. Photobiol.* **2013**, *89*, 911–918. [CrossRef]
46. Gossard, A.A.; Lindor, K.D. Autoimmune hepatitis: A review. *J. Gastroenterol.* **2012**, *47*, 498–503. [CrossRef]
47. Choi, J.S.; Han, Y.R.; Byeon, J.S.; Choung, S.Y.; Sohn, H.S.; Jung, H.A. Protective effect of fucosterol isolated from the edible brown algae, *Ecklonia stolonifera* and *Eisenia bicyclis*, on tert-butyl hydroperoxide- and tacrine-induced HepG2 cell injury. *J. Pharm. Pharmacol.* **2015**, *67*, 1170–1178. [CrossRef]
48. Hoang, M.H.; Jia, Y.; Jun, H.J.; Lee, J.H.; Lee, B.Y.; Lee, S.J. Fucosterol is a selective liver X receptor modulator that regulates the expression of key genes in cholesterol homeostasis in macrophages, hepatocytes, and intestinal cells. *J. Agric. Food Chem.* **2012**, *60*, 11567–11575. [CrossRef]
49. Chen, Z.; Liu, J.; Fu, Z.; Ye, C.; Zhang, R.; Song, Y.; Zhang, Y.; Li, H.; Ying, H.; Liu, H. 24(S)-saringosterol from edible marine seaweed *Sargassum fusiforme* is a novel selective LXRβ agonist. *J. Agric. Food Chem.* **2014**, *62*, 6130–6137. [CrossRef] [PubMed]
50. Mo, W.; Wang, C.; Li, J.; Chen, K.; Xia, Y.; Li, S.; Xu, L.; Lu, X.; Wang, W.; Guo, C. Fucosterol protects against concanavalin a-induced acute liver injury: Focus on P38 MAPK/NF-κ B pathway activity. *Gastroenterol. Res. Pract.* **2018**, *2018*. [CrossRef] [PubMed]
51. Ishwarya, M.; Narendhirakannan, R.T. The advances in neurobiology. *Adv. Neurobiol.* **2016**, *12*, 293–306. [CrossRef]
52. Oh, J.H.; Choi, J.S.; Nam, T.J. Fucosterol from an edible brown alga *Ecklonia stolonifera* prevents soluble amyloid beta-induced cognitive dysfunction in aging rats. *Mar. Drugs* **2018**, *16*, 368. [CrossRef] [PubMed]
53. Jung, H.A.; Ali, M.Y.; Choi, R.J.; Jeong, H.O.; Chung, H.Y.; Choi, J.S. Kinetics and molecular docking studies of fucosterol and fucoxanthin, BACE1 inhibitors from brown algae *Undaria pinnatifida* and *Ecklonia stolonifera*. *Food Chem. Toxicol.* **2016**, *89*, 104–111. [CrossRef]
54. Wong, C.H.; Gan, S.Y.; Tan, S.C.; Gany, S.A.; Ying, T.; Gray, A.I.; Igoli, J.; Chan, E.W.L.; Phang, S.M. Fucosterol inhibits the cholinesterase activities and reduces the release of pro-inflammatory mediators in lipopolysaccharide and amyloid-induced microglial cells. *J. Appl. Phycol.* **2018**, *30*, 3261–3270. [CrossRef]
55. Paudel, P.; Seong, S.H.; Jung, H.A.; Choi, J.S. Characterizing fucoxanthin as a selective dopamine D3/D4 receptor agonist: Relevance to Parkinson's disease. *Chem. Biol. Interact.* **2019**, *310*, 1–8. [CrossRef] [PubMed]
56. Oktaviani, D.F.; Bae, Y.; Dyah, M.; Meinita, N.; Moon, I.S. An Ethanol Extract of the Brown Seaweed *Hizikia fusiformis* and Its Active Constituent, Fucosterol, Extend the Lifespan of the Nematode Caenorhabditis elegans. *J. Life Sci.* **2019**, *29*, 1120–1125. [CrossRef]
57. Tyskiewicz, K.; Tyskiewicz, R.; Konkol, M.; Rój, E.; Jaroszuk-Scisеł, J.; Skalicka-Wozniak, K. Antifungal Properties of Fucus vesiculosus L. Supercritical Fluid Extract against *Fusarium culmorum* and *Fusarium oxysporum*. *Molecules* **2019**, *24*, 3518. [CrossRef]
58. Kumar, S.S.; Kumar, Y.; Khan, M.S.Y.; Gupta, V. New antifungal steroids from Turbinaria conoides (J. Agardh) Kutzing. *Nat. Prod. Res.* **2010**, *24*, 1481–1487. [CrossRef]
59. Rajendran, I.; Chakraborty, K.; Pananghat, V. Bioactive sterols from the brown alga *Anthophycus longifolius* (Turner) Kützing, 1849 (= *Sargassum longifolium*). *Indian J. Fish.* **2013**, *60*, 83–86.
60. Fernando, I.P.S.; Jayawardena, T.U.; Kim, H.S.; Vaas, A.P.J.P.; De Silva, H.I.C.; Nanayakkara, C.M.; Abeytunga, D.T.U.; Lee, W.W.; Ahn, G.; Lee, D.S.; et al. A keratinocyte and integrated fibroblast culture model for studying particulate matter-induced skin lesions and therapeutic intervention of fucosterol. *Life Sci.* **2019**, *233*, 116714. [CrossRef]
61. Sun, Z.; Mohamed, M.A.A.; Park, S.Y.; Yi, T.H. Fucosterol protects cobalt chloride induced inflammation by the inhibition of hypoxia-inducible factor through PI3K/Akt pathway. *Int. Immunopharmacol.* **2015**, *29*, 642–647. [CrossRef]
62. Ramos, A.A.; Almeida, T.; Lima, B.; Rocha, E. Cytotoxic activity of the seaweed compound fucosterol, alone and in combination with 5-fluorouracil, in colon cells using 2D and 3D culturing. *J. Toxicol. Environ. Heal.-Part A Curr. Issues* **2019**, *82*, 537–549. [CrossRef]
63. Ji, Y.B.; Ji, C.F.; Yue, L. Study on human promyelocytic leukemia HL-60 cells apoptosis induced by fucosterol. *Biomed. Mater. Eng.* **2014**, *24*, 845–851. [CrossRef]
64. Malhão, F.; Ramos, A.A.; Rocha, E. Cytotoxic and Anti-Proliferative Effects of Fucosterol, Alone and in Combination with Doxorubicin, in 2D and 3D Cultures of Triple-Negative Breast Cancer Cells. *Med. Sci. Forum* **2021**, *2*, 8600. [CrossRef]
65. Pacheco, B.S.; dos Santos, M.A.Z.; Schultze, E.; Martins, R.M.; Lund, R.G.; Seixas, F.K.; Colepicolo, P.; Collares, T.; Paula, F.R.; De Pereira, C.M.P. Cytotoxic activity of fatty acids from Antarctic macroalgae on the growth of human breast cancer cells. *Front. Bioeng. Biotechnol.* **2018**, *6*, 1–10. [CrossRef] [PubMed]
66. Zhen, X.H.; Quan, Y.C.; Jiang, H.Y.; Wen, Z.S.; Qu, Y.L.; Guan, L.P. Fucosterol, a sterol extracted from *Sargassum fusiforme*, shows antidepressant and anticonvulsant effects. *Eur. J. Pharmacol.* **2015**, *768*, 131–138. [CrossRef] [PubMed]
67. Moher, D.; Liberati, A.; Tetzlaff, J.; Altman, D.G.; Altman, D.; Antes, G.; Atkins, D.; Barbour, V.; Barrowman, N.; Berlin, J.A.; et al. Preferred reporting items for systematic reviews and meta-analyses: The PRISMA statement. *PLoS Med.* **2009**, *6*, e1000097. [CrossRef] [PubMed]
68. Guiry, M.D.; Guiry, G.M. *AlgaeBase. World-Wide Electronic Publication*; National University of Ireland: Galway, UK, 2021. Available online: http://www.algaebase.org (accessed on 10 September 2021).

Review

Bioactive Peptides from Algae: Traditional and Novel Generation Strategies, Structure-Function Relationships, and Bioinformatics as Predictive Tools for Bioactivity

Jack O'Connor [1,2], Marco Garcia-Vaquero [3,*], Steve Meaney [1] and Brijesh Kumar Tiwari [2]

1 School of Biological & Health Sciences, Technological University Dublin, Dublin 2, Ireland; jack.oconnor@teagasc.ie (J.O.); steve.meaney@tudublin.ie (S.M.)
2 Department of Food Chemistry and Technology, Teagasc Food Research Centre, Ashtown, Dublin 15, Ireland; brijesh.tiwari@teagasc.ie
3 Section of Food and Nutrition, School Agriculture and Food Science, University College Dublin, Belfield, Dublin 4, Ireland
* Correspondence: marco.garciavaquero@ucd.ie; Tel.: +353-(01)-716-2513

Abstract: Over the last decade, algae have been explored as alternative and sustainable protein sources for a balanced diet and more recently, as a potential source of algal-derived bioactive peptides with potential health benefits. This review will focus on the emerging processes for the generation and isolation of bioactive peptides or cryptides from algae, including: (1) pre-treatments of algae for the extraction of protein by physical and biochemical methods; and (2) methods for the generation of bioactive including enzymatic hydrolysis and other emerging methods. To date, the main biological properties of the peptides identified from algae, including anti-hypertensive, antioxidant and anti-proliferative/cytotoxic effects (for this review, anti-proliferative/cytotoxic will be referred to by the term anti-cancer), assayed in vitro and/or in vivo, will also be summarized emphasizing the structure–function relationship and mechanism of action of these peptides. Moreover, the use of in silico methods, such as quantitative structural activity relationships (QSAR) and molecular docking for the identification of specific peptides of bioactive interest from hydrolysates will be described in detail together with the main challenges and opportunities to exploit algae as a source of bioactive peptides.

Keywords: in silico; biotechnology; cryptides; anti-hypertensive; antioxidant; anti-cancer

Citation: O'Connor, J.; Garcia-Vaquero, M.; Meaney, S.; Tiwari, B.K. Bioactive Peptides from Algae: Traditional and Novel Generation Strategies, Structure-Function Relationships, and Bioinformatics as Predictive Tools for Bioactivity. *Mar. Drugs* **2022**, *20*, 317. https://doi.org/10.3390/md20050317

Academic Editor: Marialuisa Menna

Received: 13 April 2022
Accepted: 3 May 2022
Published: 10 May 2022

1. Introduction

The world's population is estimated to reach 9 billion by 2050 and there is continuing pressure on the current agricultural and food production systems to meet the increased food demands without damaging the environment [1,2]. Algae (macro- and microalgae) have become an attractive source of raw protein for the food industry [3] due to a potential protein content which often exceeds that of other common protein sources such as milk and soy (which contain approximately 40% protein on a dry weight basis) [4]. For example, the dry weight yield of protein in dried algal biomass can be up to 47% in the macroalgae *Pyropia* sp. [5] or 65% in the microalga *Chorella* sp. [6]. Thus, the production of algae offers excellent opportunities to increase food production without increasing deforestation or encroaching upon natural habitats while benefitting from an all-year harvest for algae [3]. In contrast to protein from milk (0.13 ton/acre/annum) and soybean (0.41–0.81 ton/acre/annum), macro and microalgae biomass can produce high protein yields ranging between 1.62 and 6.1, and 4.1 and 7.3 ton/acre/annum for macroalgae and microalgae, respectively [7].

Algal proteins are also a source of bioactive peptides, also named cryptides, as these compounds have the ability to exert direct physiological effects once they are released from their parent proteins where they remain inactive [8]. Similar to endogenous peptide hormones (e.g., those derived from proopiomelanocortin, insulin and angiotensin), bioactive peptides derive from parent polypeptide sequences through a series of controlled and specific proteolytic cleavages [9]. In some cases, several different sequences with a hormone-like action can be derived from the same parent polypeptide through differential cleavage [9]. Although numerous bioactive peptides with potential health benefits, e.g., anti-hypertensive properties, have been isolated from macro- and microalgae, there are technical challenges associated with the production and commercialization of bioactive peptides that still need to be addressed.

This review aims to provide a comprehensive summary of the main approaches for the generation and isolation of bioactive peptides from algae. The review will focus mainly on novel uses of pre-treatment methods for the extraction of protein from algal biomass by physical and biochemical methods, as well as the enzymatic hydrolysis and other emerging methods for the release of bioactive peptides from algae. The main biological properties of these peptides (anti-hypertensive, antioxidant, and anti-cancer) and the structure–function relationships of known peptide sequences from algae will also be discussed in relation to their hypothesized mechanisms of action. Moreover, the recent developments in bioinformatics or in silico tools helping in the identification of these structures and their health benefits will also be discussed together with the main challenges and opportunities of bioactive peptides from algae.

2. Process of Generation and Isolation of Bioactive Peptides

The initial extraction of protein from raw biomass by the use of pre-treatments is needed as the first step for further protein processing for the generation and subsequent isolation of bioactive peptides.

2.1. Pre-Treatments of the Algal Biomass

Although algae are described in the literature as biomass rich in proteins, the complex structures of the carbohydrate-rich algal cell walls prevent immediate access to these compounds, and thus, algae must be pre-treated by either physical or biochemical methods to allow the release of proteins from the biomass [10]. Amongst all the physical methods available, the application of physical methods, such as pulsed electric fields (PEF), ultrasound-assisted extraction (UAE) and microwave-assisted extraction (MAE), has shown promising results in algae. Other enzyme-based methods use the application of high specific-activity enzyme preparations to degrade the cell walls [11]. The specific pre-treatments must be tailored and optimized on the basis of the algal species studied in order to achieve high yields of proteins. For example, the macroalgae *Chondrus crispus* has a cellulose microfibril base cell wall and a carrageenan matrix and thus, the extraction of protein from this macroalgae requires an enzymatic mixture with high carrageenase and cellulase activities [12,13].

2.1.1. Physical Pre-Treatments

PEF is based on exposing cells to a high-strength electric field, inducing the formation of transient pores in the cell walls in a process known as electroporation [14]. The disruption of the cell walls by PEF can also lead to the formation of permanent or transient pores, allowing the mass transfer of small molecular weight compounds including protein extraction [15] to the outside of the cells and into the solution [16]. Pore sizes may also be further influenced by PEF duration, for instance, the pore radius for short pulses (5 μs) was approximately 0.8–1.6 nm, while when using longer pulses (10 μs) the pore radius increased up to at least approximately 5 nm, which was considered as irreversibly formed [17]. To date, PEF has been primarily used as a method to extract lipids from algae in connection with biofuel manufacture [17]. Vanthoor-Koopmans, Wijffels, Barbosa and Eppink [4] reported that small peptides and free amino acids present in the cell or created during the PEF process may be released from the biomass following PEF, while larger peptides and proteins were retained within the cell. Thus, according to the authors, adding a protein solubilization step will benefit these extraction processes. This solubilization can be achieved by adding surfactants and/or increasing the pH to untangle proteins, increasing their solubilization and extraction ratios [4,18]. The yield of the extracted protein from *Nannochloropsis* spp. doubled when using alkaline solvents (pH 8) compared to water; however, the recovery of proteins from the biomass was still low compared to other physical extraction procedures, such as UAE [19]. PEF has been described as a promising method for the extraction of proteins from the macroalga *Ulva ohnoi* [20]. The authors achieved a 3-fold increase in the yield of proteins from *U. ohnoi* of with the application of PEF. Similarly, in the case of *Arthrospira platensis*, a microalgae species with high protein content, PEF treatment followed by soaking the biomass in water increased the extraction of the intracellular proteins (c-phycocyanin) 12.7-fold compared to the control [21]. Parniakov, Barba, Grimi, Marchal, Jubeau, Lebovka and Vorobiev [19] emphasized the advantages of PEF in relation to power usage and cost-effectivity of the treatments, although the yields of protein extraction were lower compared to UAE pre-treatments. The power consumption of PEF was approximately 100 kJ/kg at 20 kV/cm for 6 ms, while in the case of UAE it was 250 kJ/kg at 200 W for 10 min [19].

UAE is based on the generation of high-frequency ultrasonic waves in a liquid, which generates a great number of bubbles that collapse (cavitation) and release a burst of energy that can disrupt the algal cell walls [22]. Advantages of this technology include the possibility to combine its application with various solvents, targeting the extraction of different compounds from the biomass, as well as an increased extraction efficiency in terms of reduced times and energy of extraction compared to conventional solvent extraction processes [23]. For example, the use of UAE with a subsequent alkaline treatment allowed the extraction of 57% of the total proteins from *Ascophyllum nodosum* [24]. Moreover, the use of ultrasounds enabled 2.29-fold higher yields of protein extraction compared to conventional methods from the microalga *A. platensis* [25].

MAE has also been used to extract protein from algae [26]. MAE uses oscillating electric fields which cause vibrational friction of polar molecules in the cells, which allows heating of the sample to occur. The main advantage of MAE is that a small amount of or even no solvents are required, as well as the fast extraction times of this technique compared to conventional methods [27,28]. Juin, Chérouvrier, Thiéry, Gagez, Bérard, Joguet, Kaas, Cadoret and Picot [26] used MAE to extract water-soluble proteins from *Porphyridium purpureum*, focusing mainly on the extraction of phycobilin proteins. The authors reported that MAE improves the extraction yields of compounds and significatively reduced the time of extraction to seconds compared to the hours needed to generate similar extraction yields when using traditional solving extraction methods. Moreover, MAE appeared to be more efficient in protecting thermolabile compounds during the process of extraction, as the application of high temperatures for prolonged periods of time needed during conventional solvent extraction may have a negative impact on these compounds once they are extracted [29].

2.1.2. Enzymatic Pre-Treatments

One of the key advantages of enzymatic pre-treatments over other mechanical methods is the relatively low temperatures required, allowing the release of both peptides and proteins from algae with minimal or no damage to their structures [30]. However, the high variability of the composition of the algal cell walls between species requires will require the customization and optimization of these enzymatic treatments [11]. For example, a mixture of trypsin, collagenase, lysozyme, and autolysin was useful as a pre-treatment to disrupt the cell walls of *Chlamydomonas reinhardtii* [31]. Autolysin was the most efficient enzyme as an enzymatic treatment and showed to be preferable to other chemical and mechanical methods such as solvents and sonication [31]. Moreover, the authors also reported that longer pre-treatments resulted in the total lysis of the cells advantageously resulting in increased extraction yields of cellular compounds measured as proteins and lipids [31]. Other pre-treatments using cellulase followed by protein hydrolysis with bromelain were effective when extracting proteinaceous concentrates from *Fucus spiralis*, increasing its extraction yields by 1.5-fold compared to those achieved by using bromelain alone [32]. The protein content of *Macrocystis pyrifera* and *Chondracanthus chamissoi* was increased by disrupting the carbohydrate matrix of the algae by using cellulase, increasing the yields of protein extraction [33]. As expected, optimum enzymatic conditions also varied between both macroalgae, achieving protein extraction yields of approximately 75% from *M. pyrifera* by an optimized enzymatic treatment with cellulase (1:10, enzyme:seaweed ratio) for 18 h and yields of 36% from *C. chamissoi* using the same enzyme:seaweed ratio for 12 h [33]. Moreover, Fleurence, Massiani, Guyader and Mabeau [13] reported that a mix of carrageenase/cellulase has a 10-fold higher extraction efficiency over the use of just carrageenase alone in *C. crispus*. Similarly, mixtures of agarase/cellulose achieved a 3-fold increase in protein extraction from *G. verrucosa*, while in the case of *P. palmata*, the combined xylanase/cellulase had similar protein extraction yields to those of control [13].

2.2. Generation of Bioactive Peptides

Once the protein is extracted from the biomass, the classical method for the generation of bioactive peptides uses proteases to break peptide bonds and generate hydrolysates containing a complex mix of peptides [34]. Enzymes exert their action by cleaving sequence motifs within a protein. The preferred cleavage sites of each of these enzymes are summarized in Table 1.

Table 1. Characteristics of proteases used for the generation of bioactive peptides.

Enzyme Name	Type of Enzyme	pH Range	Temperature	Cleavage Preference	References
Trypsin	Serine protease	7.8	37–42 °C	Positively charged amino acids; R and K	[35]
Chymotrypsin	Serine protease	7.8	37–42 °C	Hydrophobic amino acids; Y, F and W	[36]
Pepsin	Aspartic protease	1.25–2.5	37–42 °C	Positively charged amino acids; R and K	[37]
Alcalase	Serine endopeptidase	6.5–10	60–75 °C	Broad range specificity however, propensity for cleaving aromatic amino acids	[38–40]
Papain	Cysteine endopeptidase	6–7	65 °C	Broad range specificity, cleaving peptide bonds of basic amino acids, L or G. Papain will not accept V at position 1 and at position 2 prefers large hydrophobic amino acids	[41]
Bromaline	Cysteine endopeptidase	4.5–8	35–55 °C	Broad range specificity with preferred cleavage site at the C terminus of K, A, Y and G	[42]
Protamex	Mixture of endo- and exo-proteases from *Bacillus* sp.	6–9	30–65 °C	Broad cleavage range as it is a mixture of proteases	[43,44]
Elastase	Serine protease	9	37 °C	Preferred cleavage at the C-terminus of A, V, S, G, L and I	[45]
Thermolysin	Metalloproteinase	5–8.5	65–85 °C	Preferred cleavage at the N-terminus of F, V, I, L, M and A	[46]

Amongst all the proteases, trypsin, chymotrypsin and pepsin have been the most widely used enzymes for the generation of bioactive peptides. Trypsin has a specific binding affinity for positively charged side chains of the amino acids lysine and arginine. Trypsin's cleavage site is on the C-terminal side of the amino acid residues. Hydrolysis is decreased with the presence of acidic amino acids on either side of the cleavage site. Cleavage will not occur if a proline residue is present on the carboxyl side of the cleavage site [35]. Chymotrypsin is another serine protease which is itself activated by trypsin cleaving the bond at residues 15 and 16 (arginine and isoleucine). The chymotrypsin enzyme catalyzes the hydrolysis of proteins by cleaving the molecules at hydrophobic amino acid residues, such as the L-isomers of tyrosine, phenylalanine, and tryptophan. It also has the capability of acting on amides and esters of susceptible amino acids [47]. The use of these enzymes can result in the one key advantages of this approach, which is the reproducibility of the process such that similar proteins and hydrolysates products are generated.

Emerging technologies have also been used to hydrolyze protein extracts and generate bioactive peptides from algae and other food products and by-products [48,49]. Amongst them, subcritical water (SCW) processing has been gaining attention as both a green extraction and protein hydrolysis method [50–52]. SCW does not require the use of expensive and lengthy reaction times and it can be a method of extracting compounds from highly insoluble media which are ecologically damaging by-products from industries, such as poultry waste [53], hog fur [54] and fisheries waste [55]. In fact, SCW production of bioactive peptides also significantly reduces the processing time of hydrolysis of collagen by a factor of almost 300 when using enzymatic methods such as collagenase [56] to up 5 min [57]. The use of SCW to produce amino acids and peptides from waste and under-utilized by-products could give these industries a new revenue stream while mitigating the ecological and economic issues currently associated with the disposal of these by-products [58].

The SCW process maintains water in subcritical conditions inside the reaction chamber by using oven temperatures ranging from >100 °C to <374 °C and an internal pressure of <22 MPa, stimulating the formation of hydronium (H_3O^+) and hydroxide ions (HO^-) that allow water to interact as a basic or acidic catalyst [52]. The pressure applied during SCW will cause the unfolding and loss of secondary, tertiary, and quaternary structures of the protein, while the ions will interact with the amino acids [49]. The amino acids that are particularly vulnerable to hydrolysis by SCW are aspartic acid and glutamic acid, affected by the weak acidic conditions as their carboxyl group becomes a proton donor for the hydrolysis of the peptide bond next to it [52]. Ahmed, Mulla, Al-Ruwaih and Arfat [57] reported that using sequential pressure pretreatment of 300 MPa for 15 min increased the degree of hydrolysis for proteins when being hydrolyzed with alcalase [57], an enzyme that cleaves the carboxyl side of the amino acids E, L, Y, Q and E [58]. This indicates that the application of high pressure leads to a certain degree of protein unfolding, potentially increasing access of the enzyme to substrate cleavage sites [57]. A similar study using soy protein performed by Meinlschmidt et al. [59] showed similar enhanced digestibility following exposure to a pressure of 100 MPa in the presence of the enzyme flavourzyme for 15 min. Under these conditions, when the pressure exceeds 100 MPa the enzyme itself starts to become denatured by the pressure, and its activity is lost [59].

SCW appears to have some cleavage specificity for bonds adjacent to aspartyl residues, with some 44% of the peptides produced from subcritical water-mediated hydrolysis of BSA containing an N-terminal aspartic [52]. Moreover, peptide production from the microalgae *A. platensis* was optimal at 160 °C, while temperatures of over 220 °C produced an intense degradation of these proteins and the release of free amino acids rather than peptides, with no distinguishable bands when analyzing the hydrolysates by denaturing protein electrophoresis [49,55]. SCW has also been explored for the production of amino acids at temperatures of 240 °C [60,61], while temperatures reaching 260 °C will result in the degradation of amino acids to organic acids and ammonia [61]. These data illustrate the need for careful control of temperature during SCW processing to ensure the appropriate release of protein and peptides rather than terminal degradation.

After the proteins have been processed and hydrolyzed to generate bioactive peptides, one or several purification processes are frequently applied to isolate these molecules further. Overall, most authors used one or several steps of molecular weight cut-off filtration (MWCO) to fractionate the compounds of the hydrolysates based on their molecular weight [62]. Thereby, Megías et al. [63] used 5 kDa membranes to remove, concentrate and purify peptides in the hydrolysate by removing larger unhydrolyzed protein fractions and the protease enzymes themselves, as these compounds will be collected and discarded in the retentate. Further purification techniques can also be applied including chromatographic techniques, mainly reversed-phase high-performance liquid chromatography (RP-HPLC) and ultra-performance liquid chromatography (UPLC) depending on the level of purity desired in the final product. Previous studies generating bioactive peptides from the macroalga *Ulva* spp. applied MWCO followed by preparative RP-HPLC at wavelengths of 214 nm, to detect peptide bonds, and 280 nm, indicative of the presence of aromatic amino acids [64]. These and other purification strategies to isolate bioactive peptides from algae have been recently reviewed in detail by Lafarga et al. [62].

3. Biological Activities and Modes of Action of Algal Peptides

The generation of bioactive peptides is gaining momentum due to the wide range of biological properties attributed to these compounds that have been extensively reviewed [34,62]. Thus, this section will briefly mention a few examples of the anti-hypertensive, antioxidant, and anti-cancer activities from algae described in the recent scientific literature, also focusing on relating these described activities to their proposed mechanism of action and tools used for these analyses.

3.1. Antihypertensive Peptides

Cardiovascular disease (CVD) is one of the leading causes of mortality in the world today and hypertension is a significant risk factor for CVD. The regulation of blood pressure is mainly maintained by the renin angiotensin pathway. Briefly, the renin angiotensin system works by the secretion of renin into the blood system from the kidneys. Renin then binds the peptide angiotensinogen and forms angiotensin I. The angiotensin converting enzyme (ACE) binds and cleaves angiotensin I and transforms it into the highly potent vasoconstrictor angiotensin II, thus increasing blood pressure [65].

Fitzgerald et al. [66] extracted protein from the macroalgae *P. palmata* and performed an enzymatic hydrolysis with papain, identifying within the hydrolysate the peptide IRLIIVLMPILHA which potently inhibited the enzyme renin. Moreover, when this peptide sequence undergoes an in vitro digestion process, the gastrointestinal enzymes cleaved the peptide resulting in the production of the di-peptide IR with high anti-renin activity. In a follow up in vivo study using spontaneously hypertensive rats (SHR) and dosing with oral gavage, captopril reduced the blood pressure by 29 mm Hg, while the *P. palmata* hydrolysate reduced it by 34 mm Hg and IRLIIVLMPILHA peptide showed a reduction of 34 mm Hg [67].

ACE is a highly druggable target and several widely prescribed antihypertensive agents (e.g., captopril) are ACE inhibitors [68]. These inhibitors function by preventing the ACE-mediated conversion of angiotensin I into angiotensin II, preventing an increase in blood pressure. Captopril is a proline-based synthetic analog of a peptide present in snake venom that is a competitive inhibitor of ACE [69]. However, drugs like captopril, enalapril and lisinopril have serious adverse side effects that include dry cough, skin rashes, renal failure, and congenital malformations amongst others [70,71]. Thus, there is a growing interest in isolating new peptides with ACE inhibitory activity from natural sources, including those in food [72].

Multiple peptides with ACE inhibitory properties have been isolated from protein extracts from the microalgae *C. vulgaris* and *A. platensis* followed by enzymatic hydrolysis with pepsin [73]. In vivo tests evaluating the efficiency of peptides in SHR revealed that the oral administration of the tetrapeptide IAPG—isolated from *A. platensis*—resulted in a decrease in systolic blood pressure by approximately 50 mm Hg within 1 h of its ingestion [73]. The tripeptide FAL—isolated from *Chlorella*—was less potent in the SHR model, leading to a decrease of approximately 40 mm Hg within 2 h of ingestion. Moreover, the physiological effects of both IAPG and FAL in the SHR were sustained for 4 h post-ingestion [73].

Using a similar approach, Sun et al. [74] prepared hydrolysates from the macroalga *Ulva intestinalis* protein using trypsin, pepsin, papain, α-chymotrypsin and alcalase, and determined the in vitro activity of these hydrolysates when inhibiting ACE. The authors determined that trypsin-derived hydrolysates had the greatest inhibitory effect and identified the peptides FGMPLDR and MELVLR as those responsible for this effect. The authors also performed molecular docking with AutoDock 4.2 to reveal that while both peptides were bound to the active site, the mode of binding was different [74]. FGMPLDR interacted with Glu123, Ala354, Ala356, Glu384, and Arg522 and in particular with Ala354 and

Glu384 which are both present in the S1 pocket of ACE, interacting with a well-known ACE inhibitor, lisinopril. In contrast, MELVLR was predicted to interact with Asn70, Glu143, Gln281, His383, and Lys511, with Gln281 and Lys511 of particular importance and located in the S2 pocket of the active site of ACE [74].

To our knowledge, there are not many studies with bioactive peptides from algae linking their structure to a proposed mechanism of action. Zarei et al. [75] studied the ACE inhibitory mechanism of action of the bioactive peptides YLLLK, WAFS and GVQE-GAGHYALL identified from palm kernel cake. The authors noted concentration-dependent effects on enzyme inhibition, consistent with the presence of more than one binding site for the peptides and potentially multiple modes of inhibition [75]. Moreover, differences were appreciated in the way that these peptides achieved their activity, as some peptides showed variable degrees of degradation upon pre-incubation with ACE. The authors concluded that the peptide YLLLK acted as a competitive inhibitor and exhibited a higher number of total interactions with ACE compared to the other two peptides [75]. The action of the peptide YLLLK at the ACE active site visualized using molecular docking is represented in Figure 1. Ni, Li, Liu and Hu [69] determined that the ACE inhibition of the yeast peptide TPTQQS was caused by non-competitive interactions by displacing the Zn cofactor from the active site of the enzyme so the reaction cannot occur. The majority of the peptide is attached outside of the active site; however, the tail end of the peptide containing the serine 6 residue is what comes into contact and sequesters the zinc ion by forming a coordination bond with it [69].

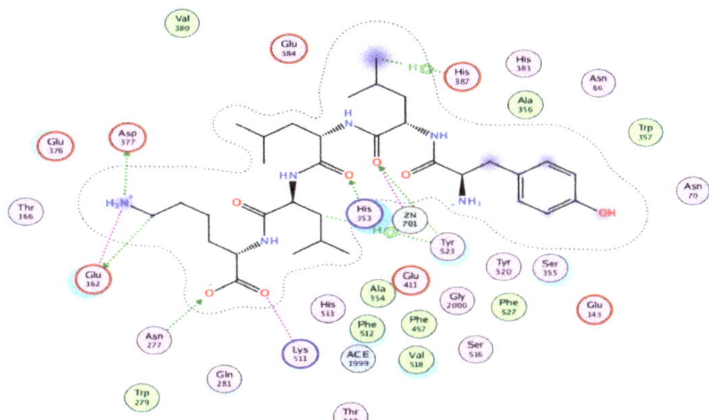

Figure 1. Automated molecular docking of the peptide YLLLK at the ACE active site. ACE hydrophobic residues are represented in green, positively charged residues in blue, and negatively charged residues in red; hydrogen bonds are purple arrows, polar residues are in turquoise color, and other residues and the zinc atom are represented automatically. Image obtained from Zarei, Abidin, Auwal, Chay, Abdul Haiyee, Md Sikin and Saari [75] originally published by MDPI.

3.2. Antioxidant Properties

Free radicals are short-lived and highly reactive chemical species that contain unpaired electrons [76]. Although reactive oxygen species (ROS) are formed during normal metabolic processes, they may also be formed due to exposure to exogenous factors such as ionizing radiation and UV light [77]. Free radicals can oxidatively modify nucleic acids, proteins, lipids, and sugars. An increased presence of these modified forms has been associated with an increased risk of various human diseases (e.g., cancer, Alzheimer's disease) [78]. Antioxidant mechanisms are present in most organisms and act to reduce or eliminate the levels of common ROS [77]. While many of the most effective mechanisms are enzyme-based (e.g., superoxide dismutase and catalase). Many small molecules (e.g., vitamin C, glutathione) also play a role in maintaining the overall redox balance in the cell [79]. While both natural and synthetic antioxidants have been added to foodstuffs, the potential toxicity of synthetic antioxidants, e.g., GHT has prompted the exploration of the use of peptides as antioxidant agents [77,79,80]. It is generally accepted that large peptides have less radical scavenging or antioxidant potential than small peptides [81]. Peptides have been reported to exert their antioxidant activity through direct metal chelation, ROS scavenging and inhibition of lipid peroxidation cascades [82].

A peptide from the microalgae *Isochrysis zhanjiangensis* has shown to have potent antioxidant capabilities towards alcohol-induced injury in cultured liver hepatoma cells (HepG2) [83]. This peptide was produced by in vitro gastro digestion using pepsin, trypsin and chymotrypsin resulting in the active sequence of NDAEYGICGF [83]. The treatment of cells with this peptide resulted in increased levels of the enzymes superoxide dismutase and glutathione. The antioxidant capabilities of the peptide appear to be related to a combination of the following factors: its molecular weight, hydrophobic amino acids (A, G, I) and aromatic amino acids (F, Y) in the sequence [83]. Other antioxidant peptides identified from *C. vulgaris* include VECYGPRPQF, which showed antioxidant capacity 26-fold higher (197 ng/mL Trolox equivalent) than trolox when tested by ORAC [84]. This peptide was found to slow oxidation by up to 10-fold compared to the control (PBS). The authors hypothesized that the Cu^{2+} chelating properties of this peptide were likely the main mechanism of action of antioxidant activity [84].

3.3. Anti-Cancer Properties

Peptides, due to their small size and chemical nature, can penetrate cell membranes without a build-up of toxic levels as seen with protein/antibodies. These compounds have shown high affinity and specificity while having low interactions with other medical treatments. However, there are limitations to overcome for their use, mainly related to the process of delivery of the peptide, as these compounds have regularly low bioavailability when taken orally, resulting in rapid clearance of the peptides. Peptides also have low levels of activity when compared to traditional drug treatments for cancer [85]. However, the peptide treatments have multiple problems mainly associated with the lack of specificity of the drugs that are not able to differentiate between carcinogenic and healthy cells. Moreover, when the chemotherapeutic agents are bound to a transport molecule, the breakage of these bounds can also reduce the efficacy of the peptides. Furthermore, the ability we have to currently treat multiple cancers is also dependent on the resistance of the cancer to the chemotherapeutic agents, which is a growing problem [86].

Limited studies are available on the anti-cancer properties of peptides derived from algae. Sheih, Fang, Wu and Lin [84] hydrolyzed protein by-products from the industrial processing of *C. vulgaris* and identified the peptide VECYGPNRPQF as an anti-proliferative. This peptide only had anti-proliferative effects on the human gastric cancer cell line AGS, but not on the other cell lines studied including human normal lung cell WI38, human colon adenocarcinoma cells C2BBel, human hepatoblastoma cell lines HepG2, human cervical epithelioid carcinoma cells Hela, and mouse BALB/c macrophage RAW 264.7 cells. The authors hypothesized that this peptide could have specific anti-cancer activity when treating certain tumor cells [84]. The peptide halted the cell cycle where the cell is given the chance to be either repaired by the TP53 mechanism or undergo apoptosis [87]. Moreover, the number of cells in the G1 phase decreased, while the Sub G1 phase category increased, indicating that the cells entered an apoptotic pathway following 48 h of incubation [84].

Anti-proliferation effects have been recorded from peptides produced by trypsin hydrolysis of proteins from *Porphyra haitanesis* [88]. The peptides generated were tested using five human cancer cell lines tested: MCF-7 (breast cancer), HepG2 (liver cancer cells), SGC-7901 (gastric cancer), A549 (lung cancer) and HT-29 (colon cancer), using the chemotherapeutic drug fluorouracil (5-FU) as a control. Four fractions (by size kDa) were obtained from the hydrolyzed peptides, and the peptide VPGTPKNLDSPR was reported as that with the highest antiproliferative activity, even significantly more potent than 5-FU in a HepG2 cell model [88]. Mechanically, this peptide appeared to interfere with the cell cycle and promoted apoptotic cell death in HepG2 and MCF7 cells [88].

The efficacy of several algal peptides in oncological treatments and their mechanisms of action have been elucidated. Kahalalides are an assortment of depsipeptides ranging in size from 31 carbon tripeptides to 75 carbon tridecapeptides. The peptide was firstly found in the mollusk *Elysia rufescens* and it was further discovered to be present in the algae *Bryopsis pennata* consumed by the mollusk and acting as a defense mechanism against predators [89]. Amongst all the Kahalalides, the one showing the most promise in cancer treatment is the largest peptide, Kahalalide F (KF) $C_{75}H_{124}N_{14}O_{16}$ [90,91]. KF has shown its potential benefits for cancer treatment in both in vitro and in vivo preclinical trials. Suárez, González, Cuadrado, Berciano, Lafarga and Muñoz [90] studied the mechanism of action of KF to determine its cytotoxic action against neoplastic cells. The authors used prostate (PC3, DU145, LNCaP) and breast cancer (SKBR-3, MCF7, BT474, MDA-MB-231) cell lines. The IC_{50} of KF for all the cell lines was around 0.3 μM, except in the case of PC3 which was 0.07 μM. Moreover, the authors also showed that the cytotoxic response of KF appears quickly, within 15 min. KF's mechanism of action differs from that of other antineoplastic drugs as it does not cause apoptosis, but generated an ATP depletion and swelling of the cells or oncosis [90,92]. KF has a similar mechanism of action to that of maitotoxin, a peptide that causes oncosis as its action linked to the function of the calcium ion channels of the cells [92]. Although maitotoxin is one of the most potent marine peptides known to date, its action is non-selective, and it is responsible for a particular human intoxication syndrome, namely ciguatera fish poisoning [93]. In this regard, KF could be a better fit for anti-cancer treatments as it has displayed tumor-selective properties in testing and has low toxicity [94].

4. Application of Novel Bioinformatic Tools

Bioinformatic tools are routinely used in peptide and protein analysis. In the context of the production of peptides via protein hydrolysis, online peptide cutters can simulate the cleavage of peptide sequences by various enzymes [95]. Toxicological and allergenic properties may also be predicted, using tools such as ToxinPred [96] and AllerTOP [97], respectively. Moreover, the BIOPEP database [98] compiles the reported activities of various peptide sequences. In silico methods being used to identify bioactivity in peptides include the use of quantitative structure–activity relationships (QSAR) and molecular docking, aiming to identify the mechanism of action underlying the biological functions of these peptides.

4.1. QSAR

One of the main challenges arising when producing a hydrolysate from a mix of proteins is to be able to elucidate which one of the multiple structures present in the hydrolysate is responsible for the biological effects appreciated during in vitro or in vivo tests and establish their mechanisms of action. When studying antioxidant peptides and only considering dipeptides, there can be potentially 400 different structural combinations accounting for all the possible combinations of 20 amino acids. However, when studying oligopeptides (2–20 amino acids in length) [99], this variability can reach levels of over 1.07×10^{39} possible structural combinations and thus, the use of bioinformatic methods, such as QSAR, can support the identification of bioactive peptides [100].

QSAR is an in silico method which takes peptides and their biological activity from pre-existing databases, such as BIOPEP, aiming to understand the link between these structures and their activity towards different biological targets [101]. The process flow of QSAR is represented in Figure 2.

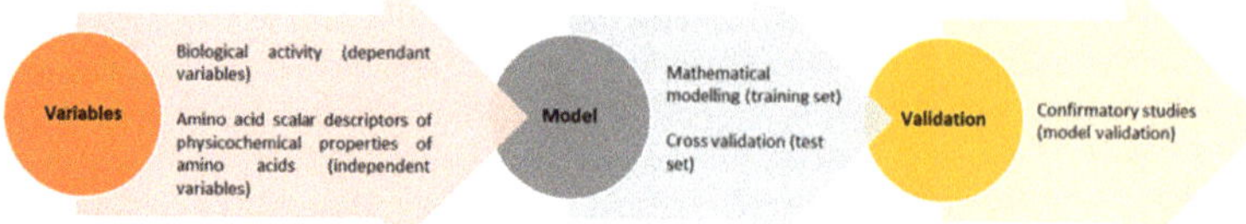

Figure 2. Process flow of QSAR applied to bioactive peptides. Content of the image adapted from Nongonierma and FitzGerald [101].

When sourcing bioactive peptides of interest in a QSAR model it is important to note that some peptides work in different ways of inhibiting their targets, such as competitive, non-competitive, and un-competitive ways. If a particular peptide library lacks the specific type of inhibitory action of these peptides then it completely skews the ability of the QSAR to estimate IC_{50} [101]. These peptides are used to create a model which aims to identify the key commonalities of the structure and composition of these peptides and link their composition to a bioactivity of interest. A portion of the peptides from the data set is randomly selected and left out of the training set; these are the test set and will be used at a later stage of the process. These peptides are used to identify the causative structure that allows for these interactions to occur, allows the identification of peptides with the most advantageous structural features, and establishes prediction scores for these structures [101]. These known peptides teach the software what to look for when unknown peptides are plugged into the equation to be identified and the peptides used to create the QSAR model should be a similar size to the peptides being analyzed [101].

Kumar et al. [102] researched novel ACE inhibitory peptides and the massive variation in these results due to the variable length of peptides with ACE inhibitory activity. This author chose the libraries where peptides with the same mechanisms of action for a particular bioactivity were classed by size, and QSAR models should be produced for each class to increase the accuracy of the results. The authors classified ACE inhibitory peptides as <3 amino acids, small peptides as 4–6 amino acids, medium peptides as 7–12 amino acids and large peptides as >12 amino acids [102].

Different scales and descriptors can be used to accurately define the features that make a certain peptide bioactive. The correct choice of descriptors is important as an excessive number of descriptors will cause background noise, causing an overfitting of data and loss of predictive accuracy [103]. These descriptors are usually physiochemical characteristics, such as the scale described by Hellberg et al. [104] which uses 29 physiochemical descriptors to analyze the amino acids. The authors grouped these descriptors into three main components known as the 3 Z approach which explains hydrophilicity (Z_1), steric properties (Z_2) and electronic properties (Z_3) [104]. This approach was improved further by Sandberg et al. [105] when characterizing 87 amino acids by adding two further components—Z_4 and Z_5—to describe other properties of the amino acids, such as heat of formation, electronegativity and electrophilicity.

From these rankings, multiple regression models can be performed to evaluate the bioactive potency of peptides on the basis of the interaction between the peptide and its target. These tests are normally performed against a positive control, such as glutathione for antioxidant peptides, and if the tested compounds score similar or even higher than the control, those compounds may have a better potential for in vitro testing [101]. After the model has been run and the IC_{50} predicted, these data require validation using the test set of peptides that were set aside at the beginning. Moreover, the models will have to be confirmed by testing the highest-ranking peptides against a laboratory-based assay and ensuring that the IC_{50} predicted by the model matches the experimental data [101]. If the model predicted the activity accurately, then the peptides which are of interest to be tested will be ranked by potency, synthesized, and experimentally tested to compare the results with those of the QSAR model [101].

The third step in the process of making a QSAR is the selection of a mathematical model to relate the physiochemical characteristics and position of the amino acids in the C and N terminus of the tested peptides with those of the peptides with known bioactivity. The models chosen are usually partial least square regression (PLSR), iterative double least square (IDLS), artificial neural networks (ANN) and multiple linear regression (MLR). When using these models, it is important that the model chosen accounts for whether the peptides being screened for activity will be of the same length as those described in the training set or if they account for peptides with a variety of different sizes [100,101].

4.2. Molecular Docking

Molecular docking is a stage to further identify the interactions between the target and substrate, complementing the QSAR modeling as it will provide three-dimensional interactions between the ligand; in this case, the peptide and the target to which the peptide is binding to help to understand further their inhibitory effects [106].

Angiotensin-converting enzyme (ACE) has been a focus of molecular docking studies in relation to cardiovascular diseases to further understand the action of peptides working within its domains. Mirzaei et al. [107] used the crystal structure of human ACE complexed with inhibitor lisinopril as a template for docking studies using the software HADDOCK (see Figure 3). The authors removed all water molecules and the inhibitor from the structure while retaining the zinc and chloride atoms in their active site before proceeding with the docking [107]. As previously stated, a disadvantage of QSAR is that it can be dependent on the amount of information granted to it by the database. An example of this is not identifying if the peptide is competitively binding to the active site or if it is having another effect on the enzyme in its entirety. Using molecular docking on the highest-ranking peptides from QSAR will show their overall binding affinity to the active site; this will hopefully mitigate any problems caused by the lack of information from these databases before synthesizing the highest-ranked peptides and final laboratory confirmatory testing.

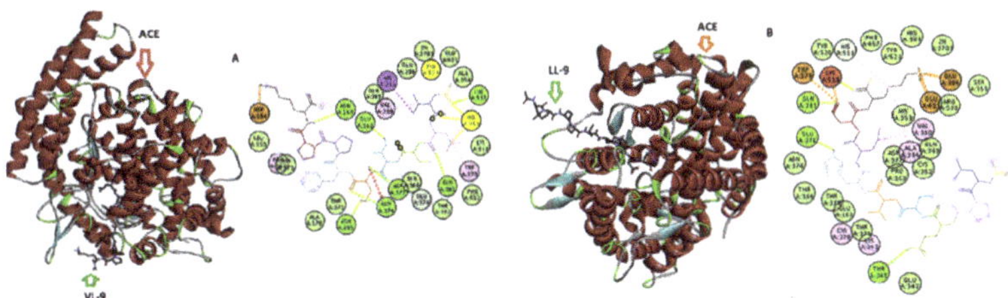

Figure 3. Representation of the molecular docking results (3D and 2D) of the ACE-inhibitory peptides VL-9 (**A**) and LL-9 (**B**). Color codes are as follows: blue (Van der Waals bonds), orange (salt bridge) and green (conventional hydrogen bond). Image originally published by Mirzaei, Mirdamadi, Ehsani and Aminlari [107] in Elsevier.

5. Opportunities and Challenges

There are huge market opportunities for algae as a source of protein due to the environmental benefits [108] associated with its production as well as their untapped potential as source of food and food ingredients for the growing world's population. However, there are still challenges, mainly related to the creation of optimum, reproducible, and sustainable protein extraction processes, mainly limited by the variable composition of the biomass as well as the presence of rigid cell walls of a variable chemical nature. Moreover, all the pre-treatments of the biomass, and the new emerging technological treatments, will have to demonstrate its economic viability in order to be adopted by industry, allowing to scale-up production and expand the use of these approaches.

In addition, further studies evaluating the activity and the chemical structure of peptides will be necessary to build upon current peptide libraries. The choice of peptide library is extremely important for the validity of the QSAR for testing unknown peptides and molecular docking studies. There are massive opportunities in the search for new peptide alternatives to be used as nutraceuticals with fewer adverse side effects than conventional treatments for multiple diseases. However, further studies and clear mechanisms of action have to be elucidated for these applications to achieve their potential.

Author Contributions: Conceptualization, M.G.-V.; writing—original draft preparation, J.O.; writing—review and editing, M.G.-V., S.M. and B.K.T.; supervision, M.G.-V., S.M. and B.K.T.; funding acquisition, B.K.T. All authors have read and agreed to the published version of the manuscript.

Funding: This research was funded by Teagasc (grant number 0018), BiOrbic SFI Bioeconomy Research Centre funded by Ireland's European Structural and Investment Programmes, Science Foundation Ireland (16/RC/3889) and the Department of Agriculture Food and the Marine (DAFM) under the umbrella of the European Joint Programming Initiative "A Healthy Diet for a Healthy Life" (JPI-HDHL) and of the ERA-NET Cofund ERA HDHL (GA No 696295 of the EU Horizon 2020 Research and Innovation Programme).

Institutional Review Board Statement: Not applicable.

Informed Consent Statement: Not applicable.

Data Availability Statement: Not applicable.

Conflicts of Interest: The authors declare no conflict of interest.

References

1. Béné, C.; Barange, M.; Subasinghe, R.; Pinstrup-Andersen, P.; Merino, G.; Hemre, G.-I.; Williams, M. Feeding 9 billion by 2050—Putting fish back on the menu. *Food Secur.* **2015**, *7*, 261–274. [CrossRef]
2. Rösch, C.; Roßmann, M.; Weickert, S. Microalgae for integrated food and fuel production. *GCB Bioenergy* **2019**, *11*, 326–334. [CrossRef]
3. Kusmayadi, A.; Leong, Y.K.; Yen, H.-W.; Huang, C.-Y.; Chang, J.-S. Microalgae as sustainable food and feed sources for animals and humans—Biotechnological and environmental aspects. *Chemosphere* **2021**, *271*, 129800. [CrossRef]
4. Vanthoor-Koopmans, M.; Wijffels, R.H.; Barbosa, M.J.; Eppink, M.H. Biorefinery of microalgae for food and fuel. *Bioresour. Technol.* **2013**, *135*, 142–149. [CrossRef] [PubMed]
5. Cian, R.E.; Drago, S.R.; De Medina, F.S.; Martínez-Augustin, O. Proteins and Carbohydrates from Red Seaweeds: Evidence for Beneficial Effects on Gut Function and Microbiota. *Mar. Drugs* **2015**, *13*, 5358–5383. [CrossRef] [PubMed]
6. Safafar, H.; Uldall Nørregaard, P.; Ljubic, A.; Møller, P.; Løvstad Holdt, S.; Jacobsen, C. Enhancement of Protein and Pigment Content in Two Chlorella Species Cultivated on Industrial Process Water. *J. Mar. Sci. Eng.* **2016**, *4*, 84. [CrossRef]
7. Van Krimpen, M.; Bikker, P.; Van der Meer, I.; Van der Peet-Schwering, C.; Vereijken, J. *Cultivation, Processing and Nutritional Aspects for Pigs and Poultry of European Protein Sources as Alternatives for Imported Soybean Products*; 1570-8616; Wageningen UR Livestock Research: Wageningen, The Netherlands, 2013.
8. Sánchez, A.; Vázquez, A. Bioactive peptides: A review. *Food Qual. Saf.* **2017**, *1*, 29–46. [CrossRef]
9. Harnedy, P.A.; FitzGerald, R.J. Bioactive proteins, peptides, and amino acids from macroalgae 1. *J. Phycol.* **2011**, *47*, 218–232. [CrossRef]
10. Geada, P.; Moreira, C.; Silva, M.; Nunes, R.; Madureira, L.; Rocha, C.M.R.; Pereira, R.N.; Vicente, A.A.; Teixeira, J.A. Algal proteins: Production strategies and nutritional and functional properties. *Bioresour. Technol.* **2021**, *332*, 125125. [CrossRef]
11. Fleurence, J. The enzymatic degradation of algal cell walls: A useful approach for improving protein accessibility? *J. Appl. Phycol.* **1999**, *11*, 313–314. [CrossRef]
12. Domozych, D.S. Algal Cell Walls. In *eLS*; John Wiley & Sons, Ltd.: Chichester, UK, 2011. [CrossRef]
13. Fleurence, J.; Massiani, L.; Guyader, O.; Mabeau, S. Use of enzymatic cell wall degradation for improvement of protein extraction from *Chondrus crispus*, *Gracilaria verrucosa* and *Palmaria palmata*. *J. Appl. Phycol.* **1995**, *7*, 393. [CrossRef]
14. Bodénès, P.; Bensalem, S.; Français, O.; Pareau, D.; Le Pioufle, B.; Lopes, F. Inducing reversible or irreversible pores in *Chlamydomonas reinhardtii* with electroporation: Impact of treatment parameters. *Algal Res.* **2019**, *37*, 124–132. [CrossRef]
15. Buchmann, L.; Brändle, I.; Haberkorn, I.; Hiestand, M.; Mathys, A. Pulsed electric field based cyclic protein extraction of microalgae towards closed-loop biorefinery concepts. *Bioresour. Technol.* **2019**, *291*, 121870. [CrossRef] [PubMed]
16. Kotnik, T.; Rems, L.; Tarek, M.; Miklavčič, D. Membrane electroporation and electropermeabilization: Mechanisms and models. *Annu. Rev. Biophys.* **2019**, *48*, 63–91. [CrossRef]
17. Bensalem, S.; Pareau, D.; Cinquin, B.; Français, O.; Le Pioufle, B.; Lopes, F. Impact of pulsed electric fields and mechanical compressions on the permeability and structure of *Chlamydomonas reinhardtii* cells. *Sci. Rep.* **2020**, *10*, 2668. [CrossRef]
18. Amorim, M.L.; Soares, J.; Vieira, B.B.; Batista-Silva, W.; Martins, M.A. Extraction of proteins from the microalga *Scenedesmus obliquus* BR003 followed by lipid extraction of the wet deproteinized biomass using hexane and ethyl acetate. *Bioresour. Technol.* **2020**, *307*, 123190. [CrossRef]

19. Parniakov, O.; Barba, F.J.; Grimi, N.; Marchal, L.; Jubeau, S.; Lebovka, N.; Vorobiev, E. Pulsed electric field and pH assisted selective extraction of intracellular components from microalgae *Nannochloropsis*. *Algal Res.* **2015**, *8*, 128–134. [CrossRef]
20. Prabhu, M.S.; Levkov, K.; Livney, Y.D.; Israel, A.; Golberg, A. High-voltage pulsed electric field preprocessing enhances extraction of starch, proteins, and ash from marine macroalgae *Ulva ohnoi*. *ACS Sustain. Chem. Eng.* **2019**, *7*, 17453–17463. [CrossRef]
21. Carullo, D.; Pataro, G.; Donsì, F.; Ferrari, G. Pulsed Electric Fields-Assisted Extraction of Valuable Compounds from Arthrospira Platensis: Effect of Pulse Polarity and Mild Heating. *Front. Bioeng. Biotechnol.* **2020**, *8*, 1050. [CrossRef]
22. Bimakr, M.; Rahman, R.A.; Taip, F.S.; Adzahan, N.M.; Sarker, M.; Islam, Z.; Ganjloo, A. Optimization of ultrasound-assisted extraction of crude oil from winter melon (*Benincasa hispida*) seed using response surface methodology and evaluation of its antioxidant activity, total phenolic content and fatty acid composition. *Molecules* **2012**, *17*, 11748–11762. [CrossRef]
23. Tiwari, B.K. Ultrasound: A clean, green extraction technology. *TrAC Trends Anal. Chem.* **2015**, *71*, 100–109. [CrossRef]
24. Kadam, S.U.; Álvarez, C.; Tiwari, B.K.; O'Donnell, C.P. Extraction and characterization of protein from Irish brown seaweed *Ascophyllum nodosum*. *Food Res. Int.* **2017**, *99*, 1021–1027. [CrossRef] [PubMed]
25. Vernes, L.; Abert-Vian, M.; El Maâtaoui, M.; Tao, Y.; Bornard, I.; Chemat, F. Application of ultrasound for green extraction of proteins from spirulina. Mechanism, optimization, modeling, and industrial prospects. *Ultrason. Sonochemistry* **2019**, *54*, 48–60. [CrossRef] [PubMed]
26. Juin, C.; Chérouvrier, J.-R.; Thiéry, V.; Gagez, A.-L.; Bérard, J.-B.; Joguet, N.; Kaas, R.; Cadoret, J.-P.; Picot, L. Microwave-assisted extraction of phycobiliproteins from *Porphyridium purpureum*. *Appl. Biochem. Biotechnol.* **2015**, *175*, 1–15. [CrossRef]
27. Grosso, C.; Valentão, P.; Ferreres, F.; Andrade, P.B. Alternative and Efficient Extraction Methods for Marine-Derived Compounds. *Mar. Drugs* **2015**, *13*, 3182–3230. [CrossRef]
28. Kaufmann, B.; Christen, P. Recent extraction techniques for natural products: Microwave-assisted extraction and pressurised solvent extraction. *Phytochem. Anal. Int. J. Plant Chem. Biochem. Tech.* **2002**, *13*, 105–113. [CrossRef]
29. Pasquet, V.; Chérouvrier, J.-R.; Farhat, F.; Thiéry, V.; Piot, J.-M.; Bérard, J.-B.; Kaas, R.; Serive, B.; Patrice, T.; Cadoret, J.-P.; et al. Study on the microalgal pigments extraction process: Performance of microwave assisted extraction. *Process Biochem.* **2011**, *46*, 59–67. [CrossRef]
30. Al-Zuhair, S.; Ashraf, S.; Hisaindee, S.; Darmaki, N.A.; Battah, S.; Svistunenko, D.; Reeder, B.; Stanway, G.; Chaudhary, A. Enzymatic pre-treatment of microalgae cells for enhanced extraction of proteins. *Eng. Life Sci.* **2017**, *17*, 175–185. [CrossRef]
31. Sierra, L.S.; Dixon, C.K.; Wilken, L.R. Enzymatic cell disruption of the microalgae *Chlamydomonas reinhardtii* for lipid and protein extraction. *Algal Res.* **2017**, *25*, 149–159. [CrossRef]
32. Paiva, L.; Lima, E.; Neto, A.I.; Baptista, J. Angiotensin I-converting enzyme (ACE) inhibitory activity, antioxidant properties, phenolic content and amino acid profiles of *Fucus spiralis* L. protein hydrolysate fractions. *Mar. Drugs* **2017**, *15*, 311. [CrossRef]
33. Vásquez, V.; Martínez, R.; Bernal, C. Enzyme-assisted extraction of proteins from the seaweeds *Macrocystis pyrifera* and *Chondracanthus chamissoi*: Characterization of the extracts and their bioactive potential. *J. Appl. Phycol.* **2019**, *31*, 1999–2010. [CrossRef]
34. Das, R.S.; Tiwari, B.K.; Garcia-Vaquero, M. Bioactive Peptides from Algae. In *Bioactive Peptides from Food: Sources, Analysis, and Functions*; CRC Press: Boca Raton, FL, USA, 2022; pp. 31–54.
35. Siepen, J.A.; Keevil, E.-J.; Knight, D.; Hubbard, S.J. Prediction of missed cleavage sites in tryptic peptides aids protein identification in proteomics. *J. Proteome Res.* **2007**, *6*, 399–408. [CrossRef] [PubMed]
36. Berg, J.; Tymoczko, J.; Stryer, L. Proteases: Facilitating a difficult reaction. *Biochemistry* **2002**.
37. Brik, A.; Wong, C.-H. HIV-1 protease: Mechanism and drug discovery. *Org. Biomol. Chem.* **2003**, *1*, 5–14. [CrossRef]
38. Available online: https://biosolutions.novozymes.com/en/plant-protein/products/alcalase (accessed on 1 May 2022).
39. Tavano, O.L. Protein hydrolysis using proteases: An important tool for food biotechnology. *J. Mol. Catal. B Enzym.* **2013**, *90*, 1–11. [CrossRef]
40. Waglay, A.; Karboune, S. Enzymatic generation of peptides from potato proteins by selected proteases and characterization of their structural properties. *Biotechnol. Prog.* **2016**, *32*, 420–429. [CrossRef]
41. Available online: www.sigmaaldrich.com/IE/en/technical-documents/technical-article/research-and-disease-areas/metabolism-research/papain (accessed on 1 May 2022).
42. Available online: http://biocon.es/wp-content/uploads/2016/12/TDS-Bromelain_eng.pdf (accessed on 1 May 2022).
43. Available online: https://biosolutions.novozymes.com/en/animal-protein/products/protamex (accessed on 1 May 2022).
44. Silva, C.M.; da Fonseca, R.d.S.; Prentice, C. Comparing the hydrolysis degree of industrialization byproducts of Withemouth croaker (*Micropogonias furnieri*) using microbial enzymes. *Int. Food Res. J.* **2014**, *21*, 1757–1761.
45. Available online: https://www.promega.com/-/media/files/resources/protocols/product-information-sheets/n/elastase-protocol.pdf?rev=996bbf57fd0441709e56af3147154437&sc_lang=en (accessed on 1 May 2022).
46. Available online: https://www.promega.com/-/media/files/resources/protocols/product-information-sheets/n/thermolysin-protocol.pdf?rev=4115cf6e496f47afb3be9168515fdc5f&sc_lang=en (accessed on 1 May 2022).
47. Gupta, S.S.; Mishra, V.; Mukherjee, M.D.; Saini, P.; Ranjan, K.R. Amino acid derived biopolymers: Recent advances and biomedical applications. *Int. J. Biol. Macromol.* **2021**, *188*, 542–567. [CrossRef]
48. Marcet, I.; Alvarez, C.; Paredes, B.; Diaz, M. Inert and oxidative subcritical water hydrolysis of insoluble egg yolk granular protein, functional properties, and comparison to enzymatic hydrolysis. *J. Agric. Food Chem.* **2014**, *62*, 8179–8186. [CrossRef]
49. Rivas-Vela, C.I.; Amaya-Llano, S.L.; Castaño-Tostado, E.; Castillo-Herrera, G.A. Protein Hydrolysis by Subcritical Water: A New Perspective on Obtaining Bioactive Peptides. *Molecules* **2021**, *26*, 6655. [CrossRef]

50. Fan, X.; Hu, S.; Wang, K.; Yang, R.; Zhang, X. Coupling of ultrasound and subcritical water for peptides production from *Spirulina platensis*. *Food Bioprod. Process.* **2020**, *121*, 105–112. [CrossRef]
51. Garcia-Moscoso, J.L.; Obeid, W.; Kumar, S.; Hatcher, P.G. Flash hydrolysis of microalgae (*Scenedesmus* sp.) for protein extraction and production of biofuels intermediates. *J. Supercrit. Fluids* **2013**, *82*, 183–190. [CrossRef]
52. Powell, T.; Bowra, S.; Cooper, H.J. Subcritical water processing of proteins: An alternative to enzymatic digestion? *Anal. Chem.* **2016**, *88*, 6425–6432. [CrossRef] [PubMed]
53. Zhu, G.-Y.; Zhu, X.; Wan, X.-L.; Fan, Q.; Ma, Y.-H.; Qian, J.; Liu, X.-L.; Shen, Y.-J.; Jiang, J.-H. Hydrolysis technology and kinetics of poultry waste to produce amino acids in subcritical water. *J. Anal. Appl. Pyrolysis* **2010**, *88*, 187–191. [CrossRef]
54. Esteban, M.B.; García, A.J.; Ramos, P.; Márquez, M.C. Sub-critical water hydrolysis of hog hair for amino acid production. *Bioresour. Technol.* **2010**, *101*, 2472–2476. [CrossRef]
55. Melgosa, R.; Marques, M.; Paiva, A.; Bernardo, A.; Fernández, N.; Sá-Nogueira, I.; Simões, P. Subcritical water extraction and hydrolysis of cod (*Gadus morhua*) frames to produce bioactive protein extracts. *Foods* **2021**, *10*, 1222. [CrossRef]
56. Banerjee, P.; Shanthi, C. Isolation of novel bioactive regions from bovine Achilles tendon collagen having angiotensin I-converting enzyme-inhibitory properties. *Process Biochem.* **2012**, *47*, 2335–2346. [CrossRef]
57. Ahmed, J.; Mulla, M.; Al-Ruwaih, N.; Arfat, Y.A. Effect of high-pressure treatment prior to enzymatic hydrolysis on rheological, thermal, and antioxidant properties of lentil protein isolate. *Legume Sci.* **2019**, *1*, e10. [CrossRef]
58. Adamson, N.J.; Reynolds, E.C. Characterization of casein phosphopeptides prepared using alcalase: Determination of enzyme specificity. *Enzym. Microb. Technol.* **1996**, *19*, 202–207. [CrossRef]
59. Meinlschmidt, P.; Brode, V.; Sevenich, R.; Ueberham, E.; Schweiggert-Weisz, U.; Lehmann, J.; Rauh, C.; Knorr, D.; Eisner, P. High pressure processing assisted enzymatic hydrolysis–An innovative approach for the reduction of soy immunoreactivity. *Innov. Food Sci. Emerg. Technol.* **2017**, *40*, 58–67. [CrossRef]
60. Asaduzzaman, A.K.M.; Chun, B.-S. Recovery of functional materials with thermally stable antioxidative properties in squid muscle hydrolyzates by subcritical water. *J. Food Sci. Technol.* **2015**, *52*, 793–802. [CrossRef] [PubMed]
61. Tavakoli, O.; Yoshida, H. Conversion of scallop viscera wastes to valuable compounds using sub-critical water. *Green Chem.* **2006**, *8*, 100–106. [CrossRef]
62. Lafarga, T.; Acién-Fernández, F.G.; Garcia-Vaquero, M. Bioactive peptides and carbohydrates from seaweed for food applications: Natural occurrence, isolation, purification, and identification. *Algal Res.* **2020**, *48*, 101909. [CrossRef]
63. Megías, C.; Yust, M.d.M.; Pedroche, J.; Lquari, H.; Girón-Calle, J.; Alaiz, M.; Millán, F.; Vioque, J. Purification of an ACE inhibitory peptide after hydrolysis of sunflower (*Helianthus annuus* L.) protein isolates. *J. Agric. Food Chem.* **2004**, *52*, 1928–1932. [CrossRef]
64. Garcia-Vaquero, M.; Mora, L.; Hayes, M. In Vitro and In Silico Approaches to Generating and Identifying Angiotensin-Converting Enzyme I Inhibitory Peptides from Green Macroalga *Ulva lactuca*. *Mar. Drugs* **2019**, *17*, 204. [CrossRef] [PubMed]
65. Daskaya-Dikmen, C.; Yucetepe, A.; Karbancioglu-Guler, F.; Daskaya, H.; Ozcelik, B. Angiotensin-I-Converting Enzyme (ACE)-Inhibitory Peptides from Plants. *Nutrients* **2017**, *9*, 316. [CrossRef]
66. Fitzgerald, C.; Mora-Soler, L.; Gallagher, E.; O'Connor, P.; Prieto, J.; Soler-Vila, A.; Hayes, M. Isolation and Characterization of Bioactive Pro-Peptides with In Vitro Renin Inhibitory Activities from the Macroalga *Palmaria palmata*. *J. Agric. Food Chem.* **2012**, *60*, 7421–7427. [CrossRef]
67. Fitzgerald, C.; Aluko, R.E.; Hossain, M.; Rai, D.K.; Hayes, M. Potential of a Renin Inhibitory Peptide from the Red Seaweed Palmaria palmata as a Functional Food Ingredient Following Confirmation and Characterization of a Hypotensive Effect in Spontaneously Hypertensive Rats. *J. Agric. Food Chem.* **2014**, *62*, 8352–8356. [CrossRef]
68. Li, E.C.; Heran, B.S.; Wright, J.M. Angiotensin converting enzyme (ACE) inhibitors versus angiotensin receptor blockers for primary hypertension. *Cochrane Database Syst. Rev.* **2014**, *2014*, CD009096. [CrossRef]
69. Ni, H.; Li, L.; Liu, G.; Hu, S.-Q. Inhibition mechanism and model of an angiotensin I-converting enzyme (ACE)-inhibitory hexapeptide from yeast (*Saccharomyces cerevisiae*). *PLoS ONE* **2012**, *7*, e37077. [CrossRef]
70. Cooper, W.O.; Hernandez-Diaz, S.; Arbogast, P.G.; Dudley, J.A.; Dyer, S.; Gideon, P.S.; Hall, K.; Ray, W.A. Major congenital malformations after first-trimester exposure to ACE inhibitors. *N. Engl. J. Med.* **2006**, *354*, 2443–2451. [CrossRef] [PubMed]
71. Pryde, P.G.; Sedman, A.B.; Nugent, C.E.; Barr, M. Angiotensin-converting enzyme inhibitor fetopathy. *J. Am. Soc. Nephrol.* **1993**, *3*, 1575–1582. [CrossRef] [PubMed]
72. Kim, S.-K.; Wijesekara, I. Development and biological activities of marine-derived bioactive peptides: A review. *J. Funct. Foods* **2010**, *2*, 1–9. [CrossRef]
73. Suetsuna, K.; Chen, J.-R. Identification of antihypertensive peptides from peptic digest of two microalgae, *Chlorella vulgaris* and *Spirulina platensis*. *Mar. Biotechnol.* **2001**, *3*, 305–309. [CrossRef] [PubMed]
74. Sun, S.; Xu, X.; Sun, X.; Zhang, X.; Chen, X.; Xu, N. Preparation and identification of ACE inhibitory peptides from the marine macroalga *Ulva intestinalis*. *Mar. Drugs* **2019**, *17*, 179. [CrossRef] [PubMed]
75. Zarei, M.; Abidin, N.B.Z.; Auwal, S.M.; Chay, S.Y.; Abdul Haiyee, Z.; Md Sikin, A.; Saari, N. Angiotensin converting enzyme (ACE)-peptide interactions: Inhibition kinetics, in silico molecular docking and stability study of three novel peptides generated from palm kernel cake proteins. *Biomolecules* **2019**, *9*, 569. [CrossRef] [PubMed]
76. Halliwell, B. Reactive species and antioxidants. Redox biology is a fundamental theme of aerobic life. *Plant Physiol.* **2006**, *141*, 312–322. [CrossRef]

77. Tadesse, S.A.; Emire, S.A. Production and processing of antioxidant bioactive peptides: A driving force for the functional food market. *Heliyon* **2020**, *6*, e04765. [CrossRef]
78. Canabady-Rochelle, L.L.; Harscoat-Schiavo, C.; Kessler, V.; Aymes, A.; Fournier, F.; Girardet, J.-M. Determination of reducing power and metal chelating ability of antioxidant peptides: Revisited methods. *Food Chem.* **2015**, *183*, 129–135. [CrossRef]
79. Wen, C.; Zhang, J.; Zhang, H.; Duan, Y.; Ma, H. Plant protein-derived antioxidant peptides: Isolation, identification, mechanism of action and application in food systems: A review. *Trends Food Sci. Technol.* **2020**, *105*, 308–322. [CrossRef]
80. Zhang, X.; Cao, D.; Sun, X.; Sun, S.; Xu, N. Preparation and identification of antioxidant peptides from protein hydrolysate of marine alga *Gracilariopsis lemaneiformis*. *J. Appl. Phycol.* **2019**, *31*, 2585–2596. [CrossRef]
81. Guo, H.; Kouzuma, Y.; Yonekura, M. Structures and properties of antioxidative peptides derived from royal jelly protein. *Food Chem.* **2009**, *113*, 238–245. [CrossRef]
82. Nimalaratne, C.; Bandara, N.; Wu, J. Purification and characterization of antioxidant peptides from enzymatically hydrolyzed chicken egg white. *Food Chem.* **2015**, *188*, 467–472. [CrossRef] [PubMed]
83. Chen, M.-F.; Zhang, Y.Y.; Di He, M.; Li, C.Y.; Zhou, C.X.; Hong, P.Z.; Qian, Z.-J. Antioxidant Peptide Purified from Enzymatic Hydrolysates of Isochrysis Zhanjiangensis and Its Protective Effect against Ethanol Induced Oxidative Stress of HepG2 Cells. *Biotechnol. Bioprocess Eng.* **2019**, *24*, 308–317. [CrossRef]
84. Sheih, I.-C.; Fang, T.J.; Wu, T.-K.; Lin, P.-H. Anticancer and antioxidant activities of the peptide fraction from algae protein waste. *J. Agric. Food Chem.* **2010**, *58*, 1202–1207. [CrossRef]
85. Marqus, S.; Pirogova, E.; Piva, T.J. Evaluation of the use of therapeutic peptides for cancer treatment. *J. Biomed. Sci.* **2017**, *24*, 21. [CrossRef]
86. Kakde, D.; Jain, D.; Shrivastava, V.; Kakde, R.; Patil, A. Cancer therapeutics-opportunities, challenges and advances in drug delivery. *J. Appl. Pharm. Sci* **2011**, *1*, 1–10.
87. Murad, H.; Hawat, M.; Ekhtiar, A.; AlJapawe, A.; Abbas, A.; Darwish, H.; Sbenati, O.; Ghannam, A. Induction of G1-phase cell cycle arrest and apoptosis pathway in MDA-MB-231 human breast cancer cells by sulfated polysaccharide extracted from Laurencia papillosa. *Cancer Cell Int.* **2016**, *16*, 39. [CrossRef]
88. Fan, X.; Bai, L.; Mao, X.; Zhang, X. Novel peptides with anti-proliferation activity from the *Porphyra haitanesis* hydrolysate. *Process Biochem.* **2017**, *60*, 98–107. [CrossRef]
89. Becerro, M.A.; Goetz, G.; Paul, V.J.; Scheuer, P.J. Chemical Defenses of the Sacoglossan Mollusk *Elysia rufescens* and Its Host Alga *Bryopsis* sp. *J. Chem. Ecol.* **2001**, *27*, 2287–2299. [CrossRef]
90. Suárez, Y.; González, L.; Cuadrado, A.; Berciano, M.; Lafarga, M.; Muñoz, A. Kahalalide F, a new marine-derived compound, induces oncosis in human prostate and breast cancer cells. *Mol. Cancer Ther.* **2003**, *2*, 863–872. [PubMed]
91. Wang, B.; Waters, A.L.; Valeriote, F.A.; Hamann, M.T. An efficient and cost-effective approach to kahalalide F N-terminal modifications using a nuisance algal bloom of *Bryopsis pennata*. *Biochim. Biophys. Acta (BBA)-Gen. Subj.* **2015**, *1850*, 1849–1854. [CrossRef] [PubMed]
92. Sewell, J.; Mayer, I.; Langdon, S.; Smyth, J.; Jodrell, D.; Guichard, S. The mechanism of action of Kahalalide F: Variable cell permeability in human hepatoma cell lines. *Eur. J. Cancer* **2005**, *41*, 1637–1644. [CrossRef] [PubMed]
93. Flores, P.L.; Rodríguez, E.; Zapata, E.; Carbó, R.; Farías, J.M.; Martínez, M. Maitotoxin Is a Potential Selective Activator of the Endogenous Transient Receptor Potential Canonical Type 1 Channel in *Xenopus laevis* Oocytes. *Mar. Drugs* **2017**, *15*, 198. [CrossRef]
94. Hamann, M.T. Technology evaluation: Kahalalide F, PharmaMar. *Curr. Opin. Mol. Ther.* **2004**, *6*, 657.
95. Available online: https://web.expasy.org/peptide_cutter/ (accessed on 29 March 2022).
96. Available online: http://crdd.osdd.net/raghava/toxinpred/ (accessed on 29 March 2022).
97. Available online: https://www.ddg-pharmfac.net/AllerTOP/method.html (accessed on 29 March 2022).
98. Minkiewicz, P.; Dziuba, J.; Iwaniak, A.; Dziuba, M.; Darewicz, M. BIOPEP Database and Other Programs for Processing Bioactive Peptide Sequences. *J. AOAC Int.* **2019**, *91*, 965–980. [CrossRef]
99. Reddy, B.; Jow, T.; Hantash, B.M. Bioactive oligopeptides in dermatology: Part I. *Exp. Dermatol.* **2012**, *21*, 563–568. [CrossRef]
100. Du, Q.S.; Ma, Y.; Xie, N.Z.; Huang, R.B. Two-level QSAR network (2L-QSAR) for peptide inhibitor design based on amino acid properties and sequence positions. *SAR QSAR Environ. Res.* **2014**, *25*, 837–851. [CrossRef]
101. Nongonierma, A.B.; FitzGerald, R.J. Learnings from quantitative structure–activity relationship (QSAR) studies with respect to food protein-derived bioactive peptides: A review. *RSC Adv.* **2016**, *6*, 75400–75413. [CrossRef]
102. Kumar, R.; Chaudhary, K.; Singh Chauhan, J.; Nagpal, G.; Kumar, R.; Sharma, M.; Raghava, G.P. An in silico platform for predicting, screening and designing of antihypertensive peptides. *Sci. Rep.* **2015**, *5*, 12512. [CrossRef]
103. Khan, A.U. Descriptors and their selection methods in QSAR analysis: Paradigm for drug design. *Drug Discov. Today* **2016**, *21*, 1291–1302.
104. Hellberg, S.; Sjoestroem, M.; Skagerberg, B.; Wold, S. Peptide quantitative structure-activity relationships, a multivariate approach. *J. Med. Chem.* **1987**, *30*, 1126–1135. [CrossRef] [PubMed]
105. Sandberg, M.; Eriksson, L.; Jonsson, J.; Sjöström, M.; Wold, S. New Chemical Descriptors Relevant for the Design of Biologically Active Peptides. A Multivariate Characterization of 87 Amino Acids. *J. Med. Chem.* **1998**, *41*, 2481–2491. [CrossRef] [PubMed]
106. Agrawal, P.; Singh, H.; Srivastava, H.K.; Singh, S.; Kishore, G.; Raghava, G.P.S. Benchmarking of different molecular docking methods for protein-peptide docking. *BMC Bioinform.* **2019**, *19*, 426. [CrossRef] [PubMed]

occurring sources with medicinal and pharmacological potentialities are ideal, owing to the ease in accessibility, free availability, natural abundance, being carbon neutral, and having fewer side effects than chemical-based synthetic formulations. Moreover, the growing technological advancement, scientific awareness, legislative authorities, and a broader variability of technological and analytical endorsement all support the exploitation of treasure of natural and underexploited sources from the marine biome [1,2]. In this context, the utilization of sustainable sources and the development of green technologies could be the key to going green. The value of "going green" has clearly transformed the pursuit of sustainable, recyclable, and socioeconomic friendlier products to fulfill the economic pressure by considering the high cost-effective ratio benefit. Therefore, the divergence from nonrenewable to renewable natural resources is fetching the center of interest in biomedical establishments. Technological legislation is a driving force and should be considered with care to develop green strategies for the cleaner obtainment of high-value biologically active products [1]. Several approaches are in practice to develop a state-of-the-art bio-based platform for various technological applications in bio- and non-bioindustries of the modern world. Green biotechnology has a noteworthy potential, of course in combination with 12 principles of green chemistry, to eradicate the generation of wasteful protection and deprotection steps [3]. The combination of green chemistry principles and modern biotechnology along with the employment of natural resources is urgently required to establish a sustainable future production and exploitation of the above-mentioned products.

MBPs are characteristically accessible from numerous natural sources, including marine biome, though in different extents with unique functionalities. Among several marine-derived compounds, MBPs are greener alternatives and have been broadly considered and utilized as pharmaceutically resourceful ingredients for numerous health determinations. Therefore, MBPs are being used to develop novel formulations in the biomedical, pharmaceutical, cosmeceutical, and nutraceutical industries. Owing to their advantageous features, such as robust affinity, functional reactivity, specificity, selectivity, etc., MBPs offer practical and positive replacements, compared with their synthetic counterpart formulations and chemical drugs [4,5]. Regardless of advantageous features, MBPs have been considered from diverse standpoints, i.e., based on (1) taxonomic sources, (2) biosynthesis pathways, (3) source of extraction, (4) ring and linear structures, (5) active fractions and compositional variations, (6) biologically active precursor molecules, etc. [2]. Irrespective of the category type, MBPs are protein-based specific fragments that solely depend on the activity based on the composition and amino acid sequence in that particular fragment [6]. However, the MBP activity indeed depends on the extraction and purification technique [2,7,8]. Thus far, an array of production, extraction, and purification approaches have been developed and exploited for MBPs [9–11]. From the above examples, an appropriately established methodology plays a significant role in the quality and quantity of the end-product of interests, specifically MBPs. It is also imperative to consider innumerable manipulating factors that can influence the complete performance of the entire isolation, extraction, and purification process and the overall product yield. For example, the source material's physiochemical composition, matrix characteristics, material's pretreatment considerations, solvent grade, type and concentration, reaction pH and temperature, pressure, reaction time, etc., are key influencing factors that should be considered before designing and running the experimental protocols [9]. The process efficacy of the entire extraction process and the end-product relies on (1) input parameters, (2) consideration of the nature of the substrate materials from the source, (3) interplay between the procedure and the substrate materials, and (4) chemistry of MBPs [2].

Considering the above critiques and opportunities herein, we sought to highlight the value, challenges, and opportunities that exist in MBPs with potent bioactivities and therapeutic potentialities. Following a brief introduction, a standardized initial methodological literature screening and inclusion/exclusion criteria were adopted to justify the scientific and literature theme. The next section concerns the marine biome and its potent MBPs with superior performance. The latter half of the paper addresses the generalized overview of bioactivities with biomedical and pharmaceutical potentialities to understand the fundamental impact and role of bioactive elements with related pathways. The FDA-approved MBPs in the market and MBPs with clinical trials, as well as sustainable development goals (SDGs) and legislation to valorize marine biome, are discussed. Finally, a detailed overview of current challenges, conclusions, and future perspectives is given to satisfy the stimulating demands of the pharmaceutical sector of the modern world.

2. Review of Methodological Approach: Inclusion/Exclusion Criteria

The summarized review contents and given examples are well justified by using a standardized methodological approach that is based on screening the initial literature and inclusion/exclusion criteria. In this study, inclusion/exclusion criteria were also considered important to justify the most recent and relevant literature as per the conceptualized theme of the research. Following the initial screening step, the studies meticulously corresponding with the title theme were included in this discussion unless otherwise excluded to avoid the literature redundancy. Careful literature search queries were made in the Scopus and PubMed literature databases using any or many of the following terms: bioactive compounds from marine sources, bioactive peptides from marine sources, and biomedical applications of marine peptides. The active search filters were article title, abstract, and keywords. The literature screening results attained from the Scopus database are shown in Figure 1. The literature search queries were executed on 21 March 2021 and 25 April 2021 at Scopus, while for PubMed database search, the simultaneous search queries were performed on 21 March 2021 and 25 April 2021 at PubMed. Table 1 summarizes the literature screening results attained from the PubMed database.

Table 1. The literature screening results attained from the PubMed database. Data were extracted from https://pubmed.ncbi.nlm.nih.gov/, access on 25 April 2021.

Search Terms	Total Articles	Total Number of Articles Published and Filtered with Best Match Term on					
		2021	2020	2019	2018	2017	All Past Years
Bioactive compounds from marine sources	995	74	172	147	97	101	404
Bioactive peptides from marine sources	307	15	41	52	36	34	129
Biomedical applications of marine peptides	76	3	10	15	11	6	31

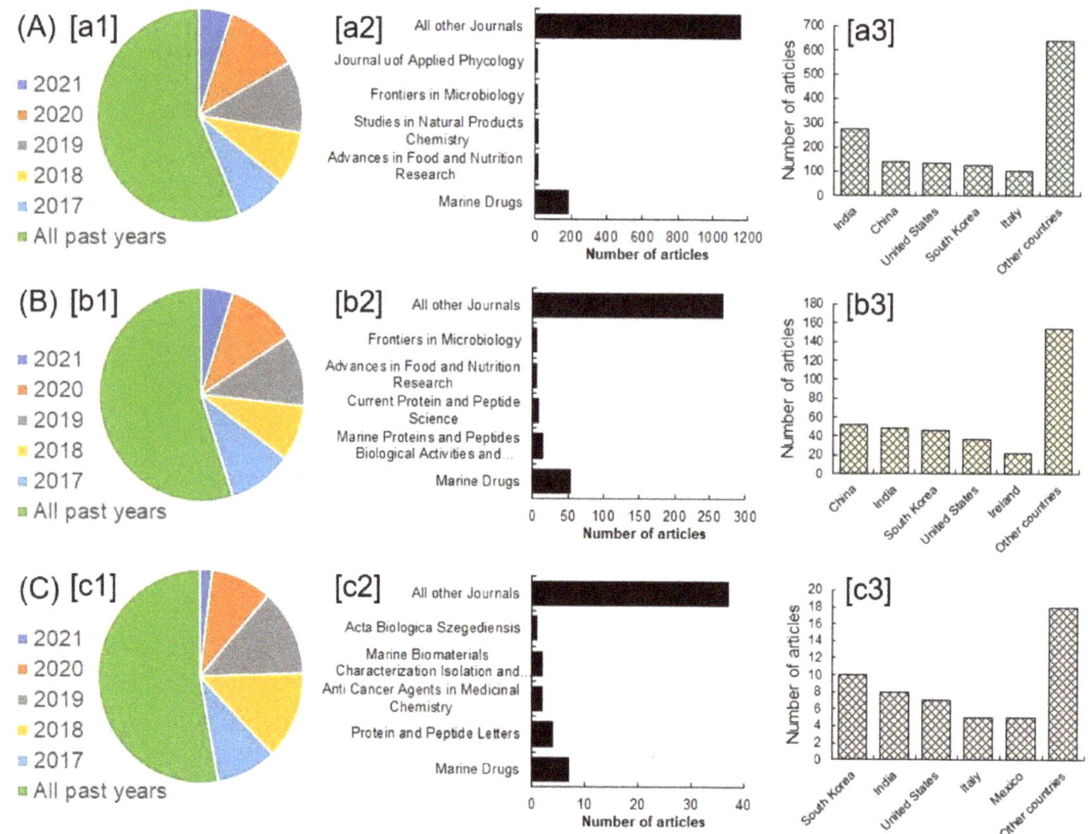

Figure 1. The literature screening results attained from the Scopus database. The letters A–C characterize the search terms: (**A**) bioactive compounds from marine sources; (**B**) bioactive peptides from marine sources; (**C**) biomedical applications of marine peptides. The letters [a1]–[c1] correspond to the number of articles from all years in that specific category of search term, [b2]–[c2] represents the number of articles published in different journals, and [a3]–[c3] represents the number of articles based on territory. Data were extracted from https://www.scopus.com, access on 25 April 2021.

3. Marine Biome—A Rich Source of MBPs

The marine biome represents the aquatic region and is considered the prime aquatic biome in the world with a plethora of untouched or underexploited sources of high-value interests, having around 500,000 to 10 million marine species [12]. Owing to the difficulty in reaching the lower/deeper zones of the marine biome, several MBPs have yet to be explored to identify and characterize marine biome; thus, marine biome constitutes a rich source of novel compounds with superior performance. As a major component of the biosphere, the aquatic biome covers almost 70% of the earth's surface [13]. The aquatic biome is broadly divided into two central regions—namely, (1) freshwater regions and (2) marine regions. The freshwater regions of the aquatic biome comprise ponds/lakes, streams/rivers, and wetlands, while the marine region comprises neritic, oceanic, and benthic biomes. Other marine biome zones include intertidal, estuaries, and coral reefs. These alone have the highest biodiversity of all marine biomes. Considering all marine biomes, marine-originated species embrace about half of the total global biodiversity, with a considerable potential to extract products of interest. Thus, several marine biome sources have been broadly discovered and utilized to obtain novel bioactive compounds, e.g., MBPs. The aquatic organisms are classified into three basic categories: (1) plankton, including phytoplankton (bacteria and algae) and zooplankton (tiny animals that feed on phytoplankton); (2) nekton (invertebrates such as shrimp and vertebrates such as fish); (3) benthos (sponges, clams, sea stars, etc.).

Owing to this massive biodiversity, the marine biome is proved to be a highly advantageous and acceptable natural source that can fulfill the growing ecological and socioeconomic demands. Moreover, marine-based natural resources and/or their extracted bioproducts are easily accessible, have high bioactive efficacy, and exert no/fewer side effects [2], compared with the similar representative from synthetic formulations. In the marine biome, the complex habitat environment and the extreme living conditions, such as ultraviolet light exposure, variation in saltiness, thermal conditions of the environment, and limited/excessive nutrient availability, are the contributory aspects to produce MBPs. In consequence, the marine resources including microorganisms such as microalgae survive these extreme environments by rapidly acclimatizing to the new and surrounding atmosphere [14]. Such quick environmental adaptions further assist them to produce more stable and highly effective metabolites, which are biologically active and not common in other similar synthetic counterparts. Owing to their high protein content, microalgae and other marine sources are considered potential sources to produce both elementary proteins and therapeutic peptides [15]. Furthermore, seeing the enormous marine biome potential for industrial sections, the principle of "going green" has scrutinized this substitute search to eco-friendly, ecological, and sustainable materials with comparable socioeconomic benefits. The expansion of distinguishing practices or approaches to improve the cutting-edge platforms also supports the green agenda. The synergistic use of marine-derived natural resources, along with green and modern biotechnology, must be considered to unveil a sustainable production of high-value-added products with highly requisite features [2].

4. Marine Bioactive Peptides (MBPs)

The amino-acid-based organic substances joined by covalent bonds (e.g., amide or peptide bonds) are termed bioactive peptides (BPs). Based on the natural source, some peptides exist freely. However, the majority of the peptides are surrounded or encoded with respective parent protein molecules. In later cases, the pretreatments in the presence of related enzymes, such as proteolytic enzymes, facilitate the release of BPs in an efficient manner [2,16,17]. Such pretreatments cause the hydrolysis of the cell walls or cell membranes, as applied. Thus, the deployment of highly effective and suitable pretreatment supports the retrieval of intracellularly captured bioactive constituents, which are mainly not easy to extract via conventional extraction procedures [2]. Following enzyme facilitated pretreatment, once the encrypted peptides are released from their respective source materials, the amino acid composition and sequence govern the activity [17]. However, owing to the similar peptide length that ranges between 2 and 20 amino acids, most of the BPs share their structural and physicochemical features [18]. The broad bioactivity spectra of MBPs, such as antimicrobial, antiviral, anticancer, anticoagulant, antidiabetic, cardioprotective, immunomodulatory, neuroregenerative, appetite suppressing, etc. [19–24], have attracted the attention of biomedical, pharmaceutical, and nutraceutical sectors, with the hope that they can be used as a new frontier to diagnose, treat, or prevent various pathologies in human. Based on the evidence derived from the literature, some of the MBPs and/or their representative biologically active byproducts have gained considerable commercial values and seized the pharmaceutical market. From the perspective of the market status and share, ziconotide and brentuximab vedotin are two important representatives of MBPs and peptide derivatives, and both have successfully reached the market. The premier, Ziconotide (Prialt®), was obtained from a marine cone snail. It was the first peptide from the marine biome approved by the Food and Drug Administration (FDA) USA in 2004 to exploit and use against analgesics [25]. Later in 2011, FDA also approved another marine-derived drug, Adcetris®, to manipulate and use against cancer. Since then, numerous other

MBPs have been evaluated for various phases of clinical related trials in the United States and Europe [13]. Figure 2 shows various marine sources along with their representative products or byproducts that have been either accepted or granted to enter the clinical trials. The commercial value of these therapeutic protein-based products was around USD 174.7 Billion in 2015. With a recent hike in interests, it is anticipated that this value will reach/cross USD 266.6 Billion in 2021 [26]. Several other marine-based peptides with different bioactivities and applications are summarized in Table 2.

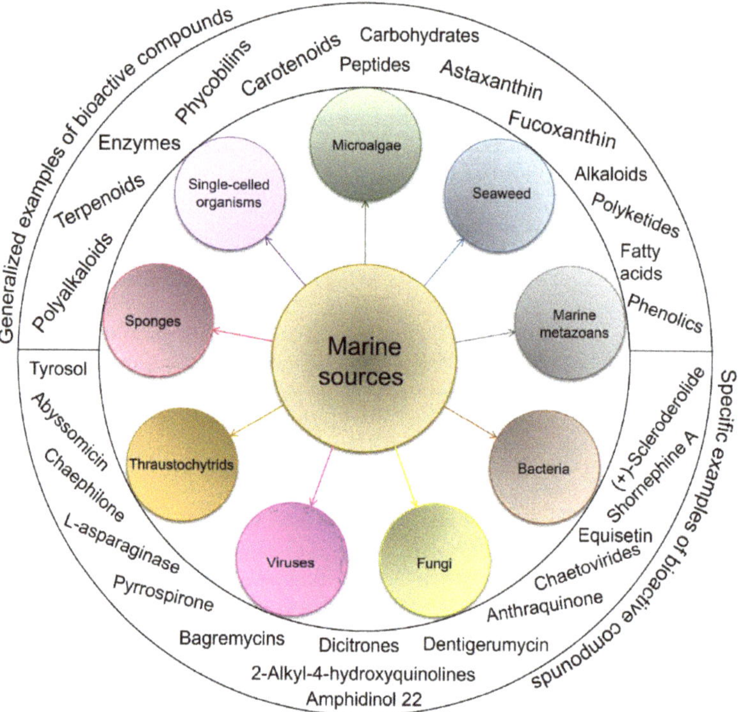

Figure 2. Various marine sources, along with their representative products or byproducts that have been either accepted or granted to enter the clinical trials.

Table 2. Marine-based peptides (MBPs) with different bioactivities and applications.

Peptide Name	Structure	Chemical Formula	Molar Mass (g/mol)	Class or Type of Chemical Compounds	Bioactivities and/or Proposed Applications	References
Astaxanthin	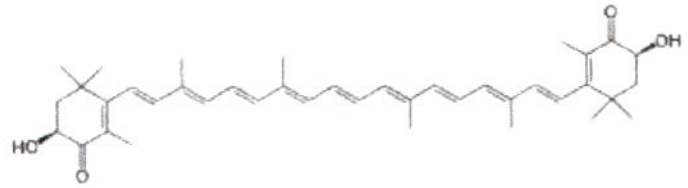	$C_{40}H_{52}O_4$	596.84	Terpenes	Nutraceutical and pharmaceutical applications	Messina et al. [27]
Cyclopolypeptide, e.g., α-amanitin, rolloamide A	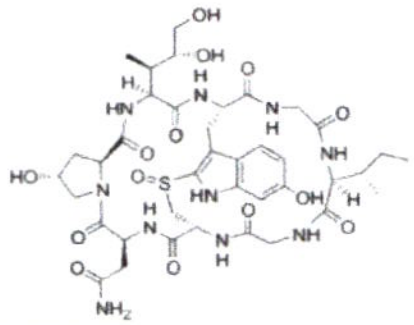	$C_{39}H_{54}N_{10}O_{14}S$	918.97	Cyclic octopeptide	Antibacterial activity, antifungal activity, anthelmintic activity	Dahiya et al. [28]
Plitidepsin	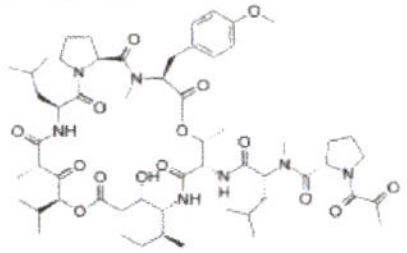	$C_{57}H_{87}N_7O_{15}$	1110.357	Cyclic depsipeptide	Effective against various cancers, e.g., breast, thyroid, lung, etc.	Leisch et al. [29]
Borophycin	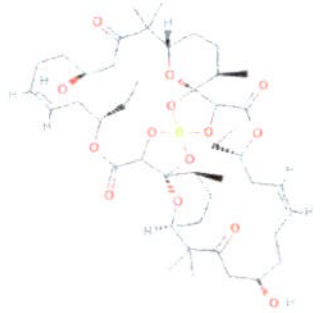	$C_{44}H_{68}BO_{14}$	831.8	Organic compounds	Various carcinoma types, e.g., epidermoid and human colorectal adenocarcinoma	Nowruzi et al. [30]
Cryptophycin-52	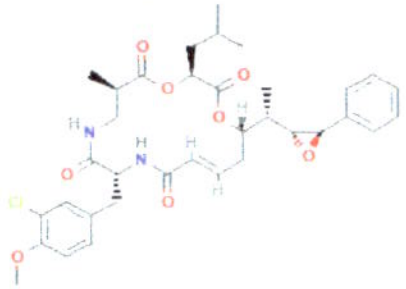	$C_{35}H_{43}ClN_2O_8$	655.2	Depsipeptide	Tumor cell lines	Nowruzi et al. [31]

Table 2. *Cont.*

Peptide Name	Structure	Chemical Formula	Molar Mass (g/mol)	Class or Type of Chemical Compounds	Bioactivities and/or Proposed Applications	References
Fucoxanthin	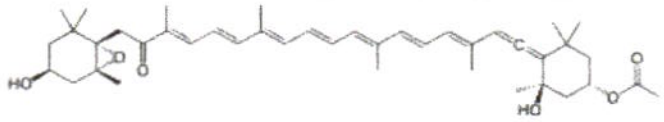	$C_{42}H_{58}O_6$	658.920	Carotenoids	Gastric cancer	Zhu et al. [32]
Pardaxin	–	$C_{154}H_{248}N_{36}O_{45}$	3323.8	Cationic peptide	Oral squamous cell carcinoma	Han et al. [33]
Hepcidin	–	$C_{113}H_{170}N_{34}O_{31}S_9$	2789.4	Cationic amphipathic peptide	Human cervical carcinoma, hepatocellular carcinoma, breast adenocarcinoma cell line	Hassana et al. [34]; Chang et al. [35]; Chen et al. [36]
Hemiasterlin	–	$C_{30}H_{46}N_4O_4$	526.7	Peptide	Inhibitory effect on microtubule assembly, cell cycle arrest, Apoptosis induction	Lai et al. [37]
Aurilide	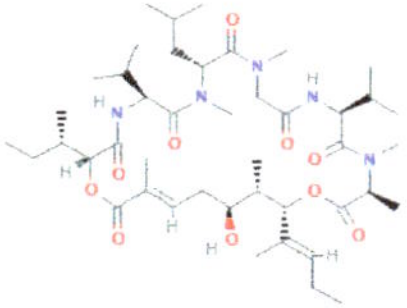	$C_{44}H_{75}N_5O_{10}$	834.1	Depsipeptides	Human lung tumor, leukemia, renal, and prostate cancer cell lines	Sato et al. [38]; Han et al. [39]; Suenaga et al. [40]
Desmethoxymajusculamide C	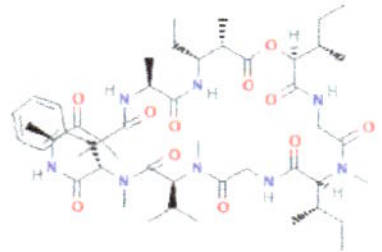	$C_{49}H_{78}N_8O_{11}$	955.2	Cyclic depsipeptide	Human colon HCT-116	Simmons et al. [41]
Diazonamide A	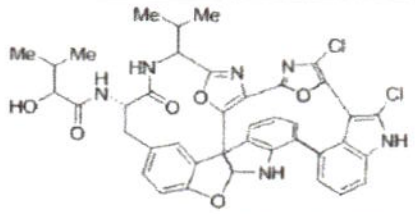	$C_{40}H_{34}Cl_2N_6O_6$	765.6	Oxazoles	Human tumor cells	Lachia and Moody [42]
PG155	–	–	15,500	Polypeptide	Potent antiangiogenic activity	Zheng et al. [43]

Table 2. *Cont.*

Peptide Name	Structure	Chemical Formula	Molar Mass (g/mol)	Class or Type of Chemical Compounds	Bioactivities and/or Proposed Applications	References
C-phycocyanin (e.g., Phycocyanobilin)	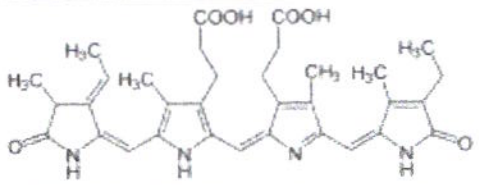	$C_{33}H_{38}N_4O_6$	586.7	Phycobiliprotein	Apoptosis induction	Li et al. [44]
Kahalalide F	–	$C_{75}H_{124}N_{14}O_{16}$	1477.9	Depsipeptide	Ovaries, breast, prostate, colon, and liver tumor cells	Sewell et al. [45]
Vitilevuamide	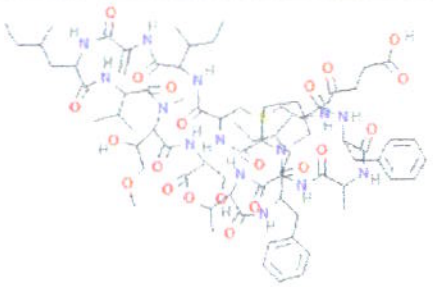	$C_{77}H_{114}N_{14}O_{21}S$	1603.9	Cyclic peptide	Lymphocytic inhibition of tubulin polymerization	Edler et al. [46]
Tachyplesin	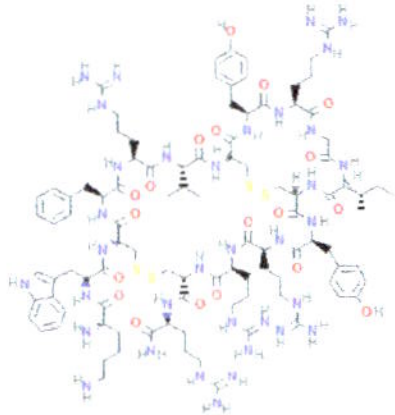	$C_{99}H_{151}N_{35}O_{19}S_4$	2263.8	Cationic Peptides	Prostate, Melanoma, and endothelial cancer cell	Chen et al. [47]
Thiocoraline	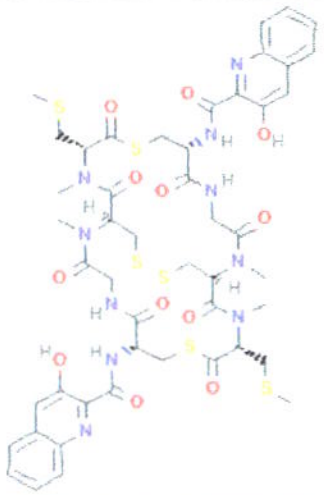	$C_{48}H_{56}N_{10}O_{12}S_6$	1157.4	Depsipeptides	Human colon cancer	Erba et al. [48]

5. Bioactivities—An Overview

As mentioned earlier, the marine biome having millions of macro- and microspecies (both animal and plants) is becoming highly significant [49,50]. This is also because the marine is considered a prolific source of structurally and functionally diverse MBPs at large, and MBPs in particular. The pharmacological activities, such as antimicrobial, antiviral, anticancer, anticoagulant, antidiabetic, cardioprotective, immunomodulatory, neuroregenerative, appetite suppressing, and many other MBPs, have been described in the literature [14–19,51,52]. Anticancer, anti-inflammatory, immunomodulating, antioxidant, hepatoprotective, and neuroprotective activities of phycobiliproteins from cyanobacteria and red algae are reported [15,53]. According to one estimate, the marine biome offers around 8000 species of red algae, which are highly enriched with several MBPs. An array of cyanobacterial-based MBPs, e.g., polypeptide or a hybrid of polyketide–polypeptide, are obtained via polyketide synthase and nonribosomal peptide synthase [54] and comprise exceptional characteristics, e.g., amino and/or hydroxy acids, heteroaromatic ring systems, and extended polyketide-derived units [55]. Likewise, several other BP-based functional entities, e.g., microcystin (a cyclic peptide), majusculamide 8 (depsipeptide), etc. have been isolated and/or extracted from marine-based blue-green algae [56,57]. Ghalamara et al. [58] proposed and extracted bioactive peptides-enriched and bioactive fractions from codfish blood, which shows notable antimicrobial properties and inhibits *Escherichia coli* growth [58].

Even though marine organisms at large, and microalgae, in particular, are highly enriched with diverse MBPs, their application as antimicrobials against human pathogens and diseases in aquaculture is still in its early stages [59]. With ever-rising heavy drug resistance and/or emergence/reemergence of new resistant microbial infections, there is a dire need to develop robust classes of natural antibiotics in combination with MBPs to treat the emerging/reemerging bacterial pathogens [60]. Diverse natural compounds with potent antimicrobial properties have been described in algae that belong to a wide range of chemical classes, including fatty acids, indoles, acetogenins, phenols, terpenes, and volatile, halogenated hydrocarbons [61]. Macro- and microalgae have grown to produce MBPs to pledge pathogenic bacteria [62], which are considered ubiquitous to their environment. Under the current scenario, effective treatment for a multi-drug-resistant *S. aureus* has become a challenge and extremely worrying concern [63]. Therefore, an alternative, novel antimicrobial agent is highly demanded.

Cancer is a multifaceted disease in which abnormal cells divide overpoweringly and abolish normal body tissue. Owing to its molecular variations and consequential cellular effects, it has expanded extensive courtesy in terms of public health. Cancer involves several types that cause adversative consequences, e.g., leading to the growth of a cell mass with the capability to penetrate adjacent normal tissues and move to new locations via a mechanism known as metastasis (Figure S1) [64,65]. Figure S2 shows a detailed metastasis to new locations in (A) lungs and (B) brain shown in Figure S1. MBPs can attack the cancerous cells membrane by the processes called either necrosis or apoptosis that may cause cell death. In necrosis, the peptides target the negatively charged molecules on the cancer cell membrane and cause cell lysis, while in the case of apoptosis (Figure 3), they cause disruption of the mitochondrial membrane [66–68]. Amphipathic peptides have shown notable bioactivities and can easily be obtained from various marine-based natural resources [5,69]. Amphipathic peptides can be used to strengthen the natural defense against invasive pathogens. The prospective therapeutic application of MBPs has been established because of their wide-ranging bioactivity spectrum and the likelihood of not inducing resistance [70]. Initiation of extrinsic apoptotic activity pathways and inhibition of angiogenesis is an excellent example of killing action of MBPs. Owing to their capability to work synergistically, most of the MBPs, expressly antitumor peptides, are considered highly active in combination with the old style, but in practice, they are chemotherapeutic agents [5].

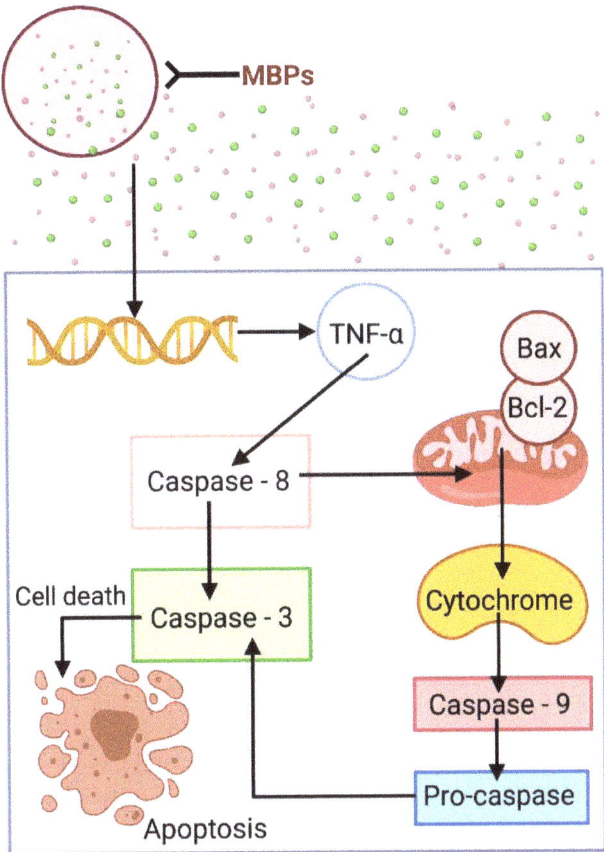

Figure 3. Anticancer potentialities of MBPs to abolish cancer cells and avoid metastasis via apoptosis mechanism. Created with BioRender.com and extracted under premium membership.

Based on the structural characteristics, they are mainly categorized into (1) cysteine-rich and β-sheet peptides, e.g., α- and β -defensins; (2) α-helices comprising peptides, e.g., LL-37 cathelicidin, cecropins, magainins, etc.; (3) peptide structures that are rich in glycine, proline, tryptophan, arginine, histidine, etc.; (4) peptides comprising a single disulfide bond, e.g., bactenecin. Further examples in the literature regarding some particular peptides having anticancerous activities are given in Table 2. Microalgal peptides with antiatherosclerotic activity have also been recognized, as described by Fan et al. [71]. Atherogenesis is an artery wall syndrome that encompasses several progressions, including cell adhesion, migration, differentiation, proliferation, and cell interaction with the extracellular matrix till the formation of atherosclerotic plaques, controlled by a complex network/cascade of cytokines and growth regulatory peptides. Besides the above-mentioned model examples, several other biological activities, such as anticoagulant, antidiabetic, cardioprotective, immunomodulatory, neuroregenerative, appetite suppressing, and others have fascinated researchers and strengthened the marine pharmacology sector [72–74]. Oxidative stress is a foremost reason for inflammatory events implicated in many of the above-mentioned diseases (e.g., neurodegenerative, cardiovascular, cancer, diabetes, etc.) [72,74].

6. FDA-Approved MBPs in the Market and Clinical Trials

Owing to the unique structural and multifunctional attributes, an array of MBPs have attained FDA-approved status, as already verified with numerous bioactivities, such as antimicrobials (antibacterial, and antifungal), antiviral, anticancer, antioxidant, antihypertensive, anticoagulant, antithrombotic, immunomodulatory, cholesterol-lowering activities, etc. The market-available MBP-based products have been developed using a unique combination of pristine counterparts and compositional alternation, as appropriate for medicinal, nutraceutical, and pharmaceutical uses. A plethora of BPs have been extracted from marine sources, treated, and purified, though using different materials and methods. However, a small fraction of those BPs have been legalized to move for clinical phase assessment, and even a few have accomplished to reach in the market. To avoid literature redundancy, selected FDA-approved MBPs are summarized in Table 3 with detailed information. In contrast, several other marine-derived biologically active compounds/products, including Pliditepsin, PM00104, Kahalalide F, Hemiasterlin, Spisulosine, Pseudopterosin A, Salinosporamide A, Tetrodotoxin, Conotoxin G, Bryostatin 1, and Plinabulin, are under clinical trials (phases I–III) [21,22,75–77].

Table 3. Selected FDA-approved MBPs.

Compound Name	Formula	Molar Mass	CAS Number	Source	Natural/Derivative	Legal Status	Bioavailability [1]	Elimination Half-Life	Applications
Ziconotide (intrathecal ziconotide)	$C_{102}H_{172}N_{36}O_{32}S_7$	2639.14 (g/mol)	107452-89-1	Cone Snail	Natural product	Prescription only	50%	2.9 to 6.5 h	Analgesics
Adcetris (Brentuximab vedotin)	$C_{6476}H_{9930}N_{1690}O_{2030}S_{40}$	149.2–151.8 (kg/mol)	914088-09-8	*Dolabella auricularia*	Derivative	Prescription only	50–80%	Approximately 4 to 6 days	Cancer treatment, treatment for patients with cutaneous T-cell lymphoma
Dactinomycin	$C_{62}H_{86}N_{12}O_{16}$	1255.438 (g/mol)	50-76-0	*Streptomyces parvullus*	Derivative	Prescription only	Not Available	36 h	Cancer treatment including for Gestational trophoblastic neoplasia, Wilms' tumor, Rhabdomyosarcoma, Ewing's sarcoma
Bacitracin (Baciim)	$C_{66}H_{103}N_{17}O_{16}S$	1422.71 (g/mol)	1405-87-4	*Bacillus subtilis*	Natural product	Prescription-only for injection and OTC	Not Available	Not Available	Acute and chronic localized skin infections
Dutasteride (Avodart)	$C_{27}H_{30}F_6N_2O_2$	528.539 (g/mol)	164656-23-9	–	Synthetic	Prescription-only	60%	4–5 Weeks	Treat enlarged prostate, Prostate cancer, hormone therapy
Curacin A	$C_{23}H_{35}NOS$	373.60 (g/mol)	155233-30-0	*Lyngbya majuscula*	Natural product	Not Available	Not Available	Not Available	Cancer treatment
Eribulin (Halaven)	$C_{40}H_{59}NO_{11}$	729.908 (g/mol)	253128-41-5	Marine Sponge	–	Prescription-only	Not Available	40 h	Cancer treatment
Trabectedin (Yondelis)	$C_{39}H_{43}N_3O_{11}S$	761.84 (g/mol)	114899-77-3	Marine Tunicate		Prescription-only	Not Available	180 h	Antitumor chemotherapy medication for the treatment of advanced soft-tissue sarcoma and ovarian cancer

[1]: In pharmacology, bioavailability refers to absorption and is the fraction (%) of an administered drug that reaches the systemic circulation.

7. Sustainable Development Goals (SDGs) and Legislation to Valorize Marine Biome

The sustainable development goals (SDGs) are blueprints that urge to attain better tomorrow and a more sustainable future. Moreover, a highly efficient transformation of innumerable marine resources into high-value entities, such as MBPs, as per SDGs, is of supreme interest. The potential of marine pharmacology pointedly subsidizes accomplishing 14 out of 17 of the United Nations (UN) SDGs (available online: https://sdgs.un.org/ -Last (accessed on 27 April 2021)). More specifically, with particular reference to SDG 14, "life below water" is a central theme for the sighting of biodiversity and the sustainable use of marine-based natural resources. Promotion of resource efficiency and technological development contribute to SDG 12 (Responsible Consumption and Production) and SDG 9 (Industry, Innovation, and Infrastructure). Therefore, in order to achieve SDG 3, with good health and well-being theme, the well-developed and effectively deployed policy frameworks and regulations or a combination of SDG measures are much needed, including the expansion of new products in the medical and pharmaceutical industries. The establishment of partnerships between governments, industry, civil society, and the scientific sector contributes to SDG 17 (Partnerships for the Goals). As confirmed most recently by Agenda 2030, the ecosystem-based approach is vital to "conserve and sustainably use the oceans, seas and marine resources for sustainable development" (SDG 14). At the sea-basin level, a close regional alliance of Member States within pertinent regional sea conventions helps synchronize the regional execution and valuation of ocean-related SDGs.

European Union (EU) legislation supports marine spatial planning (MSP) to balance the maritime economy while protecting and valorizing biodiversity. Several complementary policies have been regulated, i.e., the regulation of fisheries through the Common Fisheries Policy (CFP), EU Biodiversity Strategy to 2020, EU Regulation 1143/2014 on Invasive Alien Species, and the control of the input of nutrients and chemicals into waters through the Water Framework Directive (WFD), etc. The EU Marine Strategy Framework Directive, the environmental pillar of the EU maritime policy, introduced the principle of ecosystem-based marine spatial planning and provided a supportive framework for national initiatives toward spatial planning, designed for achieving good status for the environment. In summary, successful execution of the Marine Directive will be dynamic if the Integrated Maritime Policy is to be delivered as intended. To provide further insight on the current policy frameworks and regulations, some of the EU directives and legislation are summarized in Table S1.

8. Conclusions, Current Challenges, and Future Considerations

In summary, marine biotechnology and pharmacology have infinite potential to formulate/synthesize new MBPs that show a dynamic role in bio- and non-bioindustries of the modern world. MBPs obtained by following green chemistry agenda principles using naturally existing sources from marine biomes such as microalgae have succeeding merits, including natural abundance, ease in availability throughout the year, materials renewability and sustainability, carbon-neutral aspects, re-processibility with a zero-waste SDG agenda, facile synthesis options with possibilities to scale up, net positive and high cost-effective ratio, no or minimal consumption of harsh chemicals/reagents, and no or less toxic contaminants/byproducts, etc. Nonetheless, colossal steps have already been reserved and taken in the past years; however, focused and genetically positioned research is necessary to further strengthen the marine pharmacology consideration for a better tomorrow.

Regardless of ever-growing scientific awareness and technological advancement, several challenges still exist to address valorize MBPs from marine origins. These include (1) reachability ease to the unexplored biodiversity, (2) standardized procedural isolation (regardless of the source variability), (3) cost-effective handling of the extracted products, (4) stability maintenance under standardized environment for any or many products at the same time (regardless of the compositional variability), (5) possible scale-up probability, (6) sustainable marketing and commercialization, all of which are vital challenges and should be considered with care. A considerable amount of information is available in the literature about numerous potential aspects of marine biotechnology and pharmacology. However, substantial critiques are still unresolved, which necessitate future studies. Despite contemporary scientific advancements in marine biotechnology, extensive research with verified employability of marine sources is needed in this line of research.

Supplementary Materials: The following are available online at https://www.mdpi.com/article/10.3390/md20030208/s1, Figure S1: Schematic representation of cancer metastasis to new locations. Cancer involves several types that cause adversative consequences, e.g., leading to the growth of a cell mass with the capability to penetrate and move to adjacent normal tissues. In metastasis, cancer cells break away from where they first formed (primary cancer site), travel through the blood stream, and form new tumors (metastatic tumors) in other parts of the body (secondary cancer site). Created with BioRender.com and extracted under premium membership. Figure S2: A detailed metastasis to new locations (**A**) lungs and (**B**) brain, with reference to the model locations marked in Figure S1. Created with BioRender.com and extracted under premium membership. Table S1: EU directives and legislation that support current policy frameworks and regulations in the field of marine environmental policy.

Funding: This research received no external funding.

Institutional Review Board Statement: Not applicable.

Informed Consent Statement: Not applicable.

Data Availability Statement: Not applicable.

Acknowledgments: Consejo Nacional de Ciencia y Tecnología (CONACYT) Mexico is thankfully acknowledged for partially supporting this research under Sistema Nacional de Investigadores (SNI) program awarded to Ashutosh Sharma (CVU: 398262), Roberto Parra-Saldívar (CVU: 35753), and Hafiz M. N. Iqbal (CVU: 735340).

Conflicts of Interest: The authors declare no conflict of interest.

References

1. Centella, M.H.; Arévalo-Gallegos, A.; Parra-Saldivar, R.; Iqbal, H.M. Marine-derived bioactive compounds for value-added applications in bio-and non-bio sectors. *J. Clean. Prod.* **2017**, *168*, 1559–1565. [CrossRef]
2. Sosa-Hernández, J.E.; Escobedo-Avellaneda, Z.; Iqbal, H.; Welti-Chanes, J. State-of-the-art extraction methodologies for bioactive compounds from algal biome to meet bio-economy challenges and opportunities. *Molecules* **2018**, *23*, 2953. [CrossRef] [PubMed]
3. DeVierno Kreuder, A.; House-Knight, T.; Whitford, J.; Ponnusamy, E.; Miller, P.; Jesse, N.; Rodenborn, R.; Sayag, S.; Gebel, M.; Aped, I.; et al. A method for assessing greener alternatives between chemical products following the 12 principles of green chemistry. *ACS Sustain. Chem. Eng.* **2017**, *5*, 2927–2935. [CrossRef]
4. Wang, X.; Yu, H.; Xing, R.; Li, P. Characterization, preparation, and purification of marine bioactive peptides. *BioMed Res. Int.* **2017**, *2017*, 9746720. [CrossRef]
5. Li, D.; Liao, X.; Zhong, S.; Zhao, B.; Xu, S. Synthesis of Marine Cyclopeptide Galaxamide Analogues as Potential Anticancer Agents. *Mar. Drugs* **2022**, *20*, 158. [CrossRef]
6. Tyagi, A.; Tuknait, A.; Anand, P.; Gupta, S.; Sharma, M.; Mathur, D.; Joshi, A.; Singh, S.; Gautam, A.; Raghava, G.P. CancerPPD: A database of anticancer peptides and proteins. *Nucleic Acids Res.* **2015**, *43*, D837–D843. [CrossRef]
7. Delattre, C.; Pierre, G.; Laroche, C.; Michaud, P. Production, extraction and characterization of microalgal and cyanobacterial exopolysaccharides. *Biotechnol. Adv.* **2016**, *34*, 1159–1179. [CrossRef]
8. Al-Dhafri, K.S.; Ching, C.L. Isolation and Characterization of Antimicrobial Peptides Isolated from *Fagonia bruguieri*. *Appl. Biochem. Biotechnol.* **2022**, 1–14. [CrossRef]
9. Hernández, Y.; Lobo, M.G.; González, M. Factors affecting sample extraction in the liquid chromatographic determination of organic acids in papaya and pineapple. *Food Chem.* **2009**, *114*, 734–741. [CrossRef]
10. Hamayeli, H.; Hassanshahian, M.; Hesni, M.A. Identification of Bioactive Compounds and Evaluation of the Antimicrobial and Anti-biofilm Effect of *Psammocinia* sp. and *Hyattella* sp. Sponges from the Persian Gulf. *Thalass. Int. J. Mar. Sci.* **2021**, *37*, 357–366. [CrossRef]
11. Lim, S.M.; Agatonovic-Kustrin, S.; Lim, F.T.; Ramasamy, K. High-performance thin layer chromatography-based phytochemical and bioactivity characterisation of anticancer endophytic fungal extracts derived from marine plants. *J. Pharm. Biomed. Anal.* **2021**, *193*, 113702. [CrossRef] [PubMed]
12. Nguyen, T.T.; Heimann, K.; Zhang, W. Protein Recovery from Underutilised Marine Bioresources for Product Development with Nutraceutical and Pharmaceutical Bioactivities. *Mar. Drugs* **2020**, *18*, 391. [CrossRef] [PubMed]

13. Cheung RC, F.; Ng, T.B.; Wong, J.H. Marine peptides: Bioactivities and applications. *Mar. Drugs* **2015**, *13*, 4006–4043. [CrossRef] [PubMed]
14. De Morais, M.G.; Vaz BD, S.; de Morais, E.G.; Costa, J.A.V. Biologically active metabolites synthesized by microalgae. *BioMed Res. Int.* **2015**, *2015*, 835761. [CrossRef]
15. Talero, E.; García-Mauriño, S.; Ávila-Román, J.; Rodríguez-Luna, A.; Alcaide, A.; Motilva, V. Bioactive compounds isolated from microalgae in chronic inflammation and cancer. *Mar. Drugs* **2015**, *13*, 6152–6209. [CrossRef]
16. Rosenthal, A.; Pyle, D.L.; Niranjan, K. Aqueous and enzymatic processes for edible oil extraction. *Enzym. Microb. Technol.* **1996**, *19*, 402–420. [CrossRef]
17. Sánchez, A.; Vázquez, A. Bioactive peptides: A review. *Food Qual. Saf.* **2017**, *1*, 29–46. [CrossRef]
18. Möller, N.P.; Scholz-Ahrens, K.E.; Roos, N.; Schrezenmeir, J. Bioactive peptides and proteins from foods: Indication for health effects. *Eur. J. Nutr.* **2008**, *47*, 171–182. [CrossRef]
19. Chi, C.F.; Hu, F.Y.; Wang, B.; Li, T.; Ding, G.F. Antioxidant and anticancer peptides from the protein hydrolysate of blood clam (*Tegillarca granosa*) muscle. *J. Funct. Foods* **2015**, *15*, 301–313. [CrossRef]
20. Jose, D.T.; Uma, C.; Sivagurunathan, P.; Aswini, B.; Dinesh, M.D. Extraction and antibacterial evaluation of marine AMPs against diabetic wound pathogens. *J. Appl. Pharm. Sci.* **2020**, *10*, 087–092.
21. Ucak, I.; Afreen, M.; Montesano, D.; Carrillo, C.; Tomasevic, I.; Simal-Gandara, J.; Barba, F.J. Functional and bioactive properties of peptides derived from marine side streams. *Mar. Drugs* **2021**, *19*, 71. [CrossRef] [PubMed]
22. Zhang, Q.T.; Liu, Z.D.; Wang, Z.; Wang, T.; Wang, N.; Wang, N.; Zhang, B.; Zhao, Y.F. Recent Advances in Small Peptides of Marine Origin in Cancer Therapy. *Mar. Drugs* **2021**, *19*, 115. [CrossRef] [PubMed]
23. Safronova, V.N.; Panteleev, P.V.; Sukhanov, S.V.; Toropygin, I.Y.; Bolosov, I.A.; Ovchinnikova, T.V. Mechanism of Action and Therapeutic Potential of the β-Hairpin Antimicrobial Peptide Capitellacin from the Marine *Polychaeta Capitella teleta*. *Mar. Drugs* **2022**, *20*, 167. [CrossRef]
24. Zeng, Z.; Wang, Y.; Anwar, M.; Hu, Z.; Wang, C.; Lou, S.; Li, H. Molecular cloning and expression analysis of mytilin-like antimicrobial peptides from Asian green mussel *Perna viridis*. *Fish Shellfish Immunol.* **2022**, *121*, 239–244. [CrossRef]
25. Olivera, B.M. Biodiversity-based Discovery and Exogenomics. *J. Biol. Chem.* **2006**, *281*, 31173–31177.
26. Morlighem, J.É.R.; Radis-Baptista, G. The place for enzymes and biologically active peptides from marine organisms for application in industrial and pharmaceutical biotechnology. *Curr. Protein Pept. Sci.* **2019**, *20*, 334–355. [CrossRef]
27. Messina, C.M.; Manuguerra, S.; Arena, R.; Renda, G.; Ficano, G.; Randazzo, M.; Fricano, S.; Sadok, S.; Santulli, A. In Vitro Bioactivity of Astaxanthin and Peptides from Hydrolisates of Shrimp (*Parapenaeus longirostris*) By-Products: From the Extraction Process to Biological Effect Evaluation, as Pilot Actions for the Strategy "From Waste to Profit". *Mar. Drugs* **2021**, *19*, 216. [CrossRef]
28. Dahiya, R.; Rampersad, S.; Ramnanansingh, T.G.; Kaur, K.; Kaur, R.; Mourya, R.; Chennupati, S.V.; Fairman, R.; Jalsa, N.K.; Sharma, A.; et al. Synthesis and Bioactivity of a Cyclopolypeptide from Caribbean Marine Sponge. *Iran. J. Pharm. Res. IJPR* **2020**, *19*, 156.
29. Leisch, M.; Egle, A.; Greil, R. Plitidepsin: A potential new treatment for relapsed/refractory multiple myeloma. *Future Oncol.* **2019**, *15*, 109–120. [CrossRef]
30. Nowruzi, B.; Blanco, S.; Nejadsattari, T. Chemical and molecular evidences for the poisoning of a duck by anatoxin-a, nodularin and cryptophycin at the coast of Lake Shoormast (Mazandaran province, Iran). *Int. J. Algae* **2018**, *20*, 359–376. [CrossRef]
31. Nowruzi, B.; Haghighat, S.; Fahimi, H.; Mohammadi, E. Nostoc cyanobacteria species: A new and rich source of novel bioactive compounds with pharmaceutical potential. *J. Pharm. Health Serv. Res.* **2018**, *9*, 5–12. [CrossRef]
32. Zhu, Y.; Cheng, J.; Min, Z.; Yin, T.; Zhang, R.; Zhang, W.; Hu, L.; Cui, Z.; Gao, C.; Xu, S.; et al. Effects of fucoxanthin on autophagy and apoptosis in SGC-7901cells and the mechanism. *J. Cell. Biochem.* **2018**, *119*, 7274–7284. [CrossRef] [PubMed]
33. Han, Y.; Cui, Z.; Li, Y.H.; Hsu, W.H.; Lee, B.H. In vitro and in vivo anticancer activity of pardaxin against proliferation and growth of oral squamous cell carcinoma. *Mar. Drugs* **2016**, *14*, 2. [CrossRef] [PubMed]
34. Hassana MA, K.; Azeminb, W.A.; Dharmaraja, S.; Mohdb, K.S. Hepcidin TH1-5 Induces Apoptosis and Activate Caspase-9 in MCF-7 Cells. *J. Appl. Pharm. Sci.* **2016**, *6*, 081–086. [CrossRef]
35. Chang, W.T.; Pan, C.Y.; Rajanbabu, V.; Cheng, C.W.; Chen, J.Y. Tilapia (*Oreochromis mossambicus*) antimicrobial peptide, hepcidin 1–5, shows antitumor activity in cancer cells. *Peptides* **2011**, *32*, 342–352. [CrossRef] [PubMed]
36. Chen, J.Y.; Lin, W.J.; Lin, T.L. A fish antimicrobial peptide, tilapia hepcidin TH2-3, shows potent antitumor activity against human fibrosarcoma cells. *Peptides* **2009**, *30*, 1636–1642. [CrossRef] [PubMed]
37. Lai, W.T.; Cheng, K.L.; Baruchello, R.; Rondanin, R.; Marchetti, P.; Simoni, D.; Lee, R.M.; Guh, J.H.; Hsu, L.C. Hemiasterlin derivative (R)(S)(S)-BF65 and Akt inhibitor MK-2206 synergistically inhibit SKOV3 ovarian cancer cell growth. *Biochem. Pharmacol.* **2016**, *113*, 12–23. [CrossRef]
38. Sato, S.I.; Murata, A.; Orihara, T.; Shirakawa, T.; Suenaga, K.; Kigoshi, H.; Uesugi, M. Marine natural product aurilide activates the OPA1-mediated apoptosis by binding to prohibitin. *Chem. Biol.* **2011**, *18*, 131–139. [CrossRef]
39. Han, B.; Gross, H.; Goeger, D.E.; Mooberry, S.L.; Gerwick, W.H. Aurilides B and C, Cancer Cell Toxins from a Papua New Guinea Collection of the Marine Cyanobacterium *Lyngbya majuscula*. *J. Nat. Prod.* **2006**, *69*, 572–575. [CrossRef]

40. Suenaga, K.; Mutou, T.; Shibata, T.; Itoh, T.; Fujita, T.; Takada, N.; Hayamizu, K.; Takagi, M.; Irifune, T.; Kigoshi, H. Aurilide, a cytotoxic depsipeptide from the sea hare *Dolabella auricularia*: Isolation, structure determination, synthesis, and biological activity. *Tetrahedron* **2004**, *60*, 8509–8527. [CrossRef]
41. Simmons, T.L.; Nogle, L.M.; Media, J.; Valeriote, F.A.; Mooberry, S.L.; Gerwick, W.H. Desmethoxymajusculamide C, a cyanobacterial depsipeptide with potent cytotoxicity in both cyclic and ring-opened forms. *J. Nat. Prod.* **2009**, *72*, 1011–1016. [CrossRef] [PubMed]
42. Lachia, M.; Moody, C.J. The synthetic challenge of diazonamide A, a macrocyclic indole bis-oxazole marine natural product. *Nat. Prod. Rep.* **2008**, *25*, 227–253. [CrossRef] [PubMed]
43. Zheng, L.; Ling, P.; Wang, Z.; Niu, R.; Hu, C.; Zhang, T.; Lin, X. A novel polypeptide from shark cartilage with potent anti-angiogenic activity. *Cancer Biol. Ther.* **2007**, *6*, 775–780. [CrossRef] [PubMed]
44. Li, B.; Gao, M.H.; Zhang, X.C.; Chu, X.M. Molecular immune mechanism of C-phycocyanin from Spirulina platensis induces apoptosis in HeLa cells in vitro. *Biotechnol. Appl. Biochem.* **2006**, *43*, 155–164.
45. Sewell, J.M.; Mayer, I.; Langdon, S.P.; Smyth, J.F.; Jodrell, D.I.; Guichard, S.M. The mechanism of action of Kahalalide F: Variable cell permeability in human hepatoma cell lines. *Eur. J. Cancer* **2005**, *41*, 1637–1644. [CrossRef]
46. Edler, M.C.; Fernandez, A.M.; Lassota, P.; Ireland, C.M.; Barrows, L.R. Inhibition of tubulin polymerization by vitilevuamide, a bicyclic marine peptide, at a site distinct from colchicine, the vinca alkaloids, and dolastatin 10. *Biochem. Pharmacol.* **2002**, *63*, 707–715. [CrossRef]
47. Chen, Y.; Xu, X.; Hong, S.; Chen, J.; Liu, N.; Underhill, C.B.; Creswell, K.; Zhang, L. RGD-Tachyplesin inhibits tumor growth. *Cancer Res.* **2001**, *61*, 2434–2438.
48. Erba, E.; Bergamaschi, D.; Ronzoni, S.; Faretta, M.; Taverna, S.; Bonfanti, M.; Catapano, C.V.; Faircloth, G.; Jimeno, J.; D'incalci, M. Mode of action of thiocoraline, a natural marine compound with anti-tumour activity. *Br. J. Cancer* **1999**, *80*, 971–980. [CrossRef]
49. Li, Y.X.; Wijesekara, I.; Li, Y.; Kim, S.K. Phlorotannins as bioactive agents from brown algae. *Process Biochem.* **2011**, *46*, 2219–2224. [CrossRef]
50. Valverde, F.; Romero-Campero, F.J.; León, R.; Guerrero, M.G.; Serrano, A. New challenges in microalgae biotechnology. *Eur. J. Protistol.* **2016**, *55*, 95–101. [CrossRef]
51. De Sousa Santos AK, F.; da Fonseca, D.V.; Salgado PR, R.; Muniz, V.M.; de Arruda Torres, P.; Lira, N.S.; da Silva Dias, C.; de Morais Pordeus, L.C.; Barbosa-Filho, J.M.; de Almeida, R.N. Antinociceptive activity of Sargassum polyceratium and the isolation of its chemical components. *Rev. Bras. Farmacogn.* **2015**, *25*, 683–689.
52. Arumugam, V.; Venkatesan, M.; Ramachandran, K.; Ramachandran, S.; Palanisamy, S.K.; Sundaresan, U. Purification, Characterization and Antibacterial Properties of Peptide from Marine *Ascidian Didemnum* sp. *Int. J. Pept. Res. Ther.* **2020**, *26*, 201–208. [CrossRef]
53. El Gamal, A.A. Biological importance of marine algae. *Saudi Pharm. J.* **2010**, *18*, 1–25. [CrossRef] [PubMed]
54. Mazard, S.; Penesyan, A.; Ostrowski, M.; Paulsen, I.T.; Egan, S. Tiny microbes with a big impact: The role of cyanobacteria and their metabolites in shaping our future. *Mar. Drugs* **2016**, *14*, 97. [CrossRef] [PubMed]
55. Tan, L.T.; Okino, T.; Gerwick, W.H. Bouillonamide: A mixed polyketide–peptide cytotoxin from the marine cyanobacterium Moorea bouillonii. *Mar. Drugs* **2013**, *11*, 3015–3024. [CrossRef] [PubMed]
56. Lawrence, J.F.; Niedzwiadek, B.; Menard, C.; Lau, B.P.; Lewis, D.; Kuper-Goodman, T.; Carbone, S.; Holmes, C. Comparison of liquid chromatography/mass spectrometry, ELISA, and phosphatase assay for the determination of microcystins in blue-green algae products. *J. AOAC Int.* **2001**, *84*, 1035–1044. [CrossRef] [PubMed]
57. Abdel-Rahman, G.N.E. Cytotoxicity and antibacterial activity of the blue green alga Microcystis aeruginosa extracts against human cancer cell lines and foodborne bacteria. *Egypt. J. Chem.* **2020**, *63*, 2–3. [CrossRef]
58. Ghalamara, S.; Silva, S.; Brazinha, C.; Pintado, M. Valorization of fish by-products: Purification of bioactive peptides from codfish blood and sardine cooking wastewaters by membrane processing. *Membranes* **2020**, *10*, 44. [CrossRef]
59. Falaise, C.; François, C.; Travers, M.A.; Morga, B.; Haure, J.; Tremblay, R.; Turcotte, F.; Pasetto, P.; Gastineau, R.; Hardivillier, Y.; et al. Antimicrobial compounds from eukaryotic microalgae against human pathogens and diseases in aquaculture. *Mar. Drugs* **2016**, *14*, 159. [CrossRef]
60. Sukmarini, L. Drug Development from Peptide-derived Marine Natural Products. In *IOP Conference Series: Materials Science and Engineering*; IOP Publishing: Tangerang, Indonesia, 2021; Volume 1011, p. 012063.
61. Plaza, M.; Santoyo, S.; Jaime, L.; Reina GG, B.; Herrero, M.; Señoráns, F.J.; Ibáñez, E. Screening for bioactive compounds from algae. *J. Pharm. Biomed. Anal.* **2010**, *51*, 450–455. [CrossRef]
62. Bule, M.H.; Ahmed, I.; Maqbool, F.; Bilal, M.; Iqbal, H.M. Microalgae as a source of high-value bioactive compounds. *Front. Biosci* **2018**, *10*, 197–216.
63. Winokur, P.; Chenoweth, C.E.; Rice, L.; Mehrad, B.; Lynch, J.P. Resistant pathogens: Emergence and control. In *Ventilator-Associated Pneumonia*; Springer: Boston, MA, USA, 2001; pp. 131–164.
64. Zheng, L.H.; Wang, Y.J.; Sheng, J.; Wang, F.; Zheng, Y.; Lin, X.K.; Sun, M. Antitumor peptides from marine organisms. *Mar. Drugs* **2011**, *9*, 1840–1859. [CrossRef] [PubMed]
65. Liu, H.; Shi, Y.; Qian, F. Opportunities and delusions regarding drug delivery targeting pancreatic cancer-associated fibroblasts. *Adv. Drug Deliv. Rev.* **2021**, *172*, 37–51. [CrossRef] [PubMed]

66. Liu, F.H.; Hou, C.Y.; Zhang, D.; Zhao, W.J.; Cong, Y.; Duan, Z.Y.; Qiao, Z.Y.; Wang, H. Enzyme-sensitive cytotoxic peptide–dendrimer conjugates enhance cell apoptosis and deep tumor penetration. *Biomater. Sci.* **2018**, *6*, 604–613. [CrossRef]
67. Mirza, A.Z. Advancement in the development of heterocyclic nucleosides for the treatment of cancer-A review. *Nucleosides Nucleotides Nucleic Acids* **2019**, *38*, 836–857. [CrossRef]
68. Böhmová, E.; Pola, R.; Pechar, M.; Parnica, J.; Machová, D.; Janoušková, O.; Etrych, T. Polymer Cancerostatics Containing Cell-Penetrating Peptides: Internalization Efficacy Depends on Peptide Type and Spacer Length. *Pharmaceutics* **2020**, *12*, 59. [CrossRef]
69. Lobine, D.; Rengasamy, K.R.; Mahomoodally, M.F. Functional foods and bioactive ingredients harnessed from the ocean: Current status and future perspectives. *Crit. Rev. Food Sci. Nutr.* **2021**, 1–30. [CrossRef]
70. Reddy, P.A.; Jones, S.T.; Lewin, A.H.; Carroll, F. Synthesis of hemopressin peptides by classical solution phase fragment condensation. *Int. J. Pept.* **2012**, *2012*, 186034. [CrossRef]
71. Fan, X.; Bai, L.; Zhu, L.; Yang, L.; Zhang, X. Marine algae-derived bioactive peptides for human nutrition and health. *J. Agric. Food Chem.* **2014**, *62*, 9211–9222. [CrossRef]
72. Robertson, R.C.; Guihéneuf, F.; Bahar, B.; Schmid, M.; Stengel, D.B.; Fitzgerald, G.F.; Ross, R.P.; Stanton, C. The anti-inflammatory effect of algae-derived lipid extracts on lipopolysaccharide (LPS)-stimulated human THP-1 macrophages. *Mar. Drugs* **2015**, *13*, 5402–5424. [CrossRef]
73. Bélanger, W.; Arnold, A.A.; Turcotte, F.; Saint-Louis, R.; Deschênes, J.S.; Genard, B.; Marcotte, I.; Tremblay, R. Extraction Improvement of the Bioactive Blue-Green Pigment "Marennine" from Diatom Haslea ostrearia's Blue Water: A Solid-Phase Method Based on Graphitic Matrices. *Mar. Drugs* **2020**, *18*, 653. [CrossRef] [PubMed]
74. Bilal, M.; Iqbal, H.M. Biologically active macromolecules: Extraction strategies, therapeutic potential and biomedical perspective. *Int. J. Biol. Macromol.* **2020**, *151*, 1–18. [CrossRef] [PubMed]
75. Anjum, K.; Abbas, S.Q.; Akhter, N.; Shagufta, B.I.; Shah SA, A.; Hassan SS, U. Emerging biopharmaceuticals from bioactive peptides derived from marine organisms. *Chem. Biol. Drug Des.* **2017**, *90*, 12–30. [CrossRef] [PubMed]
76. Alves, C.; Silva, J.; Pinteus, S.; Gaspar, H.; Alpoim, M.C.; Botana, L.M.; Pedrosa, R. From marine origin to therapeutics: The antitumor potential of marine algae-derived compounds. *Front. Pharmacol.* **2018**, *9*, 777. [CrossRef] [PubMed]
77. Ghareeb, M.A.; Tammam, M.A.; El-Demerdash, A.; Atanasov, A.G. Insights about clinically approved and Preclinically investigated marine natural products. *Curr. Res. Biotechnol.* **2020**, *2*, 88–102. [CrossRef]

MDPI

Article

Evaluation of Ultrasound, Microwave, Ultrasound–Microwave, Hydrothermal and High Pressure Assisted Extraction Technologies for the Recovery of Phytochemicals and Antioxidants from Brown Macroalgae

Marco Garcia-Vaquero [1], Rajeev Ravindran [2], Orla Walsh [3], John O'Doherty [1], Amit K. Jaiswal [3], Brijesh K. Tiwari [4] and Gaurav Rajauria [1,2,*]

[1] School Agriculture and Food Science, University College Dublin, Dublin D04 V1W8, Belfield, Ireland; marco.garciavaquero@ucd.ie (M.G.-V.); john.vodoherty@ucd.ie (J.O.)

[2] Department of Biological & Pharmaceutical Sciences, Munster Technological University, Kerry Campus, Clash V92 CX88 Tralee, Co. Kerry, Ireland; rajeev.ravindran@staff.ittralee.ie

[3] School of Food Science and Environmental Health, College of Sciences and Health, Technological University Dublin, City Campus, Central Quad, Dublin D07 ADY7, Grangegorman, Ireland; C14429608@mytudublin.ie (O.W.); amit.jaiswal@tudublin.ie (A.K.J.)

[4] Teagasc Food Research Centre, Dublin D15 DY05, Ashtown, Ireland; brijesh.tiwari@teagasc.ie

* Correspondence: gaurav.rajauria@staff.ittralee.ie; Tel.: +353-(0)-66-714-5600

Citation: Garcia-Vaquero, M.; Ravindran, R.; Walsh, O.; O'Doherty, J.; Jaiswal, A.K.; Tiwari, B.K.; Rajauria, G. Evaluation of Ultrasound, Microwave, Ultrasound–Microwave, Hydrothermal and High Pressure Assisted Extraction Technologies for the Recovery of Phytochemicals and Antioxidants from Brown Macroalgae. *Mar. Drugs* **2021**, *19*, 309. https://doi.org/10.3390/md19060309

Academic Editors: Ipek Kurtboke and Bill J. Baker

Received: 5 May 2021
Accepted: 25 May 2021
Published: 27 May 2021

Publisher's Note: MDPI stays neutral with regard to jurisdictional claims in published maps and institutional affiliations.

Abstract: This study aims to explore novel extraction technologies (ultrasound-assisted extraction (UAE), microwave-assisted extraction (MAE), ultrasound–microwave-assisted extraction (UMAE), hydrothermal-assisted extraction (HAE) and high-pressure-assisted extraction (HPAE)) and extraction time post-treatment (0 and 24 h) for the recovery of phytochemicals and associated antioxidant properties from *Fucus vesiculosus* and *Pelvetia canaliculata*. When using fixed extraction conditions (solvent: 50% ethanol; extraction time: 10 min; algae/solvent ratio: 1/10) for all the novel technologies, UAE generated extracts with the highest phytochemical contents from both macroalgae. The highest yields of compounds extracted from *F. vesiculosus* using UAE were: total phenolic content (445.0 ± 4.6 mg gallic acid equivalents/g), total phlorotannin content (362.9 ± 3.7 mg phloroglucinol equivalents/g), total flavonoid content (286.3 ± 7.8 mg quercetin equivalents/g) and total tannin content (189.1 ± 4.4 mg catechin equivalents/g). In the case of the antioxidant activities, the highest DPPH activities were achieved by UAE and UMAE from both macroalgae, while no clear pattern was recorded in the case of FRAP activities. The highest DPPH scavenging activities (112.5 ± 0.7 mg trolox equivalents/g) and FRAP activities (284.8 ± 2.2 mg trolox equivalents/g) were achieved from *F. vesiculosus*. Following the extraction treatment, an additional storage post-extraction (24 h) did not improve the yields of phytochemicals or antioxidant properties of the extracts.

Keywords: seaweed; innovative technology; extraction; polyphenol; nutraceuticals

1. Introduction

Macroalgae are a diverse group of organisms with over 19,000 species identified worldwide [1], and only 5% of them are currently exploited at an industrial level for food applications [2]. Globally, the macroalgal sector was valued at USD 10.31 billion in 2018, and at an annual growth rate of 8.9% this market is predicted to be valued at USD 22.13 billion by 2024 [3]. This valuable biomass is mainly exploited as food (85%), while the remaining 15% is exploited for the generation of high value compounds with multiple applications as pharmaceuticals, nutraceuticals and cosmeceuticals that surpass in economic value any other use or application currently being considered for macroalgae [4].

The increased economic value and research interest in macroalgal compounds as pharmaceuticals, nutraceuticals and cosmeceuticals relies on the unique and wide variety of health benefits associated with their consumption. Macroalgae are able to adapt rapidly to abiotic and biotic stressors of the marine environment by producing secondary metabolites with unique chemical structures [5,6]. Thereby, abiotic stress (i.e., changes in water level, solar radiation and/or temperature) may lead to an increased production of reactive oxygen species and free radical species that may damage the structure and metabolism of algae in the long term [5,7]. As a defence mechanism, the stressed macroalgae produce high amounts of antioxidant compounds, such as polyphenols [5,6]. Polyphenols encompass a wide variety of antioxidant compounds and chemical classes that can be quantified in the biomass as total phenolic content (TPC), total flavonoid content (TFC), total tannin content (TTC), and in the case of brown macroalgae, total phlorotannin content (TPhC) [8]. These compounds have strong antioxidant properties, playing a major role in the prevention or treatment of several diseases or disorders, including obesity, diabetes, cancer and cardiovascular diseases, amongst others [5,6]. Brown macroalgae *F. vesiculosus* and *P. canaliculata* tested in this study are native species of Ireland which have reported a high amount of polyphenols and polysaccharides with potential health properties [9,10]. Moreover, Chater et al. [11] reported that ethanol extracts of selected seaweeds showed significant inhibition of lipase in a model gut system, suggesting their potential in weight management.

Polyphenols from macroalgae are typically extracted using conventional technologies, by macerating the biomass in organic solvents (methanol, ethanol, and acetone) at room temperature or heating the mixtures for several hours or days [12]. Further processing of the polyphenol extracts is also required at a later stage, aiming to eliminate carbohydrates that are normally tightly bound to these compounds using one or several purification processes [13]. The exploitation of TPC, TFC, TTC and TPhC at an industrial level requires processing the biomass to extract high yields of these compounds with low or minimal damage to their antioxidant properties, while minimising the co-extraction of carbohydrates or total sugar contents (TSC). Novel technologies, mainly UAE, MAE, UMAE, HAE and HPAE, and natural deep eutectic solvents have gained momentum as more efficient and environmentally friendly procedures for the extraction of multiple compounds from macroalgae, as seen in the recent scientific literature [8,14–17]. Novel extraction technologies and protocols are regularly described as being efficient, achieving high yields of targeted compounds, while minimising energy consumption, the time of extraction and the need for organic solvents, thus improving the sustainability of these procedures as well as the safety of the operators during these extraction processes [16,18,19]. Most of the literature available on polyphenols from macroalgae focuses on the chemical characterisation of the polyphenols extracted [20–22], on using novel extraction technologies to optimise the yields of compounds extracted [12,23,24], or comparing the efficiency of novel technologies against conventional protocols [25]. However, limited studies are available comparing the efficiency of several novel extraction technologies for the extraction of polyphenols and antioxidants from brown macroalgae [14].

The overarching objective of the study was to identify the best novel extraction technology (while keeping solvent type, concentration and extraction time constant) to recover polyphenols with high antioxidant properties from macroalgal biomass. To achieve this, the study aimed to explore (1) multiple novel extraction technologies (UAE, MAE, UMAE, HAE and HPAE) using fixed extraction conditions (solvent: 50% ethanol; extraction time: 10 min; algae/solvent ratio: 1/10) and (2) the effect of extraction time post-treatment (0 and 24 h) on the extraction of polyphenols (TPC, TFC, TTC and TPhC) and carbohydrates (TSC), as well as the antioxidant properties (DPPH and FRAP) of the extracts generated from brown macroalgae *F. vesiculosus* and *P. canaliculata*.

2. Results and Discussion

2.1. Effect of Technologies on the Yields and Phytochemical Contents of Extracts

The extraction yields, expressed as mg of dried solids extracted per g of biomass by each novel extraction technology with or without additional extraction time post-treatment

(0 and 24 h), are represented in Figure 1. In general, the yields of compounds extracted were higher from *F. vesiculosus* (yields ranging from 173.33 to 354.67 mg dried extract per g macroalgae) compared to *P. canaliculata* (147.00 to 240.33 mg dried extract per g macroalgae). Additionally, 24 h extraction of the samples following the technological treatments did not improve the extraction yields from both macroalgae. Overall, the yields of compounds extracted from both macroalgae were the highest using UMAE (354.67 ± 2.33 and 240.33 ± 1.20 mg dried extract per g macroalgae for *F. vesiculosus* and *P. canaliculata*, respectively), followed by UAE (316.33 ± 4.63 from *F. vesiculosus* and 214.33 ± 2.60 from *P. canaliculata*), while variable results were achieved for the other technologies explored. HPAE achieved the lowest yields from both *F. vesiculosus* (184.00 ± 0.58 mg dried extract/g macroalgae) and *P. canaliculata* (155.00 ± 2.31 mg dried extract/g macroalgae).

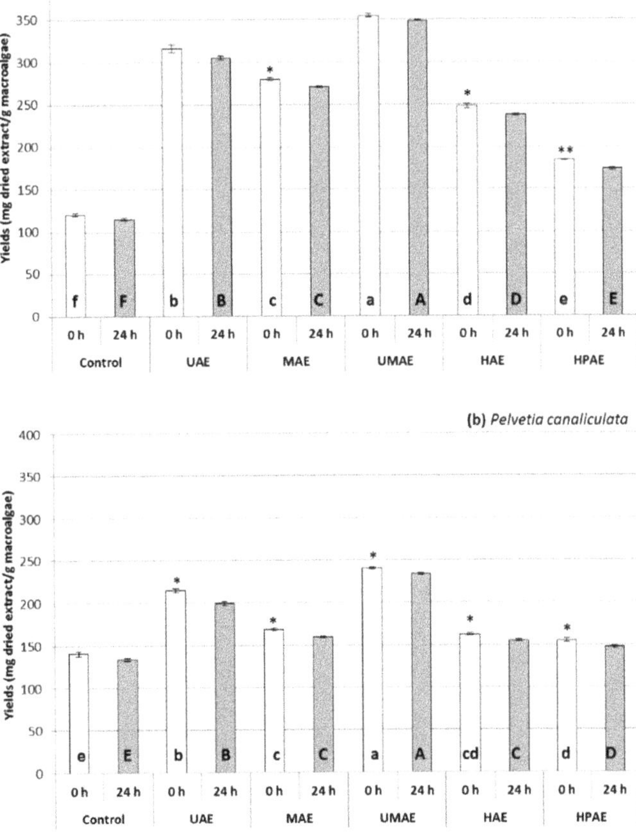

Figure 1. Yields of extract obtained from *F. vesiculosus* (**a**) and *P. canaliculata* (**b**) following control, ultrasound-assisted extraction (UAE), microwave-assisted extraction (MAE), ultrasound–microwave-assisted extraction (UMAE), hydrothermal-assisted extraction (HAE) and high-pressure-assisted extraction (HPAE) treatments. Results are expressed as average ± standard deviation ($n = 4$). Different letters indicate statistical differences ($p < 0.05$) on the yields of macroalgal extract between the different technological treatments incubated for 0 h (lower case letters) or 24 h (upper case letters). Statistical differences between the extraction yields following different extraction times post-treatment (0 or 24 h) within the same extraction treatment are reflected in the figure as ns (non-significant), * $p < 0.05$ and ** $p < 0.01$.

The effect of different extraction technologies and extraction time on the extraction of phytochemicals (TPC, TPhC, TFC, TTC and TSC) from *F. vesiculosus* and *P. canaliculata* is presented in Table 1. There was a significant variation in the extraction of each phytochemical depending on the macroalgal species, technological treatments and extraction times. In general, extracts from *F. vesiculosus* contained the highest levels of TPC, TPhC, TFC, TTC and TSC compared to *P. canaliculata*. The differences in the recovery of phytochemicals between both species could be attributed to inter-species differences; however, the effect of technological treatments on both species was similar. Garcia-Vaquero et al. [7] reported huge variation in the TPC, TSC and other relevant phytochemicals produced by the brown macroalgae *Laminaria digitata*, *Laminaria hyperborea* and *Ascophyllum nodosum* collected in Ireland every season for up to 2 years. Thus, the levels of TPC, TPhC, TFC, TTC and TSC extracted in this study may be strongly influenced by the initial concentration of these compounds in the biomass. Moreover, Harnedy and FitzGerald [26] mentioned that strong cell walls containing variable types and amounts of polysaccharides depending on the macroalgal species were one of the main obstacles hindering an efficient extraction of compounds from macroalgae. Thus, the type of polysaccharides and ionic interactions taking place with other cell wall constituents will be strongly influenced by the macroalgal species studied [27,28], affecting the yields of compounds extracted depending on the macroalgal species.

Results are expressed as average ± standard deviation (n = 3). The units of the phytochemical compounds analysed are expressed as follows: TPC (mg gallic acid equivalents (GAE)/g dried weight extract), TPhC (mg phloroglucinol equivalents (PGE)/g dried weight extract), TFC (mg quercetin equivalents (QE)/g dried weight extract), TTC (mg catechin equivalents (ChE)/g dried weight extract) and TSC (mg glucose equivalents (GlcE)/g dried weight extract). Different letters indicate statistical differences ($p < 0.05$) on the recovery of compounds at extraction times post-treatment 0 or 24 h for each individual macroalgal species. Differences between extraction times 0 and 24 h within the same technological treatment are represented in superscripts in the table: ^{ns} (not-significant), * ($p < 0.05$), ** ($p < 0.01$) and *** ($p < 0.001$).

When exploring the extraction of the samples post-treatment with different novel extraction technologies, an additional extraction of the samples for 24 h did not enhance the phytochemical contents of the extracts (see Table 1). The extraction of the samples for 24 h was performed under refrigeration conditions (4 °C) to avoid any damage to the antioxidant properties of the extracts during prolonged extraction times. An additional release of compounds may have occurred during the extraction of the samples, although the overall recovery of compounds in the extracts may even decrease due to the precipitation of solubilised compounds in ethanolic solutions during the extraction process [29].

Table 1. Effect of novel extraction technologies and extraction time post-treatment (0 and 24 h) on total phenolic content (TPC), total phlorotannin content (TPhC), total flavonoid content (TFC), total tannin content (TTC) and total sugar content (TSC) extracted from *F. vesiculosus* and *P. canaliculata*.

Macroalgae sp.	Extraction Technologies	TPC (mg GAE/g)		TPhC (mg PGE/g)		TFC (mg QE/g)		TTC (mg ChE/g)		TSC (mg GlcE/g)	
		0 h	24 h	0 h	24 h	0 h	24 h	0 h	24 h	0 h	24 h
F. vesiculosus	Control	316.4 ± 3.9 d	293.2 ± 5.9 e **	257.7 ± 3.2 d	238.7 ± 4.9 e **	180.4 ± 7.8 e	129.3 ± 6.4 e **	123.2 ± 3.4 d	91.3 ± 3.8 c ***	136.1 ± 9.1 c	116.6 ± 8.9 b ^{ns}
	UAE	445.0 ± 4.6 a	413.4 ± 5.1 a **	362.9 ± 3.7 a	337.1 ± 4.2 a **	286.3 ± 7.8 a	285.6 ± 7.7 a ^{ns}	189.1 ± 4.4 a	168.4 ± 7.1 a *	130.7 ± 6.2 c	111.2 ± 9.6 b *
	MAE	391.2 ± 6.0 c	375.9 ± 5.3 c *	318.9 ± 4.9 c	306.4 ± 4.4 c *	202.6 ± 3.4 d	198.1 ± 7.1 d ^{ns}	161.0 ± 5.6 bc	142.4 ± 4.4 b *	199.9 ± 9.2 b	162.7 ± 7.9 a **
	UMAE	431.2 ± 4.5 b	392.2 ± 7.5 b **	351.6 ± 3.7 b	319.7 ± 6.1 b **	268.5 ± 6.4 b	266.3 ± 1.3 b *	172.8 ± 7.8 b	151.3 ± 4.4 b **	194.5 ± 9.6 b	157.6 ± 6.0 a ^{ns}
	HAE	433.2 ± 6.2 ab	375.5 ± 2.1 c ***	353.3 ± 5.1 ab	306.1 ± 1.7 c ***	253.0 ± 5.6 b	250.0 ± 4.4 b ^{ns}	166.9 ± 6.7 b	146.9 ± 2.2 b **	239.4 ± 2.9 a	164.4 ± 5.0 a ***
	HPAE	387.9 ± 3.8 c	356.1 ± 2.4 d ***	316.2 ± 3.1 c	290.2 ± 2.0 d ***	231.5 ± 1.3 c	228.5 ± 7.1 c ^{ns}	148.4 ± 3.4 c	138.0 ± 9.7 b ^{ns}	123.7 ± 5.6 c	81.9 ± 2.9 c ***
P. canaliculata	Control	174.8 ± 5.5 d	158.3 ± 7.8 e *	141.9 ± 4.5 d	128.3 ± 6.4 e *	82.6 ± 8.4 b	80.4 ± 6.8 c ^{ns}	33.6 ± 2.2 d	32.1 ± 1.3 d ^{ns}	35.7 ± 5.0 d	58.4 ± 7.2 c **
	UAE	250.6 ± 6.0 a	182.3 ± 7.8 d ***	203.9 ± 4.9 a	148.0 ± 6.4 d ***	122.6 ± 3.4 a	108.5 ± 5.6 a *	79.5 ± 4.6 a	62.4 ± 2.2 a **	59.1 ± 6.1 c	42.4 ± 8.7 d ^{ns}
	MAE	205.3 ± 3.6 c	192.6 ± 7.1 bd ^{ns}	166.8 ± 2.9 c	156.4 ± 5.8 bd ^{ns}	93.7 ± 7.8 b	84.1 ± 6.4 c ^{ns}	35.8 ± 3.8 cd	32.8 ± 2.6 cd ^{ns}	123.1 ± 6.1 b	45.1 ± 8.1 d ***
	UMAE	238.4 ± 3.0 ab	210.4 ± 3.4 a ***	193.9 ± 2.5 ab	171.0 ± 2.8 a ***	113.7 ± 3.4 a	102.6 ± 6.8 ab ^{ns}	47.6 ± 4.6 b	41.7 ± 3.4 b *	109.1 ± 3.1 b	92.4 ± 7.2 b ^{ns}
	HAE	236.2 ± 1.9 b	202.3 ± 3.2 abc ***	192.1 ± 1.6 b	164.4 ± 2.7 abc ***	113.0 ± 4.6 a	87.8 ± 6.7 bc **	55.0 ± 2.6 b	41.0 ± 4.6 bc *	175.7 ± 9.0 a	123.1 ± 7.0 a **
	HPAE	226.5 ± 5.7 b	194.1 ± 3.2 cd **	184.2 ± 4.7 b	157.7 ± 2.6 cd **	110.7 ± 5.6 a	93.0 ± 6.8 abc *	46.9 ± 5.9 bc	37.3 ± 3.4 bcd ^{ns}	25.1 ± 3.1 d	15.7 ± 1.2 d *

Results are expressed as average ± standard deviation ($n = 3$). The units of the phytochemical compounds analysed are expressed as follows: TPC (mg gallic acid equivalents (GAE)/g dried weight extract), TPhC (mg phloroglucinol equivalents (PGE)/g dried weight extract), TFC (mg quercetin equivalents (QE)/g dried weight extract), TTC (mg catechin equivalents (ChE)/g dried weight extract) and TSC (mg glucose equivalents (GlcE)/g dried weight extract). Different letters indicate statistical differences ($p < 0.05$) on the recovery of compounds at extraction times post-treatment 0 or 24 h for each individual macroalgal species. Differences between extraction times 0 and 24 h within the same technological treatment are represented in superscripts in the table: ^{ns} (not-significant), * ($p < 0.05$), ** ($p < 0.01$) and *** ($p < 0.001$).

Focusing on the use of novel technologies without additional extraction time, overall, UAE achieved the highest yields of phytochemicals from both *F. vesiculosus* and *P. canaliculata*. UAE generated extracts containing the highest yields of TPC (445.0 ± 4.6 mg gallic acid equivalents (GAE)/g dried extract), TPhC (362.9 ± 3.7 mg phloroglucinol equivalents (PGE)/g dried extract), TFC (286.3 ± 7.8 mg quercetin equivalents (QE)/g dried extract), and TTC (189.1 ± 4.4 mg catechin equivalents (ChE)/g dried extract) from *F. vesiculosus*. This technology also achieved the highest yields of TPC (250.6 ± 6 mg GAE/g), TPhC (203.9 ± 4.9 mg PGE/g), TFC (122.6 ± 3.4 mg QE/g) and TTC (79.5 ± 4.6 mg ChE/g) from *P. canaliculata*. UAE has been previously used to extract polyphenols and antioxidants from brown macroalgae, although the extraction yields and antioxidant properties may vary depending on the UAE technology used, the extraction conditions and the biomass (species, season and place of collection). Vázquez-Rodríguez et al. [30] optimised UAE for the recovery of polyphenols and carbohydrates from the brown macroalgae *Silvetia compressa*. The authors reported high TPhC (10.82 mg PGE/g) using maximum ultrasound power and medium polarity solvents (62.5% ethanol solution) [30]. Ummat et al. [24] reported high yields of TPC (572.3 ± 3.2 mg GAE/g), TPhC (476.3 ± 2.2 mg PGE/g) and TFC (281.0 ± 1.7 mg QE/g) extracted from *F. vesiculosus* using UAE during 30 min and 50% ethanol as an extraction solvent. Hassan, Pham and Nguyen [23] explored UAE (60 min, 60 °C and 60% ethanol as extraction solvent) for the extraction of TPC (9.07 ± 0.49 mg GAE/g), DPPH (16.11 ± 1.69 mg TE/g) and FRAP (9.03 ± 0.58 mg TE/g) from *Padina australis*. Previous studies have reported that the algal cell wall contains variable types and amounts of complex polysaccharides depending on the type of macroalgal species, which is one of the main obstacles hindering an efficient extraction of compounds from them [26,31]. Additionally, the amount of bioactive in macroalgae is also influenced by the geographical location and harvesting season [7]. Therefore, to recover highest yield of bioactive, a specific extraction strategy focusing on extraction parameters and UAE conditions as well as harvesting season for each individual algal species is important.

MAE in the current study showed a significantly low value of TPC compared to the other novel technologies used in this study, but higher than the control. MAE exhibited 23.6% and 17.8% higher TPC than the control but up to 12% and 18% lower TPC than other technologies for *F. vesiculosus* and *P. canaliculata*, respectively. MAE in the current study did not show any advantage compared to the other novel technologies used in this study. In the case of *P. canaliculata*, it generated the extracts with the lowest levels of TFC (94 ± 8 mg QE/g) and TTC (36 ± 4 mg ChE/g), with concentrations comparable to those of the control samples (83 ± 8 mg QE/g and 34 ± 2 mg ChE/g). Amarante et al. [32], using optimised MAE conditions (75 °C, 5 min and 57% ethanol), achieved low yields of TPhC (9.8 ± 1.8 mg PGE/g extract) from *F. vesiculosus* comparable to those achieved using conventional solvent extraction (11.1 ± 1.3 mg PGE/g extract). However, Yuan et al. [12] recovered high yields of TPC from brown macroalgae (*A. nodosum*, *Laminaria japonica*, *Lessonia trabeculate* and *Lessonia nigrecens*) by MAE using a UWave-2000 reactor for microwave irradiation (2.45 GHz) at 110 °C for 15 min (5 min climbing and 10 min holding). The low extraction yields of phytochemicals in the current study may be due to operational limitations of the power of the MAE used (250 W). Similar to these results, while exploring MAE to extract phytochemicals from *A. nodosum*, the maximum yields of TPC (1790.93 ± 112.11 mg GAE/100 g biomass) were achieved by applying 600 W of microwave power for 5 min compared to treatments at 250 and 1000 W [14].

UMAE did not improve the extraction yields of phytochemicals compared to UAE in *F. vesiculosus*, while in the case of *P. canaliculata*, UMAE achieved levels comparable to those of UAE, with high yields of TPC (238 ± 3 mg GAE/g), TPhC (194 ± 2 mg PGE/g) and TFC (114 ± 3 mg QE/g). The efficiency of ultrasounds and microwave extraction forces combined seems to be influenced by the macroalgal species studied. UMAE has been previously explored for the extraction of multiple compounds from terrestrial crops, such as soluble dietary fibre from coffee silverskin [33], pectin from potato pulp [34], and lycopene from tomatoes [35], while limited literature is available on the use of UMAE in macroalgae.

To our knowledge, Garcia-Vaquero et al. [14] was the first report analysing the application of UMAE to extract phytochemicals from *A. nodosum*. The authors achieved extracts with the highest contents of TSC (10409 ± 229.11 mg glucose equivalents (GlcE)/100 g biomass) and TPC (2605.89 ± 192.97 mg GAE/100 g biomass) by using UMAE compared to UAE and MAE.

Extracts generated by HAE achieved high levels of TPC (433.2 ± 6.2 mg GAE/g) and TPhC (353.3 ± 5.1 mg PGE/g) from *F. vesiculosus*, comparable to those achieved using UAE. Moreover, HAE was the most efficient method to recover TSC from *F. vesiculosus* (239.4 ± 2.9 mg GlcE/g) and *P. canaliculata* (175.7 ± 9.0 mg GlcE/g). Similar to these results, Rajauria et al. [8] reported the efficiency of HAE (85–121 °C, 15 min using 60% methanol as a solvent) to recover TPC, TPhC and TSC from brown macroalgae *Laminaria saccharina*, *L. digitata* and *Himanthalia elongata*. In the case of carbohydrates, HAE (120 °C, 80.9 min) obtained extracts containing high yields of fucose-containing polysaccharides from *L. hyperborea* (2782 mg fucoidan/100 g biomass) [36]. The efficiency of HAE to disrupt cell walls of multiple algae and increase the extraction of compounds has also been demonstrated for multiple carbohydrates, i.e., Lemus et al. [37] used HAE (121 °C, 3 h) to extract agar from red macroalgal species *Pterocladia capillacea*, *Gelidium floridanum* and *Gelidium serrulatum*, with extraction yields ranging from 31.7 to 33% of the total biomass. HAE (103 kPa, 15 min) was also used by Lee et al. [38] to extract agar from multiple *Gracilaria* spp.

The application of HPAE did not have a significant advantage when exploring the extraction of phytochemicals from both *F. vesiculosus* and *P. canaliculata*. HPAE achieved high yields of TFC (110.7 ± 5.6 mg QE/g) from *P. canaliculata* comparable to those levels achieved by UAE and UMAE. To our knowledge, there are no data on the efficiency of this method for the extraction of TFC. However, the efficiency of HPAE has been demonstrated for the extraction of other compounds from algal matrices, such as lipids from the microalgae *Chlorella saccharophila* [39]. Moreover, Li et al. [40] used a combination of high pressure homogenisation and HAE for the extraction of sulphated polysaccharides from the brown macroalga *Nemacystus decipients*, achieving yields of sulphated carbohydrates of approximately 17%.

Overall, all the tested novel technologies extracted significantly higher amounts of phytochemicals from *F. vesiculosus* and *P. canaliculata* compared to control samples (treated by maceration), except in the case of TSC. The extraction yields of TSC from *F. vesiculosus* and *P. canaliculata* achieved in the current study may be related to the type of polysaccharides (such as laminarins and fucoidans) produced by the biomass and how these compounds interact with the polarity of the solvent during the process of extraction [11,41,42]. Variable results are available in the literature in relation to the extraction of TSC using ethanolic solutions from multiple biological matrices. Zhang et al. [43] optimised the extraction of phytochemicals and antioxidant activities from *Asparagus officinalis* L root cultivars and reported maximum recovery of TPC, TFC, and TSC when using 75.23% ethanol at 51 °C, solid:liquid ratio 1:50 and 73.02 min of extraction. In the case of macroalgae, Vázquez-Rodríguez et al. [30] optimised the UAE of TPhC and TSC from the macroalgae *Silvetia compressa*. The authors achieved extracts with the highest yields of TSC using 25% ethanol, while solvents of high polarity increased the extraction of TPhC at 50 °C, 3.8 W/cL of ultrasonic power density, and 30 mL of solvent per g of seaweed [30]. However, Foley et al. [44] reported the use of 80% ethanol as an optimum solvent for the extraction of sulphated polysaccharides from *A. nodosum* at 70 °C for 12 h of extraction. Despite the potential applications of polyphenols from macroalgae, their widespread industrial applications have been hindered as these compounds are difficult to separate, purify and characterise [13]. Carbohydrates are normally co-extracted and tightly bound to phenols, carotenoids and phytosterols [13,45], and thus, these polysaccharides need to be eliminated to improve the purification processes of polyphenols. As previously mentioned, the low co-extraction of TSC in this study for all the novel technologies explored, with exception of HAE, may indicate that the use of 50% ethanol as a solvent of extraction combined with the technologies achieving high yields of polyphenols may be a promising strategy when targeting these compounds, reducing subsequent purification steps.

2.2. Effect of Novel Technologies on the Antioxidant Capacity of the Extracts

The antioxidant properties (DPPH and FRAP) of the extracts obtained by using multiple extraction technologies and extraction times from *F. vesiculosus* and *P. canaliculata* are represented in Figure 2. There was a significant variation in the antioxidant properties of the extracts depending on the macroalgal species, novel technology and extraction time used. Overall, extracts from *F. vesiculosus* had higher DPPH (ranging from 95 to 112 mM trolox equivalents (TE)/g dried extract) and FRAP (236–285 mM TE/g) compared to those of *P. canaliculata*, ranging from 80–95 and 21–80 mM TE/g for DPPH and FRAP, respectively. Similarly to the phytochemicals analysed in this study, additional extraction of the samples for 24 h post-treatment did not improve the antioxidant properties of the extracts generated by any novel technology.

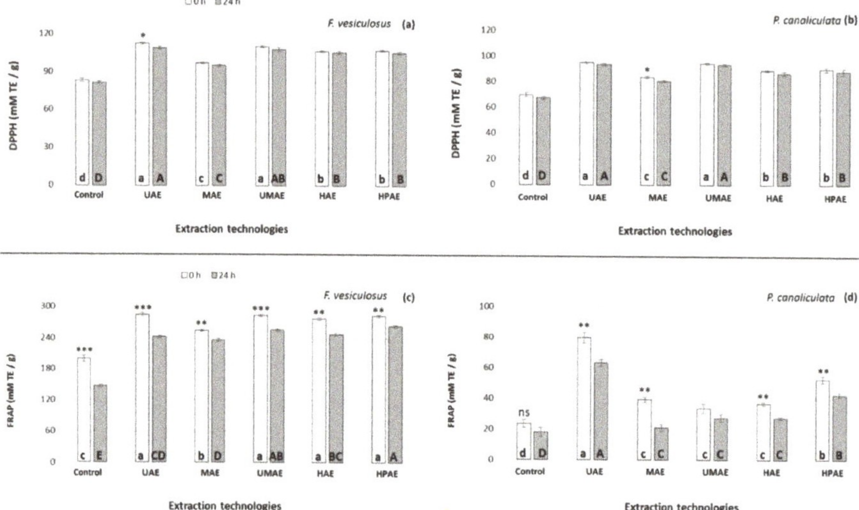

Figure 2. Antioxidant properties DPPH (**a**,**b**) and FRAP (**c**,**d**) of extracts obtained from *F. vesiculosus* and *P. canaliculata* by using novel extraction technologies and extraction times post-treatment (0 and 24 h) versus control treatment. Results are expressed as average ± standard deviation ($n = 3$). Different letters indicate statistical differences ($p < 0.05$) on the antioxidant properties of the macroalgal extracts between the different technological treatments incubated either at 0 h (lower case letters) or 24 h (upper case letters). Statistical differences between the extraction yields following different extraction times (0 and 24 h) within the same extraction treatment are reflected in the top figure as: * $p < 0.05$, ** $p < 0.01$ and *** $p < 0.001$.

All the extraction technologies generated extracts with higher antioxidant activities (DPPH and FRAP) compared to the control group. Moreover, the different novel extraction technologies applied had a similar effect on the antioxidant properties of the extracts generated from both *F. vesiculosus* and *P. canaliculata*. The highest DPPH antioxidant activities were achieved in extracts generated by UAE and UMAE (ranging from 110 to 112 mM trolox equivalents (TE)/g in *F. vesiculosus* and 94–94 mM TE/g in *P. canaliculata*) followed by HAE and HPAE (106–107 mM TE/g in *F. vesiculosus* and 89–90 mM TE/g in *P. canaliculata*), and the lowest DPPH radical scavenging activities were achieved in the extracts obtained by MAE (97 ± 0.1 mM TE/g in *F. vesiculosus* and 83 ± 0.6 mM TE/g in *P. canaliculata*).

In the case of FRAP, the extracts obtained by different technological treatments did not behave in the same way in both macroalgae. In *F. vesiculosus*, the extracts generated by MAE had the lowest FRAP activities (255 ± 3 mM TE/g), with no differences in FRAP activity between the other extracts achieved by any the other novel technologies, ranging from 277 to 285 mM TE/g, despite the fact that the phytochemicals analysed in these samples significantly varied amongst the treatments. In the case of *P. canaliculata*, the extracts containing maximum FRAP activity were those generated by UAE (80 ± 3 mM TE/g), followed by HPAE (53 ± 2 mM TE/g), and the lowest FRAP activity (ranging from 34–39 mM TE/g) was achieved in extracts generated by MAE, UMAE and HAE without statistical differences amongst the treatments despite their different phytochemical composition.

To understand the relationship between the phytochemical composition of the extracts and their antioxidant properties in *F. vesiculosus* and *P. canaliculata*, correlation matrices were analysed in Figure 3. Both antioxidant activities (DPPH and FRAP) were positively and significantly correlated with each other in both macroalgal species, as well as with the levels of TPC, TPhC, TFC and TTC. In the case of *F. vesiculosus*, the antioxidant properties of the extracts were not correlated with TSC, and in *P. canaliculata* TSC was only correlated with the levels of DPPH. These results are in agreement with previous studies associating the polyphenol contents of macroalgae with high antioxidant properties [7,46–48]. Thereby, macroalgae react to environmental stressors by increasing the production of antioxidant phytochemicals to avoid any structural and metabolic changes due to the increased levels of oxidising agents (free radicals and other reactive species) [46]. Several studies have linked an increased production of TPC and antioxidant activities of macroalgae during spring and summer seasons as a defence mechanism against oxidative stress [7,47,48]. Thus, the amount of phytochemicals and associated antioxidant activity in macroalgae are strongly influenced by the environmental conditions and the seasons they are harvested. Moreover, studies extracting TPC, TPhC, TFC and TTC have also focused on the antioxidant activity of these compounds, as it is essential for the potential use of polyphenols for high value applications as pharmaceuticals, nutraceuticals and cosmeceuticals [5,6,49].

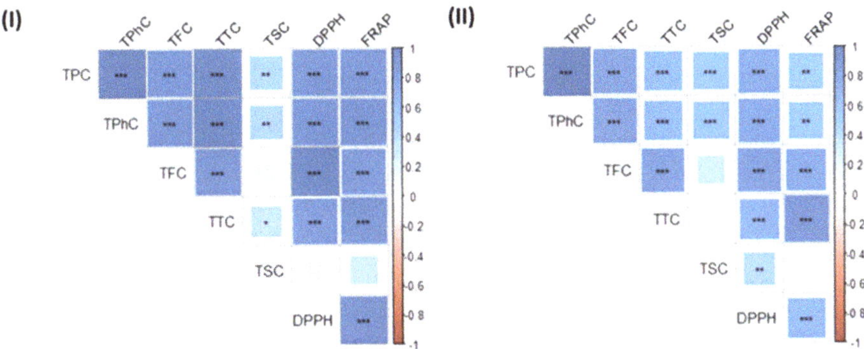

Figure 3. Correlation matrices of the composition and antioxidant properties in extracts obtained from (**I**) *F. vesiculosus* and (**II**) *P. canaliculata*. The sign of the correlations is colour coded (blue = + and red = −) and the strength of the correlations (0–1) relates to the size of each square and depth of each colour. Abbreviations in the figure are as follows: total phenolic content (TPC), total phlorotannin content (TPhC), total flavonoid content (TFC), total tannin content (TTC), total sugar content (TSC), 1,1-diphenyl-2-picryl-hydrazil radical scavenging activity (DPPH) and ferric reducing antioxidant power (FRAP). The statistical significance of the correlations is indicated in the figure as * $p < 0.05$, ** $p < 0.01$, *** $p < 0.001$.

Principal component analysis (PCA) was also performed to obtain an overview of the similarities and differences in the phytochemical and antioxidant composition of the extracts produced from *F. vesiculosus* and *P. canaliculata* following different extraction technological treatments and extraction times (see Figure 4). The two principal components (PC) extracted from the data set explained a combined 68.44% of the variation of the data. PC1 explained 52.99% of this variation and clustered all phytochemical and antioxidant properties from both macroalgae together with the application of UAE in the right side of

PC1. The second component explained the remaining variability of the data set and clearly separated the recovery of TSC from both macroalgae together with HAE treatment. These results agree with the previous literature, emphasising the efficiency of multiple HAE protocols for the extraction of carbohydrates from multiple macroalgal species [8,36,37]. Thus, the use of methods achieving high yields of phytochemicals (TPC, TPhC, TFC and TTC) and antioxidant properties, such as UAE combined with 50% ethanolic solutions, as indicated by the results of this study, may be the most promising strategy to target the extraction of polyphenols from *F. vesiculosus* and *P. canaliculata* while decreasing the co-extraction of undesirable carbohydrates, reducing the use of further purification strategies for the commercial exploitation of these compounds.

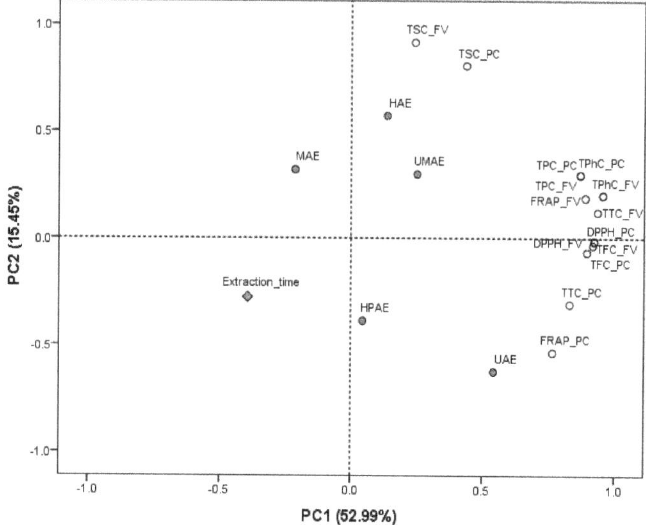

Figure 4. Principal component analysis (PCA) scatter plot representing the scores for the extraction technologies and extraction time applied for the extraction of phytochemicals (total phenolic content (TPC), total phlorotannin content (TPhC), total flavonoid content (TFC), total tannin content (TTC) and total sugar content (TSC)) and antioxidants (DPPH and FRAP) from macroalgae, FV (*F. vesiculosus*) and PC (*P. canaliculata*).

3. Materials and Methods

3.1. Macroalgal Biomass and Processing

Brown macroalgae *F. vesiculosus* and *P. canaliculata* were collected during November 2017 by Quality Sea Veg in Co Donegal (Ireland). The biomass was washed to remove epiphytes, sand and other debris, chopped and oven dried (55 °C, 5 days). Oven-dried samples were ground and sieved to a uniform particle size (1 mm), vacuum-packed and stored in a cool and dry place for further extraction experiments.

3.2. Extraction Procedures

The macroalgal biomass was mixed with ethanol solutions (50% *v/v*) at a biomass/solvent ratio of 1:10 *w/v*, and the mixtures were stirred for 10 min prior to the extraction processes. The mixtures were then subjected to novel extraction techniques including ultrasound-assisted extraction (UAE), microwave-assisted extraction (MAE), ultrasound–microwave-assisted extraction (UMAE), hydrothermal-assisted extraction (HAE) and high-pressure-assisted extraction (HPAE). All the extraction procedures were performed in duplicate and repeated twice (n = 4). UAE treatments were performed using a semi-industrial UIP500hdT 26 kHz (Hielscher Ultrasound Technology, Teltow, Germany) at 100%

ultrasonic amplitude. MAE was performed in a microwave oven (Panasonic NN-CF778S0, Bracknell, UK; 2450 MHz) at 250 W. UMAE was performed by coupling the aforementioned devices, UIP500hdT 26 kHz and Panasonic NN-CF778S0 2450 MHz, keeping the ultrasonic power (100%) and microwave power (250 W) constant through the extraction process. HAE was performed using an autoclave (Tomy SS-325, Tomy Seiko, Tokyo, Japan) at 100 °C and 15 psi. HPAE was performed in the HPP Tolling facility (Dublin, Ireland) using a 200 L Hiperbaric HPP (Hiperbaric, Burgos, Spain) at 600 MPa. All the extraction procedures were performed at fixed extraction time (10 min), solvent (50% ethanol solution) and macroalgae:solvent ratio (1:10 *w/v*) based on previous optimisation protocols for the recovery of polyphenols from brown macroalgae [24]. Selection of 50% ethanol as an extraction solvent in this study was based upon previous findings wherein 50% ethanolic solution resulted the highest recovery of total phenols from macroalgae compared to 30% and 70% ethanol solutions [24]. Control samples were performed at room temperature using a magnetic stirring plate (IKA® C-MAG HS 7, Staufen, Germany) for the same duration as the technological treatments (10 min).

After the process of 10 min extraction (fixed for each extraction method), the samples were either processed immediately (termed as—0 h extraction samples) or stored for an additional 24 h (termed as—24 h extraction samples) in closed containers under refrigeration at 4 °C. The objective of storing the extracts for 24 h after the extraction process was to evaluate if more phytochemicals leach out from treated algal cell wall or if they degrade, which would have an impact on total antioxidant activity of the extracts. After each extraction period (0 or 24 h), the samples were filtered through Whattman® number 3 (GE Healthcare, Buckinghamshire, UK) and the ethanol of the extracts was evaporated under vacuum (Rotavapor R-100, Büchi Labortechnik AG, Flawil, Switzerland). The extracts were freeze-dried (FreeZone 6, Labconco Corporation, Kansas City, MO, USA) until constant dry weight (94.82 ± 0.14% dry weight). The extracts were then vacuum packed and stored at −20 °C for further analyses. The extraction yields for each treatment and extraction time were calculated using the formula: Extraction yield (mg/g) = weight of dry extract (mg)/weight of dry sample (g).

3.3. Phytochemical Analyses

All the phytochemical analyses of the extracts were performed in triplicate (n = 3). All the freeze-dried extracts were dissolved in ethanol, a stock concentration of 1 mg/mL was prepared and used in analyses. All standards used in this study were purchased from Sigma-Aldrich (Arklow, Co. Wicklow, Ireland).

3.3.1. Total Phenolic Content (TPC) and Total Phlorotannin Content (TPhC) Analyses

TPC and TPhC were determined following the Folin–Ciocalteu reagent method as described by Rajauria et al. [50]. A total of 100 µL of sample/standard was mixed with 2 mL of sodium carbonate solution (Na_2CO_3 solution, 2% *w/v*). Following 2 min extraction, 100 µL of Folin–Ciocalteu's solution (1 M) was added to all the mixtures and the reaction was incubated for 30 min at room temperature in dark conditions. The absorbance of the reactions was read at 720 nm in a spectrophotometer (UVmini-1240, Shimadzu, Kyoto, Japan). Distilled water (instead of extract or standard) along with other reagents was used as blank and gallic acid (>97.5% purity) and phloroglucinol (>99% purity) at concentrations of 25–300 mg/L were used as standards for TPC and TPhC, respectively. TPC results were expressed as mg GAE/g and TPhC as mg PGE/g.

3.3.2. Total Flavonoid Content (TFC)

TFC was determined following the protocol described by Liu et al. [51] with slight modifications. Briefly, 250 µL of sample/standard was mixed with 1.475 mL of distilled water and 75 µL of sodium nitrite solution ($NaNO_2$ solution 5% *w/v*) and the reaction was allowed to stand for 6 min. A total of 150 µL of aluminium chloride hexahydrate solution ($AlCl_3{\cdot}6H_2O$ solution 10% *w/v*) was added, mixed thoroughly and allowed to stand for

5 min. Following extraction, 0.5 mL of sodium hydroxide solution (NaOH solution 1 M) was added and the absorbance of the reactions was read at 510 nm in a spectrophotometer. Distilled water (instead of extract or standard) along with other reagents was used as blank and quercetin (>95% purity) was used as standard at concentrations of 30–150 mg/L. TFC results were expressed as mg QE/g.

3.3.3. Total Sugar Content (TSC)

TSC was determined by the phenol–sulphuric acid method as described by [52] with minor modifications. Briefly, 100 µL of sample/standard was mixed with 100 µL of phenol solution (0.8% *w/v*), followed by 2 mL of concentrated sulphuric acid (H_2SO_4 95–98%). The mixtures were allowed to stand for 10 min at room temperature and later incubated at 30 °C in a water bath for 20 min. The absorbance of the reaction was read in a spectrophotometer at 490 nm. Distilled water was used as blank and D-glucose (>99.5% purity) as a standard at concentrations of 50–250 mg/L. TSC results were expressed as mg GlcE/g.

3.3.4. Total Tannin Content (TTC)

TTC was analysed following the protocol as described by Liu et al. [51]. A total of 50 µL of sample was mixed with 1.5 mL of a methanolic solution of vanillin (4% *w/v*) (99% purity, Sigma-Aldrich, Arklow, Co. Wicklow, Ireland) and 750 µL of hydrochloric acid (HCl 37% *w/v*). The solutions were mixed thoroughly and incubated in dark conditions at room temperature for 20 min. The absorbance of the reaction was read in a spectrophotometer at 500 nm. Distilled water was used as blank and (+)-catechin hydrate (>98% purity) was used as standard at concentrations of 15–150 mg/L. TTC results were expressed as mg ChE/g.

3.4. *Antioxidant Analyses*

All the antioxidant analyses of the extracts were performed in triplicate ($n = 3$).

3.4.1. 1,1-Diphenyl-2-Picryl-Hydrazil (DPPH) Radical Scavenging Activity

DPPH assay was performed following the protocol as described by Sridhar and Charles [53]. A total of 700 µL of a methanolic DPPH solution (100 µM) was added to 700 µL of sample/standard, incubating the mixtures at room temperature in the dark for 20 min. The absorbance of the reactions was measured against blank (methanol without DPPH solution) at a wavelength of 515 nm in a spectrophotometer. Trolox (97% purity) was used as standard at concentrations ranging from 10–150 mM/L. DPPH results were expressed as mM TE/g.

3.4.2. Ferric Reducing Antioxidant Power (FRAP) Assay

FRAP assay was performed following the method described by Benzie and Strain [54]. Fresh FRAP working solutions were prepared by mixing 300 mM acetate buffer, pH 3.6; 10 mM 2,4,6-Tri(2-pyridyl)-s-triazine in 40 mM hydrochloric acid and 20 mM Iron(III) chloride hexahydrate in the ratio of 10:1:1, *v/v/v*. A total of 2.5 mL of pre-heated FRAP working solution (37 °C) was added to 83 µL of sample/standard and the mixtures were incubated 10 min in dark conditions at room temperature. The absorbance of the reactions was measured at 593 nm in a spectrophotometer against a reagent blank (FRAP working solution). Trolox was used as standard at concentrations of 10–150 mM/L. FRAP results were expressed as mM TE/g.

3.5. *Statistical Analyses*

The influences of the macroalgal species, extraction technology and extraction time post-treatment on the extracted phytochemicals (TPC, TPhC, TFC, TSC and TTC) and antioxidant activities (DPPH and FRAP) were analysed by multivariate general linear model in SPSS version 24.0. The differences were further analysed by either Tukey's HSD post hoc tests or Student's *t*-test. Differences were considered to be significant at p-values < 0.05. The relationships between the composition of the extracts and their antioxidant properties were explored in R version 4.0.2 [55]. The statistical packages "ggplot2" and "corrplot" were used to generate the correlation matrix [56] and "cor.mtest" produced the *p*-values for the Pearson's correlation matrix. The variance in the data set was further analysed by principal component analysis (PCA) using Equamax rotation with Kaiser normalisation extracting the weight for each component with eigenvalues higher than 1 using SPSS version 24.0.

4. Conclusions

The yields of compounds extracted from *F. vesiculosus* (173.33 to 354.67 mg dried extract per g macroalgae) were higher when compared to *P. canaliculata* (147.00 to 240.33 mg dried extract/g macroalgae). The yields were influenced by the technology used and, overall, the highest were achieved using UMAE (354.67 ± 2.33 for *F. vesiculosus* and 240.33 ± 1.20 for *P. canaliculata*), followed by UAE (316.33 ± 4.63 and 214.33 ± 2.60 for *F. vesiculosus* and *P. canaliculata*, respectively). Following the extraction treatment, additional extraction (24 h) did not improve the yields of phytochemicals or the antioxidant properties of the extracts, indicating possible precipitation or degradation loss of the compounds dissolved for prolonged periods of time in ethanolic solutions. When analysing the extraction of phytochemicals using multiple novel technologies without additional extraction after extraction, overall UAE achieved the highest yields of most phytochemicals from both *F. vesiculosus* (445.0 ± 4.6 mg GAE/g, 362.9 ± 3.7 mg PGE/g, 286.3 ± 7.8 mg QE/g, and 189.1 ± 4.4 mg ChE/g) and *P. canaliculata* (250.6 ± 6 mg GAE/g, 203.9 ± 4.9 mg PGE/g, 122.6 ± 3.4 mg QE/g and 79.5 ± 4.6 mg ChE/g). Moreover, UMAE also achieved high yields of TPC, TPhC and TFC from *P. canaliculata*. Overall, the extraction procedures using 50% ethanol as solvent achieved low TSC yields, except when using HAE, from both macroalgae. In the case of the antioxidant properties, the extracts with the highest DPPH radical scavenging activities were extracted by UAE and UMAE and the lowest by MAE from both *F. vesiculosus* and *P. canaliculata*. Meanwhile, strong variations were appreciated depending on the extraction technology and macroalgal species considered in the case of FRAP activity. Both DPPH and FRAP were positively correlated with the levels of TPC, TPhC, TFC and TTC in both macroalgal species. This study evaluated the effectiveness of multiple extraction technologies for the recovery of phytochemicals (keeping a fixed solvent type, concentration and extraction time) from macroalgae. Further, studies are needed optimising the use of novel extraction technologies, varying extraction time and the polarity of the extraction solvents targeting multiple compounds from macroalgae in order to establish efficient extraction protocols to allow the future commercialisation of these compounds. Moreover, further chemical characterisation of the extracted compounds as well as confirmation of their biological effects in vivo will also be needed in order to commercialise these compounds. Based on the results of this study, the use of UAE combined with 50% ethanolic solution as an extraction solvent could be a promising strategy targeting the extraction of TPC, TPhC, TFC and TTC, while reducing the co-extraction of undesirable carbohydrates from both *F. vesiculosus* and *P. canaliculata*, with promising applications when using these compounds as pharmaceuticals, nutraceuticals and cosmeceuticals.

Author Contributions: M.G.-V. collaborated in designing the work, performed the extraction experiments and the statistical treatment of the data, and wrote and reviewed the manuscript. O.W. and R.R. performed the chemical analyses of the extracts and helped in writing the first draft of the manuscript. J.O. and B.K.T. contributed to the revision of the manuscript and provided funds for the work. A.K.J. and G.R. designed and supervised the work and revised the manuscript. All authors have read and agreed to the published version of the manuscript.

Funding: This research work was supported by Science Foundation Ireland (SFI). Grant numbers: [16/RC/3889] and [14/IA/2548].

Institutional Review Board Statement: Not applicable.

Informed Consent Statement: Not applicable.

Data Availability Statement: Not applicable as all the data is contained in the manuscript.

Conflicts of Interest: The authors declare no conflict of interest.

References

1. Guiry, M.; Guiry, G.; AlgaeBase. World-Wide Electronic Publication. Available online: www.algaebase.org (accessed on 27 April 2021).
2. Chojnacka, K.; Saeid, A.; Witkowska, Z.; Tuhy, L. Biologically active compounds in seaweed extracts-the prospects for the application. *Open Conf. Proc. J.* **2012**, *3*, 20–28. [CrossRef]
3. Fatima, F.; Løvstad, H.S.; Rohan, S.; Pedro, M.; Zhengyong, Y. *The Global Statusof Seaweed Production, Trade and Utilization*; Food and Agriculture Organization of the United Nations: Rome, Italy, 2018.
4. Zhu, X.; Healy, L.; Zhang, Z.; Maguire, J.; Sun, D.; Tiwari, B.K. Novel postharvest processing strategies for value-added applications of marine algae. *J. Sci. Food Agric.* **2021**. [CrossRef]
5. Cotas, J.; Leandro, A.; Monteiro, P.; Pacheco, D.; Figueirinha, A.; Gonçalves, A.M.M.; Da Silva, G.J.; Pereira, L. Seaweed phenolics: From extraction to applications. *Mar. Drugs* **2020**, *18*, 384. [CrossRef] [PubMed]
6. Mekinić, I.G.; Skroza, D.; Šimat, V.; Hamed, I.; Čagalj, M.; Perković, Z.P. Phenolic content of brown algae (pheophyceae) species: Extraction, identification, and quantification. *Biomolecules* **2019**, *9*, 244. [CrossRef] [PubMed]
7. Garcia-Vaquero, M.; Rajauria, G.; Miranda, M.; Sweeney, T.; Lopez-Alonso, M.; O'Doherty, J. Seasonal variation of the proximate composition, mineral content, fatty acid profiles and other phytochemical constituents of selected brown macroalgae. *Mar. Drugs* **2021**, *19*, 204. [CrossRef]
8. Rajauria, G.; Jaiswal, A.K.; Abu-Ghannam, N.; Gupta, S. Effect of hydrothermal processing on colour, antioxidant and free radical scavenging capacities of edible Irish brown seaweeds. *Int. J. Food Sci. Technol.* **2010**, *45*, 2485–2493. [CrossRef]
9. Qin, Y. Health benefits of bioactive seaweed substances. In *Bioactive Seaweeds for Food Applications: Natural Ingredients for Healthy Diets*; Qin, Y., Ed.; Academic Press Elsevier: Cambridge, MA, USA, 2018; pp. 179–200.
10. Tierney, M.S.; Smyth, T.J.; Hayes, M.; Soler-Vila, A.; Croft, A.K.; Brunton, N. Influence of pressurised liquid extraction and solid–liquid extraction methods on the phenolic content and antioxidant activities of I rish macroalgae. *Int. J. Food Sci. Technol.* **2013**, *48*, 860–869. [CrossRef]
11. Chater, P.I.; Wilcox, M.; Cherry, P.; Herford, A.; Mustar, S.; Wheater, H.; Brownlee, I.; Seal, C.; Pearson, J. Inhibitory activity of extracts of Hebridean brown seaweeds on lipase activity. *Environ. Boil. Fishes* **2016**, *28*, 1303–1313. [CrossRef]
12. Yuan, Y.; Zhang, J.; Fan, J.; Clark, J.; Shen, P.; Li, Y.; Zhang, C. Microwave assisted extraction of phenolic compounds from four economic brown macroalgae species and evaluation of their antioxidant activities and inhibitory effects on α-amylase, α-glucosidase, pancreatic lipase and tyrosinase. *Food Res. Int.* **2018**, *113*, 288–297. [CrossRef]
13. Tierney, M.S.; Smyth, T.J.; Rai, D.K.; Soler-Vila, A.; Croft, A.K.; Brunton, N. Enrichment of polyphenol contents and antioxidant activities of Irish brown macroalgae using food-friendly techniques based on polarity and molecular size. *Food Chem.* **2013**, *139*, 753–761. [CrossRef]
14. Garcia-Vaquero, M.; Ummat, V.; Tiwari, B.; Rajauria, G. Exploring ultrasound, microwave and ultrasound-microwave assisted extraction technologies to increase the extraction of bioactive compounds and antioxidants from brown macroalgae. *Mar. Drugs* **2020**, *18*, 172. [CrossRef]
15. Getachew, A.T.; Jacobsen, C.; Holdt, S.L. Emerging technologies for the extraction of marine phenolics: Opportunities and challenges. *Mar. Drugs* **2020**, *18*, 389. [CrossRef]
16. Tiwari, B.K. Ultrasound: A clean, green extraction technology. *Trends Anal. Chem.* **2015**, *71*, 100–109. [CrossRef]
17. Obluchinskaya, E.D.; Daurtseva, A.V.; Pozharitskaya, O.N.; Flisyuk, E.V.; Shikov, A.N. Natural deep eutectic solvents as alternatives for extracting phlorotannins from brown algae. *Pharm. Chem. J.* **2019**, *53*, 243–247. [CrossRef]
18. Anastas, P.T.; Warner, J.C. *Green Chemistry: Theory and Practice*; Oxford University Press Oxford: Oxford, UK, 2000; Volume 30.
19. Kadam, S.U.; Tiwari, B.K.; O'Donnell, C. Application of novel extraction technologies for bioactives from marine algae. *J. Agric. Food Chem.* **2013**, *61*, 4667–4675. [CrossRef] [PubMed]
20. Li, Y.; Fu, X.; Duan, D.; Liu, X.; Xu, J.; Gao, X. Extraction and identification of phlorotannins from the brown alga, sargassum fusiforme (Harvey) setchell. *Mar. Drugs* **2017**, *15*, 49. [CrossRef]
21. Montero, L.; Sánchez-Camargo, A.D.P.; García-Cañas, V.; Tanniou, A.; Stiger-Pouvreau, V.; Russo, M.; Rastrelli, L.; Cifuentes, A.; Herrero, M.; Ibáñez, E. Anti-proliferative activity and chemical characterization by comprehensive two-dimensional liquid chromatography coupled to mass spectrometry of phlorotannins from the brown macroalga Sargassum muticum collected on North-Atlantic coasts. *J. Chromatogr. A* **2016**, *1428*, 115–125. [CrossRef] [PubMed]
22. Puspita, M.; Déniel, M.; Widowati, I.; Radjasa, O.K.; Douzenel, P.; Marty, C.; Vandanjon, L.; Bedoux, G.; Bourgougnon, N. Total phenolic content and biological activities of enzymatic extracts from Sargassum muticum (Yendo) Fensholt. *Environ. Boil. Fishes* **2017**, *29*, 2521–2537. [CrossRef] [PubMed]
23. Hassan, I.H.; Pham, H.N.T.; Nguyen, T.H. Optimization of ultrasound-assisted extraction conditions for phenolics, antioxidant, and tyrosinase inhibitory activities of Vietnamese brown seaweed (Padina australis). *J. Food Process. Preserv.* **2021**, *45*, e15386. [CrossRef]

24. Ummat, V.; Tiwari, B.K.; Jaiswal, A.K.; Condon, K.; Garcia-Vaquero, M.; O'Doherty, J.; O'Donnell, C.; Rajauria, G. Optimisation of ultrasound frequency, extraction time and solvent for the recovery of polyphenols, phlorotannins and associated antioxidant activity from brown seaweeds. *Mar. Drugs* **2020**, *18*, 250. [CrossRef]
25. Magnusson, M.; Yuen, A.K.; Zhang, R.; Wright, J.T.; Taylor, R.B.; Maschmeyer, T.; De Nys, R. A comparative assessment of microwave assisted (MAE) and conventional solid-liquid (SLE) techniques for the extraction of phloroglucinol from brown seaweed. *Algal Res.* **2017**, *23*, 28–36. [CrossRef]
26. Harnedy, P.; FitzGerald, R.J. Extraction of protein from the macroalga Palmaria palmata. *LWT Food Sci. Technol.* **2013**, *51*, 375–382. [CrossRef]
27. Mabeau, S.; Kloareg, B. Isolation and analysis of the cell walls of brown algae: *Fucus spiralis, F. ceranoides, F. vesiculosus, F. serratus, Bifurcaria bifurcata* and *Laminaria digitata*. *J. Exp. Bot.* **1987**, *38*, 1573–1580. [CrossRef]
28. Deniaud-Bouët, E.; Kervarec, N.; Michel, G.; Tonon, T.; Kloareg, B.; Hervé, C. Chemical and enzymatic fractionation of cell walls from Fucales: Insights into the structure of the extracellular matrix of brown algae. *Ann. Bot.* **2014**, *114*, 1203–1216. [CrossRef]
29. Tai, Y.; Shen, J.; Luo, Y.; Qu, H.; Gong, X. Research progress on the ethanol precipitation process of traditional Chinese medicine. *Chin. Med.* **2020**, *15*, 84. [CrossRef] [PubMed]
30. Vázquez-Rodríguez, B.; Gutiérrez-Uribe, J.A.; Antunes-Ricardo, M.; Santos-Zea, L.; Cruz-Suárez, L.E. Ultrasound-assisted extraction of phlorotannins and polysaccharides from *Silvetia compressa* (Phaeophyceae). *J. Appl. Phycol.* **2020**, *32*, 1441–1453. [CrossRef]
31. Deniaud-Bouët, E.; Hardouin, K.; Potin, P.; Kloareg, B.; Hervé, C. A review about brown algal cell walls and fucose-containing sulfated polysaccharides: Cell wall context, biomedical properties and key research challenges. *Carbohydr. Polym.* **2017**, *175*, 395–408. [CrossRef] [PubMed]
32. Amarante, S.J.; Catarino, M.D.; Marçal, C.; Silva, A.M.S.; Ferreira, R.; Cardoso, S.M. Microwave-assisted extraction of phlorotannins from fucus vesiculosus. *Mar. Drugs* **2020**, *18*, 559. [CrossRef]
33. Wen, L.; Zhang, Z.; Zhao, M.; Senthamaraikannan, R.; Padamati, R.B.; Sun, D.; Tiwari, B.K. Green extraction of soluble dietary fibre from coffee silverskin: Impact of ultrasound/microwave-assisted extraction. *Int. J. Food Sci. Technol.* **2019**, *55*, 2242–2250. [CrossRef]
34. Yang, J.-S.; Mu, T.-H.; Ma, M.-M. Optimization of ultrasound-microwave assisted acid extraction of pectin from potato pulp by response surface methodology and its characterization. *Food Chem.* **2019**, *289*, 351–359. [CrossRef] [PubMed]
35. Lianfu, Z.; Zelong, L. Optimization and comparison of ultrasound/microwave assisted extraction (UMAE) and ultrasonic assisted extraction (UAE) of lycopene from tomatoes. *Ultrason. Sonochemistry* **2008**, *15*, 731–737. [CrossRef]
36. Garcia-Vaquero, M.; O'Doherty, J.V.; Tiwari, B.K.; Sweeney, T.; Rajauria, G. Enhancing the extraction of polysaccharides and antioxidants from macroalgae using sequential hydrothermal-assisted extraction followed by ultrasound and thermal technologies. *Mar. Drugs* **2019**, *17*, 457. [CrossRef]
37. Lemus, A.; Bird, K.; Kapraun, D.F.; Koehn, F. Agar yield, quality and standing crop biomass of Gelidium serrulatum, Gelidium floridanum and Pterocladia capillacea in Venezuela. *Food Hydrocoll.* **1991**, *5*, 469–479. [CrossRef]
38. Lee, W.-K.; Namasivayam, P.; Ho, C.-L. Effects of sulfate starvation on agar polysaccharides of Gracilaria species (Gracilariaceae, Rhodophyta) from Morib, Malaysia. *Environ. Boil. Fishes* **2013**, *26*, 1791–1799. [CrossRef]
39. Mulchandani, K.; Kar, J.R.; Singhal, R.S. Extraction of lipids from chlorella saccharophila using high-pressure homogenization followed by three phase partitioning. *Appl. Biochem. Biotechnol.* **2015**, *176*, 1613–1626. [CrossRef] [PubMed]
40. Li, G.-Y.; Luo, Z.-C.; Yuan, F.; Yu, X.-B. Combined process of high-pressure homogenization and hydrothermal extraction for the extraction of fucoidan with good antioxidant properties from Nemacystus decipients. *Food Bioprod. Process.* **2017**, *106*, 35–42. [CrossRef]
41. Graiff, A.; Ruth, W.; Kragl, U.; Karsten, U. Chemical characterization and quantification of the brown algal storage compound laminarin—A new methodological approach. *Environ. Boil. Fishes* **2015**, *28*, 533–543. [CrossRef]
42. Michalak, I.; Chojnacka, K. Algal extracts: Technology and advances. *Eng. Life Sci.* **2014**, *14*, 581–591. [CrossRef]
43. Zhang, H.; Birch, J.; Xie, C.; Yang, H.; Dias, G.; Kong, L.; Bekhit, A.E.-D. Optimization of extraction parameters of antioxidant activity of extracts from New Zealand and Chinese Asparagus officinalis L root cultivars. *Ind. Crop. Prod.* **2018**, *119*, 191–200. [CrossRef]
44. Foley, S.A.; Szegezdi, E.; Mulloy, B.; Samali, A.; Tuohy, M.G. Correction to an unfractionated fucoidan from ascophyllum nodosum: Extraction, characterization, and apoptotic effects in vitro. *J. Nat. Prod.* **2012**, *75*, 1674. [CrossRef]
45. Calixto, F.D.S. Dietary fiber as a carrier of dietary antioxidants: An essential physiological function. *J. Agric. Food Chem.* **2011**, *59*, 43–49. [CrossRef]
46. Dang, T.T.; Bowyer, M.C.; Van Altena, I.A.; Scarlett, C.J. Comparison of chemical profile and antioxidant properties of the brown algae. *Int. J. Food Sci. Technol.* **2017**, *53*, 174–181. [CrossRef]
47. Parys, S.; Kehraus, S.; Pete, R.; Küpper, F.C.; Glombitza, K.-W.; König, G.M. Seasonal variation of polyphenolics in *Ascophyllum* nodosum (Phaeophyceae). *Eur. J. Phycol.* **2009**, *44*, 331–338. [CrossRef]
48. Schiener, P.; Black, K.D.; Stanley, M.S.; Green, D. The seasonal variation in the chemical composition of the kelp species Laminaria digitata, Laminaria hyperborea, Saccharina latissima and Alaria esculenta. *Environ. Boil. Fishes* **2015**, *27*, 363–373. [CrossRef]
49. Fraga-Corral, M.; García-Oliveira, P.; Pereira, A.G.; Lourenço-Lopes, C.; Jimenez-Lopez, C.; Prieto, M.A.; Simal-Gandara, J. Technological application of tannin-based extracts. *Molecules* **2020**, *25*, 614. [CrossRef] [PubMed]

50. Rajauria, G.; Jaiswal, A.K.; Abu-Ghannam, N.; Gupta, S. Antimicrobial, antioxidant and free radical-scavenging capacity of brown seaweed Himanthalia elongata from western coast of Ireland. *J. Food Biochem.* **2013**, *37*, 322–335. [CrossRef]
51. Liu, S.-C.; Lin, J.-T.; Wang, C.-K.; Chen, H.-Y.; Yang, D.-J. Antioxidant properties of various solvent extracts from lychee (Litchi chinenesis Sonn.) flowers. *Food Chem.* **2009**, *114*, 577–581. [CrossRef]
52. Brummer, Y.; Cui, S.W. Understanding carbohydrate analysis. In *Food Carbohydrates: Chemistry, Physical Properties and Applications*; CRC Press: Boca Raton, FL, USA, 2005; pp. 1–38.
53. Sridhar, K.; Charles, A.L. In vitro antioxidant activity of Kyoho grape extracts in DPPH and ABTS assays: Estimation methods for EC50 using advanced statistical programs. *Food Chem.* **2019**, *275*, 41–49. [CrossRef]
54. Benzie, I.F.F.; Strain, J.J. The ferric reducing ability of plasma (FRAP) as a measure of "antioxidant power": The FRAP assay. *Anal. Biochem.* **1996**, *239*, 70–76. [CrossRef]
55. R Core Team. *R: A Language and Environment for Statistical Computing*; R Foundation for Statistical Computing: Vienna, Austria, 2020. Available online: http://www.R-project.org/ (accessed on 11 March 2021).
56. Friendly, M. Corrgrams: Exploratory displays for correlation matrices. *Am. Stat.* **2002**, *56*, 316–324. [CrossRef]

Article

Anticancer Effects of New Ceramides Isolated from the Red Sea Red Algae *Hypnea musciformis* in a Model of Ehrlich Ascites Carcinoma: LC-HRMS Analysis Profile and Molecular Modeling

Sameh S. Elhady [1,†], Eman S. Habib [2,†], Reda F. A. Abdelhameed [3], Marwa S. Goda [2], Reem M. Hazem [4], Eman T. Mehanna [5], Mohamed A. Helal [6,7], Khaled M. Hosny [8], Reem M. Diri [9], Hashim A. Hassanean [2], Amany K. Ibrahim [2], Enas E. Eltamany [2], Usama Ramadan Abdelmohsen [10,11] and Safwat A. Ahmed [2,*]

1 Department of Natural Products, Faculty of Pharmacy, King Abdulaziz University, Jeddah 21589, Saudi Arabia; ssahmed@kau.edu.sa
2 Department of Pharmacognosy, Faculty of Pharmacy, Suez Canal University, Ismailia 41522, Egypt; emy_197@hotmail.com (E.S.H.); marwa_saeed@pharm.suez.edu.eg (M.S.G.); hashem_omar@pharm.suez.edu.eg (H.A.H.); amany_mohamed@pharm.suez.edu.eg (A.K.I.); enastamany@gmail.com (E.E.E.)
3 Department of Pharmacognosy, Faculty of Pharmacy, Galala University, New Galala 43713, Egypt; reda.fouad@gu.edu.eg
4 Department of Pharmacology and Toxicology, Faculty of Pharmacy, Suez Canal University, Ismailia 41522, Egypt; reem_ahmed@pharm.suez.edu.eg
5 Department of Biochemistry, Faculty of Pharmacy, Suez Canal University, Ismailia 41522, Egypt; eman.taha@pharm.suez.edu.eg
6 Biomedical Sciences Program, University of Science and Technology, Zewail City of Science and Technology, October Gardens, 6th of October, Giza 12578, Egypt; mohamed_hilal@pharm.suez.edu.eg
7 Department of Medicinal Chemistry, Faculty of Pharmacy, Suez Canal University, Ismailia 41522, Egypt
8 Department of Pharmaceutics, Faculty of Pharmacy, King Abdulaziz University, Jeddah 21589, Saudi Arabia; kmhomar@kau.edu.sa
9 Department of Pharmacy Practice, Faculty of Pharmacy, King Abdulaziz University, Jeddah 21589, Saudi Arabia; rdiri@kau.edu.sa
10 Department of Pharmacognosy, Faculty of Pharmacy, Minia University, Minia 61519, Egypt; usama.ramadan@mu.edu.eg
11 Department of Pharmacognosy, Faculty of Pharmacy, Deraya University, New Minia 61111, Egypt
* Correspondence: safwat_aa@yahoo.com or safwat_ahmed@pharm.suez.edu.eg; Tel.: +20-010-92638387
† These authors equally contributed to this work.

Citation: Elhady, S.S.; Habib, E.S.; Abdelhameed, R.F.A.; Goda, M.S.; Hazem, R.M.; Mehanna, E.T.; Helal, M.A.; Hosny, K.M.; Diri, R.M.; Hassanean, H.A.; et al. Anticancer Effects of New Ceramides Isolated from the Red Sea Red Algae *Hypnea musciformis* in a Model of Ehrlich Ascites Carcinoma: LC-HRMS Analysis Profile and Molecular Modeling. *Mar. Drugs* **2022**, *20*, 63. https://doi.org/10.3390/md20010063

Academic Editors: Marco García-Vaquero and Brijesh K. Tiwari

Received: 21 December 2021
Accepted: 7 January 2022
Published: 10 January 2022

Abstract: Different classes of phytochemicals were previously isolated from the Red Sea algae *Hypnea musciformis* as sterols, ketosteroids, fatty acids, and terpenoids. Herein, we report the isolation of three fatty acids—docosanoic acid **4**, hexadecenoic acid **5**, and alpha hydroxy octadecanoic acid **6**—as well as three ceramides—**A** (**1**), **B** (**2**), and **C** (**3**)—with 9-methyl-sphinga-4,8-dienes and phytosphingosine bases. Additionally, different phytochemicals were determined using the liquid chromatography coupled with electrospray ionization high-resolution mass spectrometry (LC-ESI-HRMS) technique. Ceramides **A** (**1**) and **B** (**2**) exhibited promising in vitro cytotoxic activity against the human breast adenocarcinoma (MCF-7) cell line when compared with doxorubicin as a positive control. Further in vivo study and biochemical estimation in a mouse model of Ehrlich ascites carcinoma (EAC) revealed that both ceramides **A** (**1**) and **B** (**2**) at doses of 1 and 2 mg/kg, respectively, significantly decreased the tumor size in mice inoculated with EAC cells. The higher dose (2 mg/kg) of ceramide **B** (**2**) particularly expressed the most pronounced decrease in serum levels of vascular endothelial growth factor -B (VEGF-B) and tumor necrosis factor-α (TNF-α) markers, as well as the expression levels of the growth factor midkine in tumor tissue relative to the EAC control group. The highest expression of apoptotic factors, p53, Bax, and caspase 3 was observed in the same group that received 2 mg/kg of ceramide **B** (**2**). Molecular docking simulations suggested that ceramides **A** (**1**) and **B** (**2**) could bind in the deep grove between the H2 helix and the Ser240-P250 loop of p53, preventing its interaction with MDM2 and leading to its accumulation. In conclusion, this study reports the cytotoxic, apoptotic, and antiangiogenic effects of ceramides isolated from the Red Sea algae *Hypnea musciformis* in an experimental model of EAC.

Keywords: ceramides; MCF-7; Ehrlich ascites carcinoma; VEGF-B; TNF-α; midkine; apoptosis; p53; LC-ESI-HRMS; *Hypnea musciformis*

1. Introduction

Of the three classifications of macroalgae, the chemistry of red algae is more diverse than that of green or brown algae. Therefore, red algae are considered the most important source of many biologically active metabolites in comparison with other algal classes [1,2]. The genus *Hypnea* is one of the widest spread red algae, with economic importance as a source of carrageenan [3]. The *Hypnea* species were extensively assessed for their biological activities. Methanolic extract of *Hypnea flagelliformis* was subjected to the 1,1-diphenyl-2-picrylhydrazyl (DPPH) free radical scavenging assay, and it showed a stronger antioxidant activity compared with the standard quercetin [4]. Likewise, the ethyl acetate fraction showed a significantly higher total phenolic content, DPPH·scavenging activity, H_2O_2 scavenging activity, and lipid peroxidation inhibition than its crude extract, fractions of *n*-hexane and dichloromethane, and other methanolic fractions of *Hypnea valentiae*. This study introduced the red seaweeds *Hypnea* sp. to be used as food supplements for increasing shelf-life in the food industry and combating carcinogenesis [5]. Methanol extract of *Hypnea valentiae* inhibited acetylcholinesterase (AChE), and this neuroprotective action is considered a first line in the treatment of dementia [6]. Bitencourt and his colleagues observed that a lectin isolated from the red marine alga *Hypnea cervicornis* possessed antinociceptive and anti-inflammatory activity via interaction with the lectin carbohydrate-binding site. Additionally, the lectin did not show visible signs of toxicity at effective doses [7]. The methanolic extract of the algea *Hypnea esperi* from the Suez Canal region showed potent antibacterial activity toward Gram-positive bacteria that correlated to long chain fatty acids of more than 10 carbon atoms in length, which induced lysis of bacterial protoplasts. In addition, *H. esperi* also exhibited anticoagulant activity by delaying the blood clotting to 120 s in comparison with the control blood's 40-s clotting time [8]. Moreover, several fatty acids such as palmitic, oleic, pentacosanoic, and hexacosenoic acids, as well as sesquiterpene and sterols, were reported in *Hypnea* [9]. The previously isolated compounds from the genus *Hypnea* can be classified into three categories; sterols and ketosteroids, terpenoids, and polymers as polypeptides and polysaccharides [10–15]. Although many biological studies were performed, less research work was performed for the isolation of these pure active compounds. Our study was oriented toward finding out other classes of bioactive compounds that attributed to the previously mentioned pharmacological activities of *Hypnea* sp.

The sphingolipid-signaling pathway is a novel anticancer target system. It has been suggested that sphingolipids play fundamental roles in the regulation of cancer pathogenesis and development [16]. Ceramide serves as a central mediator in sphingolipid metabolism and signaling pathways, regulating many essential cellular responses [17]. Many drugs used in the treatment of cancer are themselves ceramide generators. This property contributes in part to their apoptosis-inducing effects [18]. Consequently, targeting the ceramide-signaling pathway by activating ceramide downstream receptors, inhibiting ceramide-metabolizing enzymes, or exogenously increasing the ceramide levels comprise the novel targets for cancer treatment [19]. In the current work, we aimed to assess the potential antitumor and apoptotic activities of two novel ceramides isolated from *Hypnea musciformis*.

2. Results and Discussion

2.1. Metabolic Analysis Profile

The metabolic analysis profile produced by the LC-HR-ESI-MS technique (Figures S1 and S2) manifested different metabolites (Table 1 and Figure 1) that were detected by comparing their detected masses with those recorded in some databases (e.g., the Dictionary of Natural Products (DNP) and Metabolite and Chemical Entity (METLIN)). The mass accuracy was calculated by ((detected mass − expected mass)/expected mass) $\times 10^6$ and expressed in parts per million (ppm) error [20]. Sterols with cholesterol nucleui were the most common chemical class isolated from *Hypnea musciformis*. Other bioactive metabolites were identified as ptilodene, an antimicrobial and anti-inflammatory icosanoid [21], agardhilactone, an oxylipin with epoxide ring showing anticancer activity [22,23], and oxytocic prostaglandin-E2, which induces labor [24]. Additionally, a phytosphingosine base that exhibited antiphlogistic and antimicrobial activity against Gram-positive and Gram-negative bacteria, viruses, and fungi [25,26] was detected. Therefore, the above-mentioned biological activity may have been related to the identified metabolites.

Table 1. Metabolic profiling (LC-ESI-HRMS) of methanolic crude extract of *Hypnea* sp.

No.	Polarity Mode	Ret. Time (min)	*m/z*	MZmine ID	Detected Mass	Expected Mass	Mass Error (ppm)	Name	Source	Ref.
1	Positive	12.70	399.3255	292	398.3182	398.3185	−0.75	Diketosteroid	Red alga *Hypnea musciformis*	[10]
2	Positive	12.70	399.3255	292	398.3182	398.3185	−0.75	6α-Hydroxy-cholesta-4,22-diene-3-one	Red alga*Hypnea musciformis*	[11]
3	Positive	12.70	399.3255	292	398.3182	398.3185	−0.75	Cholesta-5,22-diene-3β -ol-7-one	Red alga *Hypnea flagelliformis*	[13]
4	Positive	12.70	399.3255	292	398.3182	398.3185	−0.75	5β -Cholest-3-ene-7,11-dione	Red alga*Hypnea musciformis*	[10]
5	Positive	12.49	385.3465	326	384.3392	384.3392	0.00	Cholesta-5,22-dien-3β-ol	Red alga *Hypnea flagelliformis*	[13]
6	Positive	12.49	385.3465	326	384.3392	384.3392	0.00	22-Dehydrocholesterol	Red alga *Hypnea flagelliformis*	[13]
7	Positive	8.44	318.3005	108	317.2932	317.2930	0.63	Phytosphingosine	Fungi	[25]
8	Negative	6.33	349.2013	1839	350.2085	350.2093	−2.28	PGE2, Prostaglandin-E2	Red alga *Gracilaria verrucosa*	[24]
9	Positive	6.99	333.2064	1326	332.1991	332.1988	0.90	Ptilodene	Red alga*Ptilota filicina*	[21]
10	Negative	6.78	331.1914	1751	332.1987	332.1988	−0.30	Agardhilactone	Red alga *Agardhiella subulata*	[22]

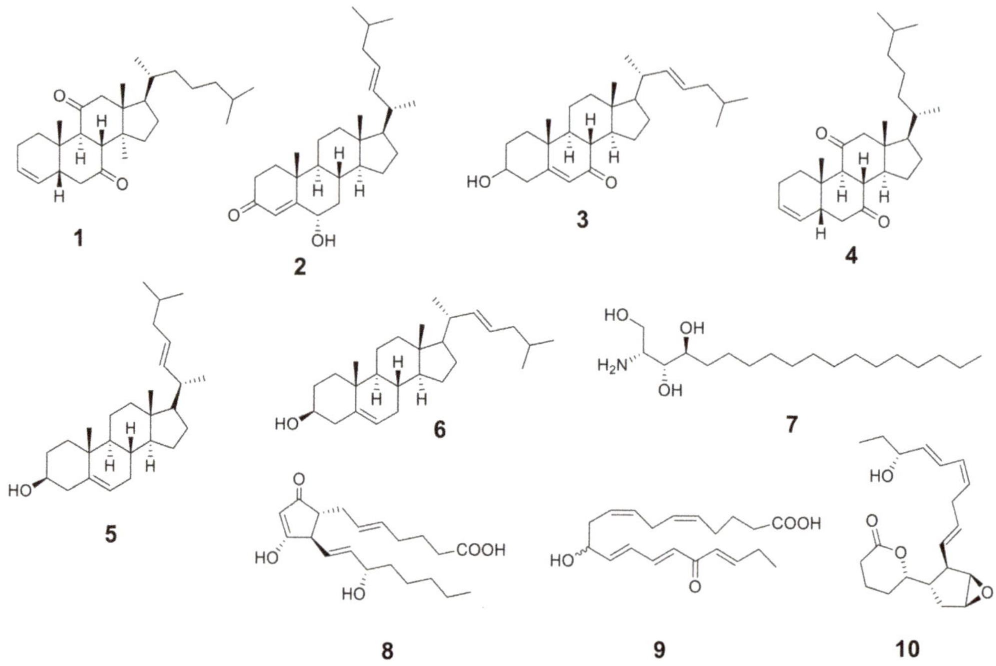

Figure 1. Structures of the identified metabolites listed in Table 1.

2.2. Identification of Isolated Compounds

Ceramide **A** (**1**) (Figure 2) was obtained as a white powder, and its molecular formula was determined to be $C_{43}H_{83}NNaO_4$ by ESI-HRMS with m/z 700.5921 $[M + Na]^+$ (Figure S3), calculated as 700.6220, representing 3 degrees of unsaturation. The ^1H NMR and ^{13}C NMR spectral data of ceramide **A** (**1**) are listed in Table 2 (Figures S4 and S5). The backbone of a ceramide nucleus was recognized by the presence of an amide group at δ_H 7.26 ppm/δ_C 175.74 ppm and multiplet peaks of a long methylene chain at δ_H 1.12–1.32 ppm/δ_C 29.2–29.7 ppm. An oxygenated methylene was determined at δ_H 3.75, 4.08 ppm/δ_C 61.9 ppm. Two oxygenated methine groups as well as a nitrogen-bearing methine were determined at δ_H 4.08 ppm/δ_C 74 ppm, δ_H 4.23 ppm/δ_C = 72.5 ppm, and δ_H 3.9 ppm/δ_C 54.4 ppm, respectively. Four olefinic peaks were detected at δ_C 129, 134.1, 123.1, and 136.3 ppm and δ_H 5.51, 5.67, and 5.08 ppm. The length of the fatty acid chain was analyzed by HRMS after performing a protocol of methanolysis [27]. The HRMS showed a molecular ion peak at m/z 369.3231 $[M + H]^+$ (Figure S6), calculated as m/z 369.3290, indicating a C_{22} fatty acid with a molecular formula of $C_{23}H_{45}O_3$, recognized as 2-hydroxy docosanoic acid methyl ester, while the long chain base was recognized as 1,3-dihydroxy-2-amino-9-methyl-icosene-4,8-diene. Straight chains of both the fatty acid and sphingosine base ended with terminal methyl groups of a normal form at δ_H 0.88 ppm/δ_C 14 ppm. Finally, ceramide **A** (**1**) could be identified as a ceramide with alpha-hydroxylated unsaturated fatty acid and long chain of 9-methyl-sphinga-4,8-diene base possessing 2*S*, 2′*R*, 3*R* relative configurations. The configuration of the ceramide moieties was assigned by comparing its physical data, optical rotation $[\alpha]_D^{22}$ +6.7 *c* 0.23, $CHCl_3$), ^{1}H-NMR, and ^{13}C-NMR (measured in $CDCl_3$) with analogs, using deuterated chloroform as an NMR solvent as reported in the literature [28,29]. The structure of compound **1** was determined to be 2′-hydroxy-N-[(2*S*,2′*R*,3*R*,4*E*,8*E*)-1,3-dihydroxy-9-methyl-icosene-4,8-diene-2-yl]-docosanamide which, to the best of our knowledge, is a new compound.

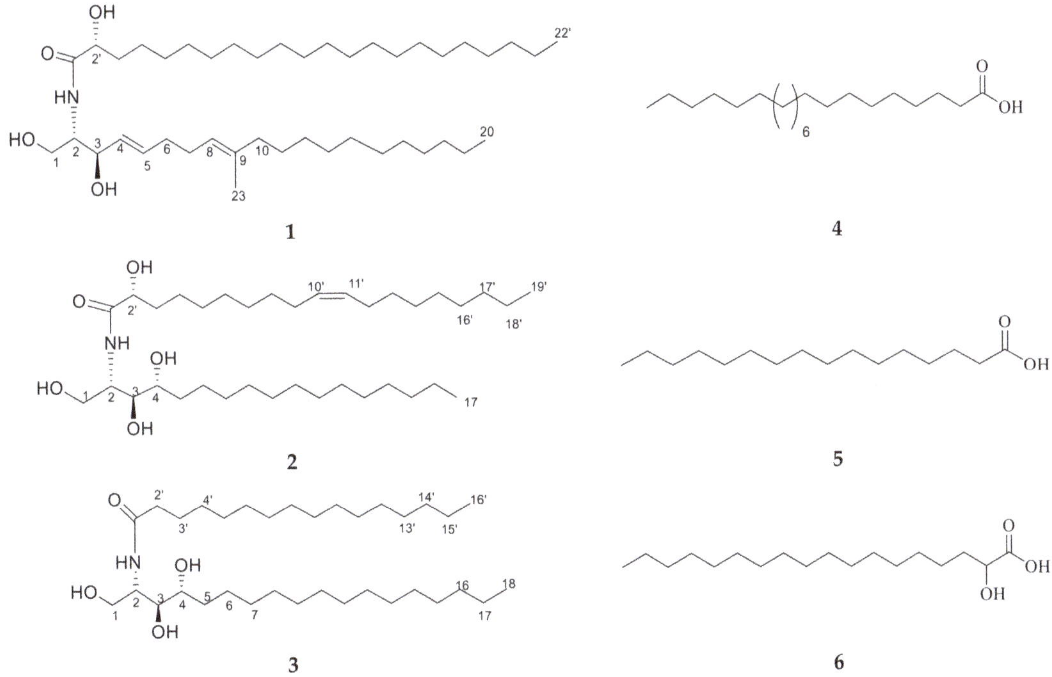

1 **4**

2 **5**

3 **6**

Figure 2. Chemical structures of isolated compounds **1–6**.

Ceramide **B** (**2**) (Figure 2) was obtained as a white powder, and its molecular formula was determined to be $C_{36}H_{72}NO_5$ by ESI-HRMS with m/z 584.4410 [M + H]$^+$ (Figure S7), calculated as 584.5176, representing two degrees of unsaturation. The ^{1}H NMR and ^{13}C NMR spectral data are listed in Table 2 (Figures S8 and S9). The backbone structure of compound **2** was identified as a ceramide. The core of a ceramide nucleus was confirmed by the presence of an amide group at δ_C 175.1 ppm/δ_H 8.58 ppm and an overlapped long methylene chain at δ_C 29.6 ppm/δ_H 1.22–1.29 ppm. An oxygenated methylene group as well as a nitrogen-bearing methine group were determined at δ_H 4.40, 4.5 ppm/δ_C 61.7 ppm and δ_H 5.11 ppm/δ_C 52.6 ppm, respectively. Three groups of oxygenated methine were detected at δ_C 72.1, 72.7, and 76.4 ppm and δ_H 4.61, 4.28, and 4.35 ppm. Additionally, two olefinic peaks were determined at δ_H 5.47 ppm/δ_C 130.0 ppm. Terminal methyl groups of a normal type were detected at δ_C 13.9 ppm/δ_H 0.85 ppm. The length of the fatty acid chain was determined on the basis of the results of its oxidative methanolysis followed by peak detection by HRMS, which showed a molecular ion peak at m/z 313.2709 [M + H]$^+$ (Figure S10), calculated as m/z 313.2743, indicating a methyl ester of $C_{18:1}$ fatty acid with a molecular formula of $C_{19}H_{37}O_3$. Therefore, the fatty ester methyl ester moiety was recognized as 2-hydroxy-10-nonadecenoic acid methyl ester, which was confirmed by GC-MS analysis (Figure S11). At last, ceramide **B** (**2**) could be identified as a ceramide with 2′-hydroxy monounsaturated fatty acid and a long chain phytosphingosine base possessing 2*S*,2′*R*,3*S*,4*R*,9′*Z* relative configurations. The configuration of the ceramide moieties was assigned by comparing its physical data, optical rotation $[\alpha]_D^{22}$ +7.70 (*c* 0.27, pyridine), ^{1}H-NMR, and ^{13}C-NMR (measured in C_5D_5N) with the analogs using deuterated pyridine as an NMR solvent, as reported in the literature [27]. The structure of ceramide **B** (**2**) was determined to be 2′-hydroxy-N-[(2*S*,2′*R*,3*S*,4*R*,9′*Z*)-1,3,4-trihydroxy-nonadecan-2-yl]-10-heptadecenamide which, to the best of our knowledge, is a new compound.

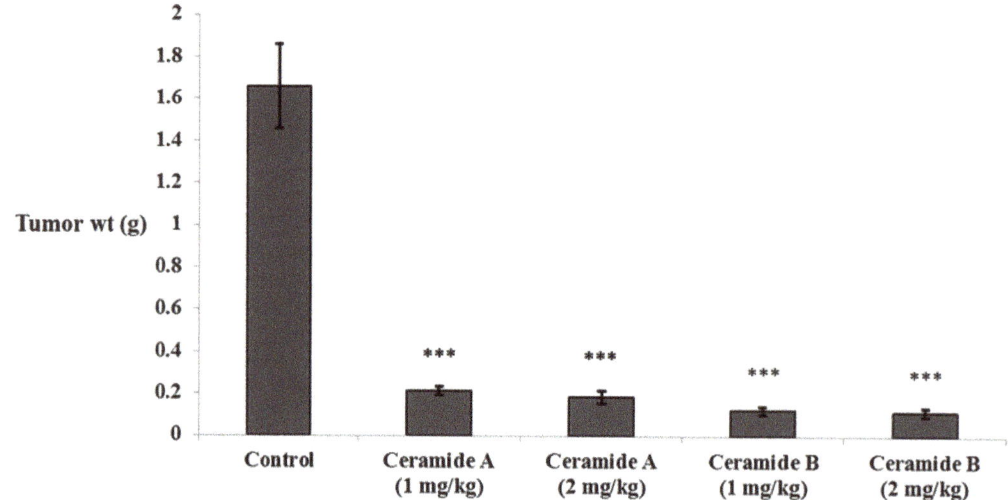

Figure 3. Effect of treatment with ceramide **A** and ceramide **B** at two dose levels (1 mg/kg and 2 mg/kg I.P) on the tumor weight in EAC-bearing mice. Values are expressed as mean ± SD. All data were analyzed using ANOVA followed by a Bonferroni post hoc test. *** Significantly different compared with the EAC control group at $p < 0.001$.

2.4.3. Effect of the Investigated Ceramides on the Serum Levels of Vascular Endothelial Growth Factor B (VEGF-B) and Tumor Necrosis Factor (TNF-α) and the Expression of Midkine (MDK) in the Tumor Tissue

The serum levels of vascular endothelial growth factor B (VEGF-B) and the tumor necrosis factor (TNF-α) were assessed by ELISA (Figure 4). Both markers were significantly increased in the EAC control group compared with the normal group ($p < 0.001$). The levels of VEGF-B were significantly decreased upon treatment by ceramide **A** (**1**) (1 and 2 mg/kg) and ceramide **B** (**2**) (1 mg/kg) ($p < 0.01$). The most significant decrease in the levels of VEGF-B in comparison with the EAC control group was recorded in the group treated by 2 mg/kg of ceramide **B** (**2**) ($p < 0.001$) (Figure 4A). Similarly, the levels of TNF-α showed a significant decrease in all treated groups: $p < 0.05$ in groups 3, 4, and 5 (1 and 2 mg/kg of ceramide **A** (**1**) and 1 mg/kg of ceramide **B** (**2**)) and $p < 0.01$ in group 6 (2 mg/kg of ceramide **B** (**2**)) (Figure 4B).

The levels of expression of midkine (MDK) in the tumor tissue were determined by real-time PCR. MDK was significantly upregulated in the EAC control group compared with the normal level ($p < 0.001$). The expression levels were significantly decreased in the groups treated with both doses of ceramide **A** (**1**) (1 and 2 mg/kg) and the lower dose of ceramide **B** (**2**) (1 mg/kg) ($p < 0.01$). The group treated with the higher dose of ceramide **B** (**2**) (2 mg/kg) showed the most significant downregulation of MDK compared with the EAC control group ($p < 0.001$) (Figure 5).

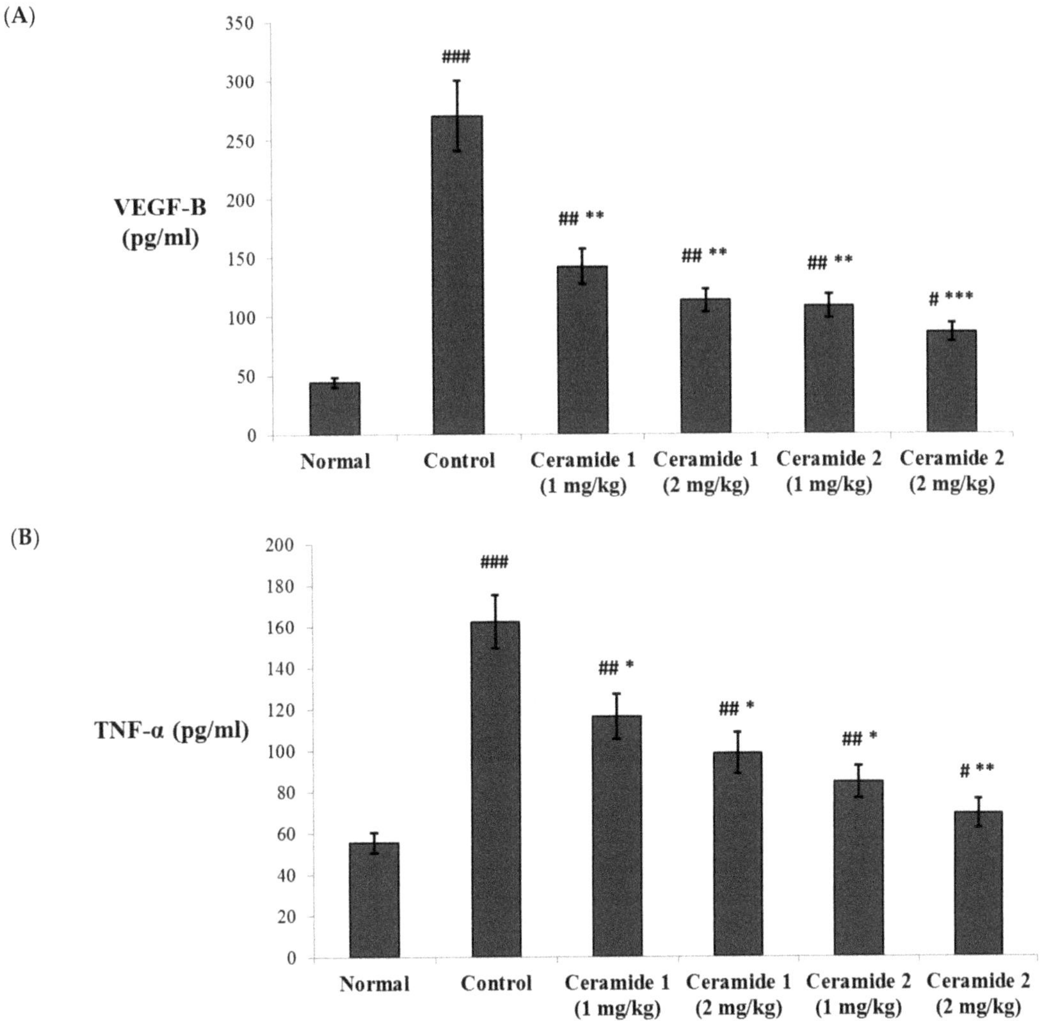

Figure 4. Effect of treatment with ceramide **A** and ceramide **B** at two dose levels (1 mg/kg and 2 mg/kg I.P) on the serum levels of (**A**) vascular endothelial growth factor B (VEGF-B) and (**B**) tumor necrosis factor-α (TNF-α). Values are expressed as mean ± SD. All data were analyzed using ANOVA followed by a Bonferroni post hoc test. # Significantly different compared with the normal group at $p < 0.05$. ## Significantly different compared with the normal group at $p < 0.01$. ### Significantly different compared with the normal group at $p < 0.001$. * Significantly different compared with the EAC control group at $p < 0.05$. ** Significantly different compared with the EAC control group at $p < 0.01$. *** Significantly different compared with the EAC control group at $p < 0.001$.

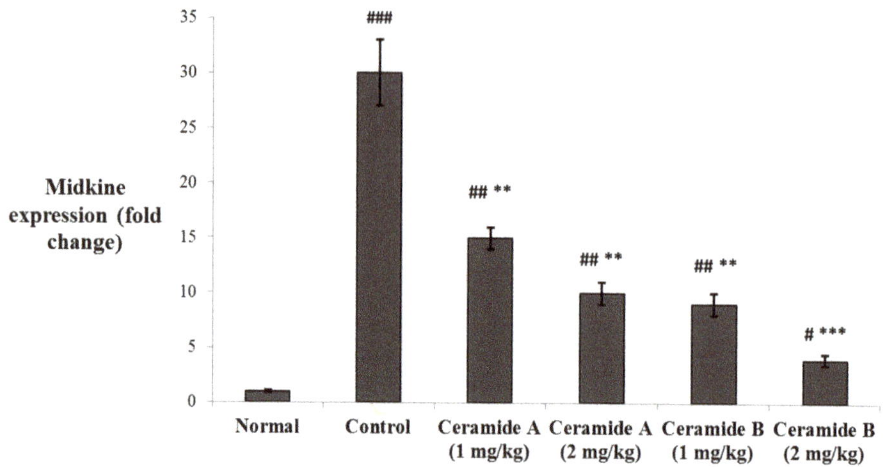

Figure 5. Effect of treatment with ceramide **A** and ceramide **B** at two dose levels (1 mg/kg and 2 mg/kg I.P) on the expression of midkine (MDK) in the tumor tissue. Values are expressed as mean ± SD. All data were analyzed using ANOVA followed by a Bonferroni post hoc test. [#] Significantly different compared with the normal group at $p < 0.05$. [##] Significantly different compared with the normal group at $p < 0.01$. [###] Significantly different compared with the normal group at $p < 0.001$. ** Significantly different compared with the EAC control group at $p < 0.01$. *** Significantly different compared with the EAC control group at $p < 0.001$.

The VEGF members are key promotors of angiogenesis and lymphangiogenesis in malignancies. The level of VEGF-B in plasma was reported as a sensitive marker in breast cancer [38]. Overexpression of VEGF-B was found to promote metastasis in patients with pulmonary [39] and bladder [40] cancers. A higher expression of VEGF-B was also correlated with multiple tumors and positive vascular invasion in hepatocellular carcinoma patients [41]. Zhu et al. [42] suggested that downregulation of VEGF-B signaling may enhance the antitumor effect of resveratrol against pancreatic cancer. VEGF-B acts through binding to vascular endothelial growth factor receptor-1 (VEGFR1), leading to downstream activation of the angiogenetic and proliferative pathways including P38 mitogen-activated protein kinase (p38 MAPK), extracellular signal-regulated kinase (ERK)/MAPK, protein kinase B/serine threonine protein kinase (PKB/AKT), and phosphoinositide 3-kinase (PI3K) [43,44]. On the other hand, the role of TNF-α in cancer has been extensively investigated. It is known to be a double player that has a marked effect in tumor progression on one hand but also may act as a pro-apoptotic agent through activation of the c-Jun N-terminal kinase (JNK) pathway [45,46]. TNF-α is a major pro-inflammatory cytokine secreted by tumor-associated macrophages (TAMs) and by breast cancer cells. It is involved in all stages of the development of breast cancer, including tumor cell proliferation, epithelial-to-mesenchymal transition, metastasis, and recurrence [47]. Higher serum levels of TNF-α were reported in breast cancer patients compared with healthy individuals [48]. Additionally, the levels of TNF-α showed a correlation with the tumor stage in breast cancer patients [49–51]. TNF-α also activates nuclear factor kappa B (NFkβ) and induces Jagged1 expression, leading to activation of Notch signaling [52]. In a recent study conducted on a mammary carcinoma model, downregulation of TNF-α was associated with reduced VEGF, interleukin 6 (IL-6), interferon γ (IFN-γ), Jagged1, and shutting the Notch signaling pathway associated with enhanced apoptosis and declined angiogenesis [53].

MDK is a heparin-binding growth factor that is abnormally expressed in many human cancers, and it was found to mediate several tumor physiological processes involving cell growth, metastasis, migration, and angiogenesis. MDK is considered a key player in cancer progression and is proposed as a potential therapeutic target [54]. MDK expression is induced by cytokines and growth factors, mainly by TNF-α [55], and it acts through activating the NFkβ and MAPK/PI3K proliferative pathway [54]. In clinical studies, MDK was considered a potential prognostic biomarker in solid tumors [56]. In breast cancer patients, it was suggested as both a diagnostic and a prognostic biomarker [57,58]. Interestingly, MDK is a potent proangiogenic factor that promotes tumor angiogenesis [59] and is thought to play a role in controlling the plasma bioavailability of VEGF-A [60]. MDK was also suggested to be implicated in the hypoxia-induced tumor angiogenesis [54,61]. The metastatic effects of MDK are most probably mediated by its combined proinflammatory, angiogenic, and mitogenic functions [62–64].

It is noteworthy that the group treated with the higher dose (2 mg/kg) of phytosphingosine ceramide **B** (**2**) expressed the most pronounced decrease in all the biochemically determined markers relative to the EAC control group, as mentioned above. The serum levels of VEGF-B and TNF-α, as well as the expression levels of MDK in the mice treated with 2 mg/kg of ceramide **B** (**2**), revealed the least significant difference compared with the normal (negative control) group ($p < 0.05$). This agrees with the findings of Kwon et al. [65], who reported that a phytosphingosine derivative exhibited an anti-angiogenic effect through markedly decreasing VEGF-induced proteolytic enzyme production, VEGF-induced chemotactic migration, and capillary-like tube formation.

2.4.4. Effect of the Investigated Ceramides on the Expression of the Apoptotic Markers p35, Bax, and Caspase 3 as Determined by Immunohistochemistry in the Tumor Tissue

Treatment with ceramide **A** (**1**) and **B** (**2**) at both doses augmented the expression of p53, with a significant difference from the control EAC group ($p < 0.001$) (Figure 6). Similarly, the expression of Bax was raised after treatment with both ceramides at the given doses (1 at 2 mg/kg), with a significant difference compared with the control EAC group ($p < 0.001$). The highest expression was observed in the group that received ceramide **B** (**2**) (2 mg/kg) (Figure 7).

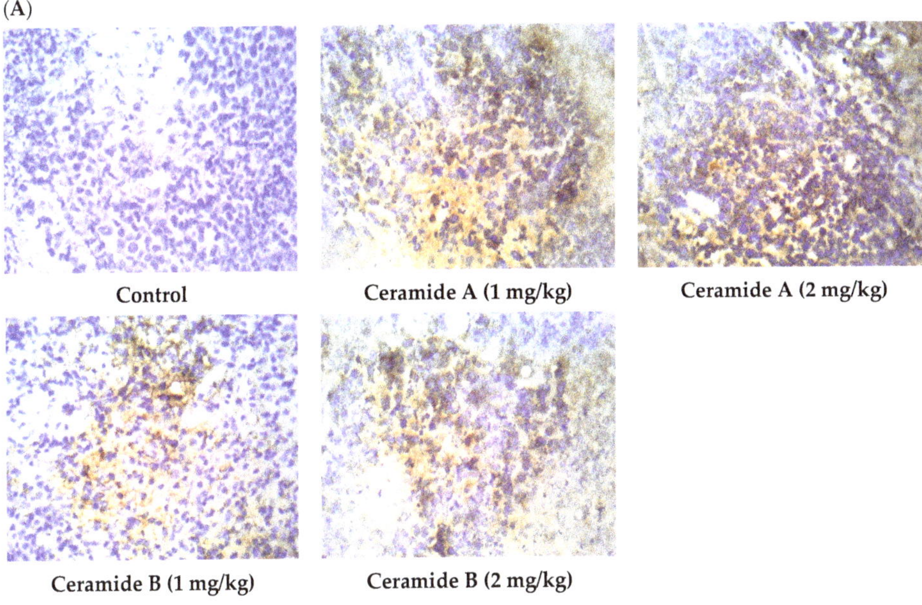

Figure 6. *Cont.*

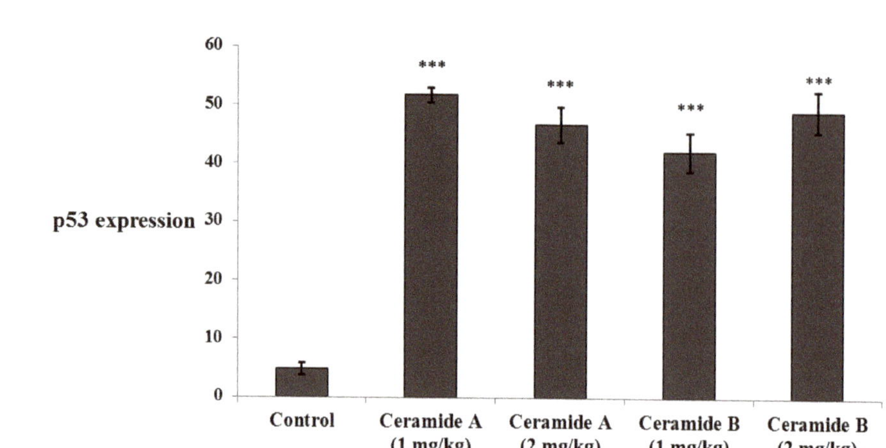

Figure 6. Effect of treatment with ceramide **A** and ceramide **B** at two dose levels (1 mg/kg and 2 mg/kg I.P) on the expression of p53. (**A**) Representative photomicrographs of p53 assessed immunohistochemically on day 21 in EAC-bearing female mice (40× magnification). (**B**) Optical density of positive immunohistochemical reactions (brown), determined using ImageJ. Values are expressed as mean ± SD. All data were analyzed using ANOVA followed by a Bonferroni post hoc test. *** Significantly different compared with the EAC control group at $p < 0.001$.

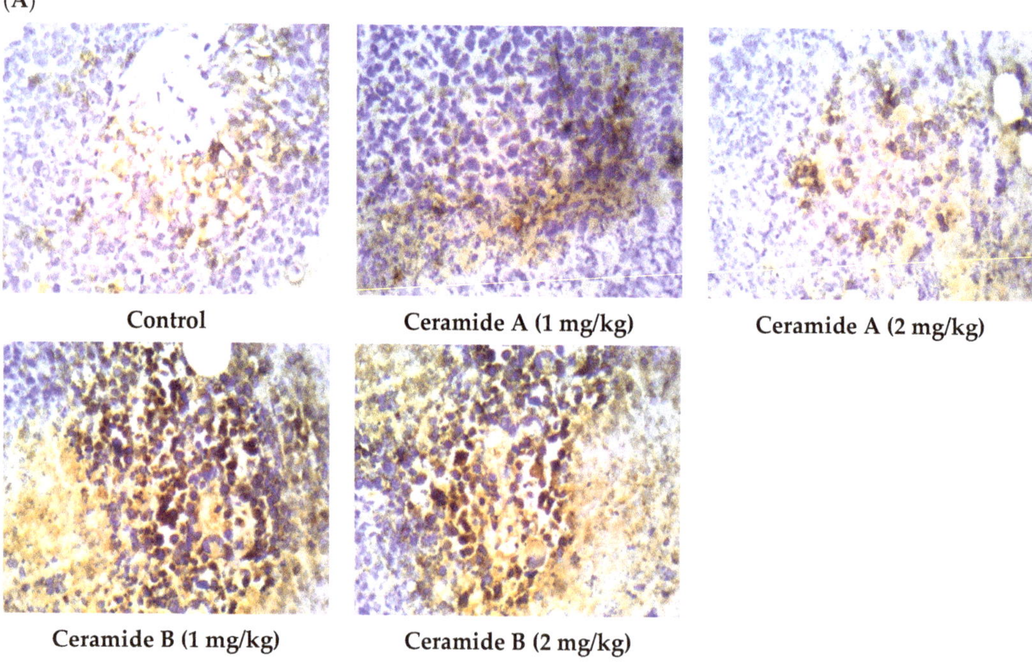

Figure 7. *Cont.*

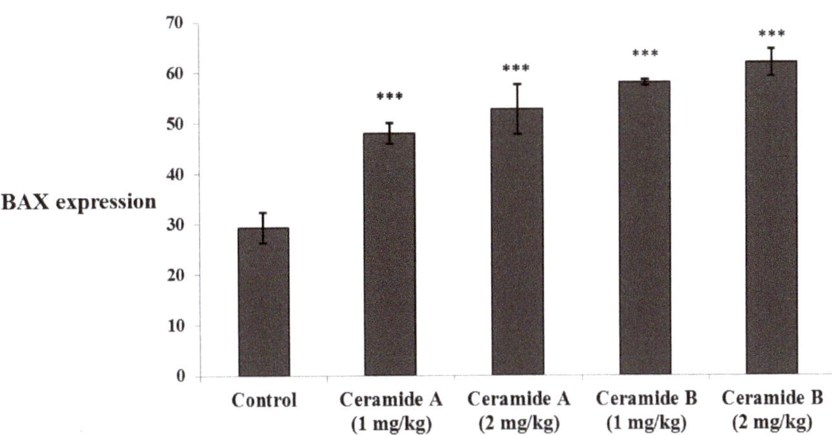

Figure 7. Effect of treatment with ceramide **A** and ceramide **B** at two dose levels (1 mg/kg and 2 mg/kg I.P) on the expression of p53. (**A**) Representative photomicrographs of Bax, assessed immunohistochemically on day 21 in EAC-bearing female mice (40× magnification). (**B**) Optical density of positive immunohistochemical reactions (brown) determined using ImageJ. Values are expressed as mean ± SD. All data were analyzed using ANOVA followed by a Bonferroni post hoc test. *** Significantly different compared with the EAC control group at $p < 0.001$.

The relationship between ceramide and p53 is very complex, and the mechanisms underlying their coregulation are diverse and not fully characterized [66]. The p53 protein was established as a tumor suppressor [67]. It has been assumed that p53 exerts its effect by inducing apoptosis [68]. Cancer research was concerned with both the p53 and ceramide pathways in the regulation of cell growth, cell cycle arrest, and apoptosis [69]. Therefore, the present study investigated the connection between exogenous ceramide uptake and the levels of p53.

An association between p53 and ceramide was observed upon investigation of the cellular response to folate stress. Stressing the A549 cells caused p53-dependent activation of de novo ceramide biosynthesis and C16-ceramide elevation followed by apoptosis [70]. Coadministration of C6-ceramide with vincristine caused apoptosis in multiple cell lines. Significant activation of p53 was detected in these cells, leading to apoptosis [71]. C2-ceramide was shown to induce cell death via elevation of p53, a subsequent increase in the Bax/Bcl-2 ratio, and caspase activation [72].

Likewise, the expression of caspase 3 showed significant elevation compared with the control EAC ($p < 0.001$) in all treatment groups (Figure 8). This finding is in agreement with a previous study which reported that phytosphingosine can potently induce apoptotic cell death in human cancer cells via activation of caspase 3, 8, and 9, mitochondrial translocation of Bax, and the subsequent release of cytochrome c into the cytoplasm, providing a potential mechanism for the anticancer activity of phytosphingosine [73]. Additionally, sphingosine was reported to mediate apoptosis in various cancer cell lines through a caspase-dependent mechanism as well as truncation of Bax, which promotes pro-death activity [74].

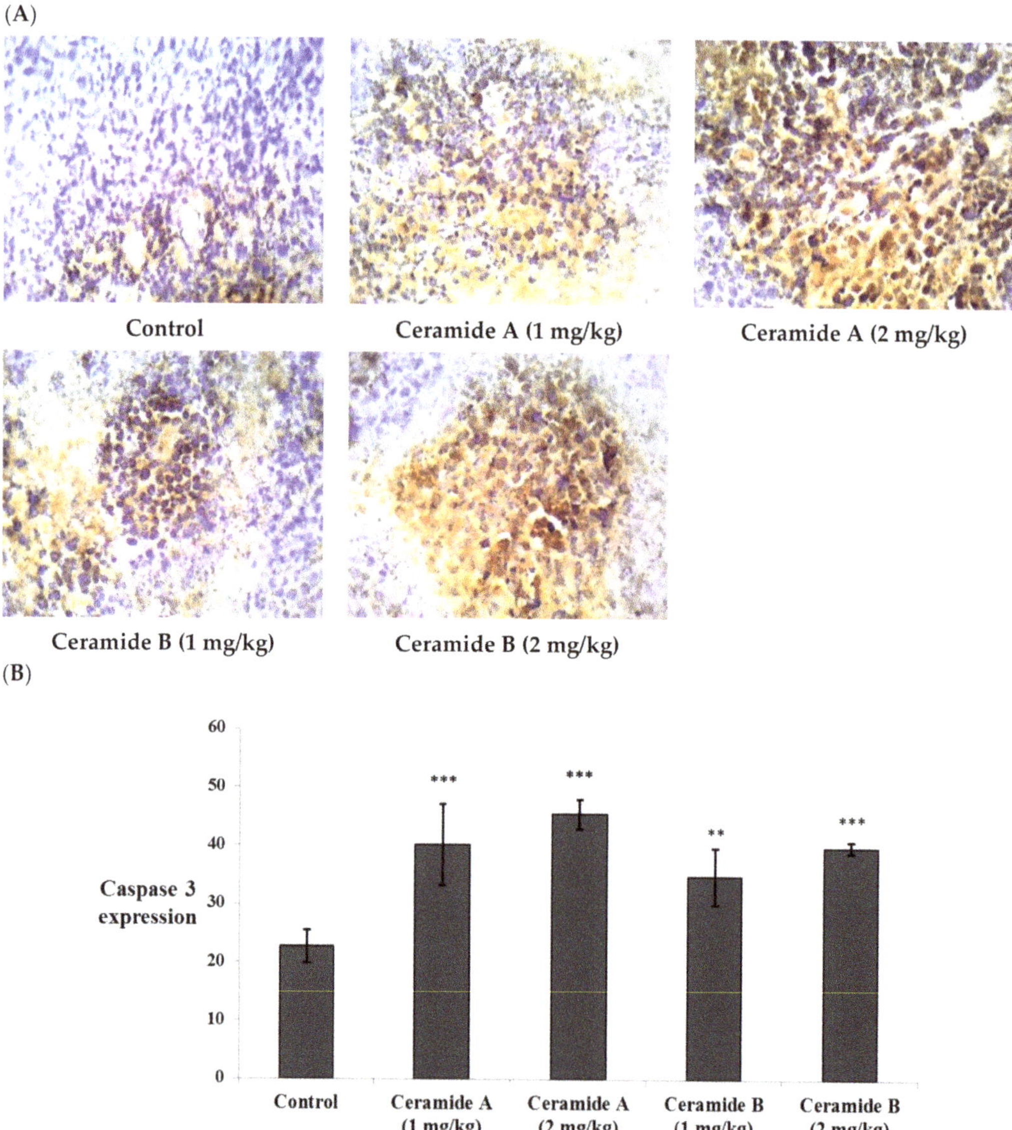

Figure 8. Effect of treatment with ceramide **A** and ceramide **B** at two dose levels (1 mg/kg and 2 mg/kg I.P) on the expression of caspase 3. (**A**) Representative photomicrographs of caspase 3, assessed immunohistochemically on day 21 in EAC-bearing female mice (40× magnification). (**B**) Optical density of positive immunohistochemical reactions (brown) determined using ImageJ. Values are expressed as mean ± SD. All data were analyzed using ANOVA followed by a Bonferroni post hoc test. ** Significantly different compared with the EAC control group at $p < 0.01$. *** Significantly different compared with the EAC control group at $p < 0.001$.

Ceramide is involved in the induction of apoptosis and growth arrest in breast cancer [75]. The mechanism of ceramide-induced apoptosis involves elevated ceramide levels in the mitochondria, resulting in mitochondrial dysfunction, including a loss of cytochrome c [76]. Mitochondrial apoptosis is dependent on the increased mitochondrial outer membrane permeability. This permeability is enhanced by proteins such as Bax [77]. Therefore, channel formation by ceramide is an upstream event to the induction of apoptosis [78]. Just after the passage of ceramides into the mitochondria, many pro-apoptotic proteins are released into the cytoplasm, primarily cytochrome c [79]. Cytochrome c binds to Apaf-1 (apoptotic protease-activating factor-1). As a result, inactive procaspase-9 is cleaved into active caspase-9. Caspase-9 stimulates caspase-3, which is the crucial step for the caspase cascade in intrinsic apoptosis [80]. The start of intrinsic apoptosis is associated with a rise in mitochondrial ceramide levels [81]. Moreover, exogenous ceramide addition to cells induces apoptosis and DNA fragmentation [82]. Ceramide has been considered a crucial performer in the extrinsic and intrinsic pathways of apoptosis. The extrinsic pathway is activated by the death receptors through interaction with their ligands or by inducing receptor clusterization [83]. Acid sphingomyelinase catalyzes the hydrolysis of sphingomyelin into ceramide, consequently generating ceramide-rich stages and the subsequent clusterization of death receptors that enables the formation of a death-inducing signaling complex and caspase activation [84].

2.5. Molecular Modeling

Ceramides have been considered a group of endogenous sphingolipids with the ability to interfere with the signaling pathways both upstream and downstream of p53 [85]. Under normal non-stressed conditions, the p53 transcription factor undergoes ubiquitination by the E3-ligase mouse double minute 2 (MDM2). This process prepares p53 for the proteasomal degradation. Disruption of the complex between p53 and MDM2 leads to p53 accumulation, nuclear translocation, and activation of the downstream apoptotic pathways [86]. The role of ceramides in the regulation of p53 has remained controversial for a long time, with studies suggesting direct and indirect cross-talk. Recently, in 2018, Fekry et al. from the University of North Carolina showed that C_{16}-ceramides could directly interact with p53 [87]. They provided experimental evidence that ceramides bind in close proximity to the BOX V motif of p53, which is a part of the p53–MDM2 interface. Consequently, this binding disrupts the p53-MDM2 binding, reducing the ubiquitination and proteasomal degradation of p53. To gain insights into the molecular determinants of the binding of our ceramides **A** (**1**) and **B** (**2**) into the putative binding site of p53, we decided to perform a molecular docking simulation into the crystal structure of the human p53 (2MEJ). Both molecules showed very similar docked poses to the one proposed by Fekry et al. (Figure 9). In both cases, the central polar part of the compounds was oriented close to the polar residues Arg280 and Asp281 at the base of the H2 helix, probably stabilized by forming water-mediated H bonds. The two long hydrophobic tails of both compounds fit in a complementary fashion into the deep groove between the H2 helix and the Ser240-P250 loop. To our delight, the C10 atom of the acyl chain of both ceramides was docked in close proximity to Ser240 and Ser241. These residues have been proven by the aforementioned research group, using MS-proteomics experiments, to participate in C_{16}-ceramide binding to p53 [87]. Moreover, they used modified C_{16}-ceramide with a diaziridine group to show that the ceramide C10 atom binds proximal to Ser240 and Ser241. Interestingly, a very similar orientation of C10 was noticed in the top-ranked docking poses of both ceramides **A** (**1**) and **B** (**2**).

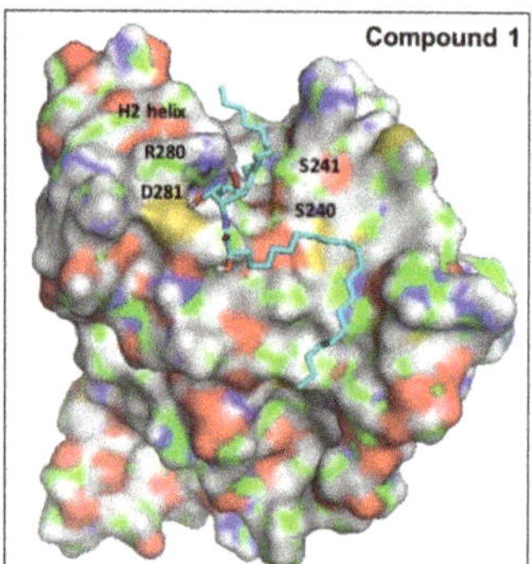

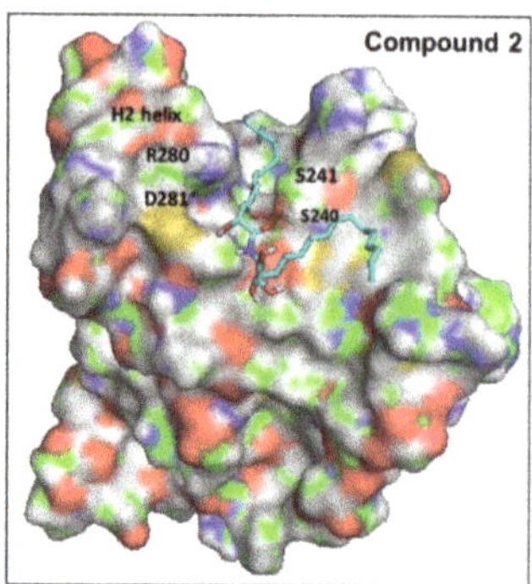

Figure 9. Proposed binding mode of ceramides **1** and **2** into the putative binding site of p53. The surface of the protein is colored by atom type, with carbons in green. Important residues of the binding site are labeled. Docked ligands are shown as cyan sticks.

3. Material and Methods

3.1. Plant Material

The red algae *Hypnea musciformis* was collected from Safaga at the Egyptian Red Sea in August 2017, air-dried, and stored at low temperature (−24 °C) until further processing. The plant was identified by Dr. Tarek Temraz of the Marine Science Department in the Faculty of Science at Suez Canal University in Ismailia, Egypt. A voucher specimen was deposited in the herbarium section of the Pharmacognosy Department of the Faculty of Pharmacy at Suez Canal University in Ismailia, Egypt under registration number SAA-130.

3.2. Metabolic Profiling

Metabolic profiling was performed using the liquid chromatography coupled with high-resolution electron spray ionization mass spectrometry technique (LC-HR-ESI-MS) as previously mentioned [20]. The obtained metabolites were described by comparison with several databases (e.g., DNP and METLIN).

3.3. Extraction and Isolation of Pure Compounds

The red algae *Hypnea musciformis* (430 g fresh weight) was freeze-dried, ground, and soaked in methanol (3 × 2 L) at room temperature. Then, the combined methanol extracts were evaporated under a vacuum, resulting in 32 g of green residue. The total methanol extract was then fractionated by partitioning with different solvents in increasing order of polarity several times to afford an *n*-hexane fraction (3.67 g), a chloroform fraction (2.33 g), an ethyl acetate fraction (7.43 g), an *n*-butanol fraction (3.07 g), and finally an aqueous fraction. The ethyl acetate fraction (Hp-EA, 7.43 g) was chromatographed with SiO_2 column chromatography using hexane:EtOAc:methanol (80:20:0 up to 0:75:25) gradient elution, giving seven subfractions, Hp-EA-(1-7). Subfraction Hp-EA-4 (437 mg) was rechromatographed on silica gel using hexane:EtOAc:methanol (50:50:0 up to 0:90:10) gradient elution, giving five subfractions, Hp-EA-4-(1′-5′). One of the resultant subfractions, Hp-EA-4-2′, was applied to a sephadex LH-20 column eluted with $CHCl_3$-MeOH (1:1), giving 3 subfractions, Hp-EA-4-2′-(1-3). Among them, ceramide **A** (**1**) (13 mg, white powder) was afforded from subfraction Hp-EA-4-2′-2. Ceramide **A** (**1**) was finally purified on an open ODS column using MeOH/H_2O (8:2). Another subfraction, Hp-EA-4-2′-3, was subjected to silica gel column chromatography using EtOAc:methanol (95:5) to afford ceramide **B** (**2**) (15 mg, white powder) and ceramide **C** (**3**) (6 mg, white powder). Regarding the chloroform fraction (Hp-C, 2.33 g), it was chromatographed with SiO_2 column chromatography using hexane:EtOAc (100:0 up to 25:75) gradient elution, giving 3 subfractions, Hp-C-(1-3). Subfraction Hp-C-1 (223 mg) was rechromatographed on silica gel using hexane:EtOAc (90:10 up to 70:30) gradient elution, giving two subfractions, Hp-C-1-(1′-2′). The first one,

Hp-C-1-1′, afforded compound **4**, white powder 10 mg in weight, while the second subfraction (Hp-C-1-2′) was applied to a sephadex LH-20 column eluted with $CHCl_3$-MeOH (1:1), giving compound **5** (12 mg, white powder) and compound **6** (13 mg, white powder).

3.4. *In Vitro Cytotoxic Activity of Ceramides* **A** (**1**), **B** (**2**), *and* **C** (**3**)

3.4.1. In Vitro Cell Culture

The human breast adenocarcinoma (MCF-7) cell line was purchased from the American Type Culture Collection (ATCC HTB-22), Minnesota, USA). The tumor cell lines were maintained at the National Cancer Institute in Cairo, Egypt by serial subculturing. The cells were subcultured on RPMI 1640 medium supplemented by 1% penicillin/streptomycin and 10% fetal bovine serum [20].

3.4.2. Sulforhodamine-B Assay

The cytotoxic activity of three isolated ceramides **A** (**1**), **B** (**2**), and **C** (**3**) was determined by a Sulforhodamine-B (SRB) assay. SRB is able to bind to the intracellular proteins, providing a sensitive index of the cellular protein content. It was assessed as previously described in detail [20]. The experiment was repeated 3 times, and the IC_{50} values (concentration that caused a 50% decrease in cell viability) were calculated. Doxorubicin was used as a positive control with the same concentration range.

3.5. *In Vivo Study*

Two ceramides were chosen for further investigation in vivo: ceramide **A** (**1**) and ceramide **B** (**2**). EAC is a spontaneous murine mammary adenocarcinoma model that has extensively been used as a study model of breast cancer [88]. The effect of these ceramides against EAC and their role in regulating vascular and apoptotic factors were tested.

3.5.1. Tumor Induction

Human breast cancer cell line MCF-7 and EAC cells were purchased from the Tumor Biology Department at the National Cancer Institute of Cairo University. MCF-7 cells were cultured in Dulbecco's Modified Eagle's Medium (DMEM) supplemented with 10% heat-inactivated fetal bovine serum, 1% L-glutamine, HEPES buffer, and 50 µg/mL gentamycin. All cells were maintained at 37 °C in a fully humidified air atmosphere containing 5% CO_2 and were subcultured twice a week. EAC served as the original tumor from which an ascites variant was obtained. EAC cells were suspended in normal saline (2.5×10^6 cells/100 µL). Cells were counted by a hemocytometer under the microscope. The mice were inoculated intradermally with a 100-µL EAC suspension on the lower ventral side after shaving [89].

3.5.2. Study Design

The study involved 48 Swiss albino mice obtained from the Egyptian Organization for Biological Products and Vaccines (Vacsera, Giza, Egypt) weighing 25–30 g. The mice were housed in plastic cages at a 25 °C temperature under a normal light/dark cycle, with water and food provided ad libitum. The mice were left for 1 week before the experiments to adjust. They were randomly separated into 6 groups (8 mice each). The first group was considered the normal group (negative control) and received Tween 80. The tumor cells were injected in all 5 other groups (groups 2–6). Group 2 was considered the EAC control group. Groups 3 and 4 received 1 mg/kg and 2 mg/kg of ceramide **A** (**1**), respectively. Groups 5 and 6 were treated by 1 mg/kg and 2 mg/kg of ceramide **B** (**2**), respectively. Ceramide **A** (**1**) and ceramide **B** (**2**) were dissolved in Tween 80 as a vehicle. The day of tumor cell injection was considered day zero (0). Groups 3–6 were treated daily from day 7 until day 21. The in vivo study was in agreement with the *Guide for the Care and Use of Laboratory Animals*. The study protocol was approved by the ethical committee of the Faculty of Pharmacy at Suez Canal University (201605PHDA1).

3.5.3. Sample Collection

At the 21st day of EAC cell inoculation, the mice were anesthetized with thiopental sodium (50 mg/kg). Blood samples were collected from the orbital sinus (retro-orbital plexus). The blood samples were left to clot for 20 min, followed by separation of the serum by centrifugation at 2000× *g* for 15 min. The mice were sacrificed by cervical dislocation, and the tumor discs were separated, weighed, and separated into two portions; one portion of each tumor disc was fixed in 10% neutral buffered formalin for immunohistochemical investigations, while the other portion was kept at −80 °C for PCR analysis.

3.5.4. Determination of Endothelial Growth Factor B (VEGF-B) and Tumor Necrosis Factor-α (TNF-α) in the Serum by ELISA

The serum samples were stored at −20 °C and used for determination of the levels of VEGF-B and TNF-α by ELISA kits ab213897 and ab181421, respectively (Abcam, Cambridge, UK), according to the manufacturer's instructions. The serum levels of the liver function enzymes ALT and AST were assessed by colorimetric kits AL1031 and AS1061, respectively (Biodiagnostic, Giza, Egypt). Similarly, the serum levels of the kidney markers urea and creatinine were also determined calorimetrically via UR2110 and CR1250, respectively (Biodiagnostic, Giza, Egypt).

3.5.5. Quantitative Real-Time PCR (q RT-PCT) for Assessment of the Expression of Midkine (MDK) in Tumor Tissue

The total RNA was isolated from the tumor tissue by an SV total RNA isolation system (Promega, Madison, WI, USA) according to the manufacturer's instructions. The extracted RNA was stored at −80 °C. The concentration and purity of the isolated RNA were measured by a Nanodrop NA-1000 UV/Vis spectrophotomter (Thermo Fisher Scientific Inc., Wilmington, DE, USA). A GoTaq® 1-Step RT-qPCR System (Promega, Madison, WI, USA) and the two primers, 5′-GTCAATCACGCCTGTCCTCT-3′ (forward) and 5′-CAAGTATCAGGGTGGGGAGA-3′ (reverse), were used for determination of the MDK expression. β-actin was marked as the housekeeping gene and was amplified using two primers: 5′-ACGGCCAGGTCATCACTATTG-3′ (forward) and 5′-CAAGAAGGAAGGCT GGAAAAGA-3′ (reverse). The 20-µL reaction mixture of each sample was composed of 4 µL of the RNA template, 1 µL of each of the forward and reverse primers, 0.4 µL of GoScript™ reverse transcriptase (RT) mix for 1-step RT-qPCR, 10 µL of GoTaq® qPCR master mix, 0.31 µL of supplemental CXR reference dye, and 3.29 µL of nuclease-free water. The reaction was carried out in a StepOnePlus™ Real-Time PCR thermal cycler (Applied Biosystems, Waltham, MA, USA). The program was formed of reverse transcription at 37 °C for 15 min, inactivation of the reverse transcriptase enzyme, and initial denaturation at 10 min at 95 °C, followed by 40 cycles of denaturation at 95 °C for 10 s, annealing at 52 °C for 30 s, and extension at 72 °C for 30 s. The cycle threshold (Ct) of each reaction was recorded, and the ΔCt was calculated against β-actin. The fold change of each sample was calculated to be $2^{-\Delta\Delta Ct}$.

3.5.6. Immunohistochemical Assessment of the Expression of Apoptotic Markers in the Tumor Tissue

The tumor discs were fixed in 10% neutral buffered formalin overnight and then embedded in paraffin. Deparaffinization was performed by adding xylene and ethyl alcohol in decreasing concentrations from 100% to 70%. Antigen retrieval was performed according to the Tris/EDTA buffer (pH = 9) antigen retrieval protocol. The EnVision™ FLEX HRP-labeled high-pH method was used for staining according to the manufacturer's protocol (Dako, Glostrup, Denmark). The primary polyclonal antibodies for p53, Bax, and caspase 3 (Bioss Inc., Woburn, MA, USA) were diluted in PBS (normal phosphate buffered saline) at a ratio of 1:250. Finally, Mayer's hematoxylin was used for counterstaining.

ImageJ was used for the semiquantitative analysis of the immunohistochemical reactions. The images were captured by an optical microscope with a 40× objective (Optika B-352A, OPTICA, Via Rigla, Italy) coupled to a camera (HDCE30C) using its software and quantified using the ImageJ MacBiophotonics (National Institutes of Health, Bethesda, MD, USA) software package developed by McMaster University (Hamilton, Ontario, Canada). The expressions of p53, Bax, and caspase 3 were all assessed, and the percentages of stained areas were calculated using the color deconvolution plugin.

3.5.7. Statistical Analysis

The values of the determined parameters were expressed as mean ± standard deviation (SD). Comparisons were performed by one-way ANOVA followed by a Bonferroni post hoc test for multiple comparisons. Differences at $p < 005$ were considered statistically significant.

3.6. Molecular Modeling Study

Molecular Operating Environment (MOE) was used to dock both ceramides **A** (**1**) and **B** (**2**) into the transcription factor P53 [90]. First, the crystal structures of P53 (2MEJ) were imported into the MOE graphical interface, and the protein was then prepared for docking using the default parameters of the Protein Preparation module and the Protonate 3D tool. Ligands were also sketched using MOE and minimized with the MMFF94 force field to a gradient of 0.001 kcal/mol·Å^2. They were then docked into the putative binding site of the P53 protein using the induced-fit protocol and the default parameters of the MOE Dock module with the Triangle Matcher method. Residues Ser240 and Ser241 were used to specify the binding site as reported in the literature. The default values of 30 docked structures for each ligand were used. Poses were arranged according to the docking scores and inspected visually.

4. Conclusions

From the methanolic extract of the Red Sea red algae *Hypnea musciformis*, two new ceramides and another first reported one were isolated. Moreover, other chemical metabolites were detected by using the LC-ESI-HRMS technique. Both new metabolites, ceramides **A** (**1**) and **B** (**2**), exhibited significant in vitro cytotoxic activity against the human breast adenocarcinoma (MCF-7) cell line. The activity of ceramides **A** (**1**) and **B** (**2**) was investigated in an EAC mouse model, where both ceramides at doses of 1 and 2 mg/kg significantly decreased the tumor size, serum levels of VEGF-B and TNF-α, and expression of the biomarker midkine growth factor in the tumor tissue, with significant upregulation of the apoptotic factors p53, Bax, and caspase-3. Ceramide **B** (**2**), at a therapeutic dose of 2 mg/kg, showed the most potent antiangiogenic activity and the highest expression of the investigated apoptotic factors. Molecular docking suggested the interaction of these ceramides with the p53 transcription factor, leading to its accumulation and activation of the downstream apoptotic pathways. The current study introduced two potentially effective anti-cancer ceramides isolated from the Red Sea algae *Hypnea musciformis*, which exhibited antiangiogenic and apoptotic effects in the experimentally induced mammary tumor. The major limitation of this work is that the effects of the isolated ceramides were not investigated on healthy cell lines, suggesting that further research is required to evaluate the toxicity of those compounds and to determine their therapeutic indices.

Supplementary Materials: The following are available online at https://www.mdpi.com/article/10.3390/md20010063/s1, Figures S1 and S2: LC-ESI-HRMS chromatogram of crude extract of *Hypnea musciformis* (positive and negative modes); Figure S3: ESI-HRMS chromatogram of ceramide **A** (**1**); Figures S4 and S5: ^{1}H NMR and ^{13}C NMR spectra of ceramide **A** (**1**); Figure S6: ESI-HRMS chromatogram of fatty acid methyl ester of ceramide **A** (**1**); Figure S7: ESI-HRMS chromatogram of ceramide **B** (**2**); Figures S8 and S9: ^{1}H NMR and ^{13}C NMR spectra of ceramide **B** (**2**); Figure S10: ESI-HRMS chromatogram of fatty acid methyl ester of ceramide **B** (**2**); Figure S11: GC-MS analysis of fatty acid methyl ester of ceramide **B** (**2**); Figure S12: ESI-HRMS chromatogram of ceramide **C** (**3**); Figures S13 and S14: ^{1}H NMR and ^{13}C NMR spectra of ceramide **C** (**3**); Figure S15: ESI-HRMS chromatogram of fatty acid methyl ester of ceramide **C** (**3**); Figure S16: ESI-HRMS chromatogram of compound **4**; Figures S17 and S18: ^{1}H NMR and ^{13}C NMR spectra of compound **4**; Figure S19: ESI-HRMS chromatogram of compound **5**; Figures S20 and S21: ^{1}H NMR and ^{13}C NMR spectra of compound **5**; Figure S22: ESI-HRMS chromatogram of compound **6**; Figure S23: ^{1}H NMR spectra of compound **6**; Table S1: Liver enzymes and kidney markers in the study groups.

Author Contributions: Conceptualization, R.F.A.A., S.A.A., S.S.E., E.E.E. and M.S.G.; methodology, R.F.A.A., E.E.E., R.M.H., E.T.M., M.S.G., E.S.H. and S.S.E.; software, M.S.G. and M.A.H.; validation, K.M.H., E.E.E., A.K.I. and M.S.G.; formal analysis, K.M.H., E.E.E., R.M.H., E.T.M., S.S.E., M.S.G. and A.K.I.; investigation, R.F.A.A., S.S.E., A.K.I., E.E.E., E.S.H., M.S.G., R.M.H., E.T.M. and U.R.A.; resources, K.M.H., S.S.E., R.M.D. and S.A.A.; funding acquisition, R.M.D., S.S.E., K.M.H. and S.A.A.; data curation, M.S.G., E.E.E., A.K.I., H.A.H., S.S.E. and M.A.H.; writing—original draft preparation, R.F.A.A., E.S.H., S.S.E., E.E.E., R.M.H., E.T.M. and M.S.G.; writing—review and editing, R.F.A.A., S.A.A., E.E.E., E.S.H., U.R.A. and S.S.E.; supervision, S.A.A., R.F.A.A., E.E.E., H.A.H., E.S.H. and A.K.I. All authors have read and agreed to the published version of the manuscript.

Funding: The Deanship of Scientific Research (DSR) at King Abdulaziz University in Jeddah, Saudi Arabia funded this project under grant no. D-029-166-1440.

Institutional Review Board Statement: The study protocol was approved by the ethical committee of the Faculty of Pharmacy at Suez Canal University (201605PHDA1).

Informed Consent Statement: Not applicable.

Data Availability Statement: Data are contained within the article and the Supplementary Materials.

Acknowledgments: This work was supported by the Deanship of Scientific Research (DSR) at King Abdulaziz University in Jeddah, Saudi Arabia under grant no. D-029-166-1440. The authors, therefore, gratefully acknowledge the DSR for technical and financial support. The authors are grateful to Tarek A. Temraz of the Marine Science Department in the Faculty of Science at Suez Canal University for his taxonomical identification of the Red Sea red algae *Hypnea musciformis*.

Conflicts of Interest: The authors declare no conflict of interest.

References

1. El Gamal, A.A. A review article: Biological importance of marine algae. *Saudi Pharm. J.* **2010**, *18*, 1–25. [CrossRef]
2. Kasanah, N.; Triyanto, T.; SarwoSeto, D.; Amelia, W.; Isnansetyo, A. Antibacterial compounds from red seaweeds (*Rhodophyta*). *Indones. J. Chem.* **2015**, *15*, 201–209. [CrossRef]
3. Dhanalakshmi, S.; Jayakumari, S. A Review on The Pharmacognostical, Ecology and Pharmacological Studies on Marine Red Algae—*Hypnea valentiae*. *Int. J. Pharm. Sci. Res.* **2018**, *10*, 1065–1071. [CrossRef]
4. Jassbia, A.R.; Mohabatia, M.; Eslamia, S.; Sohrabipourb, J.; Miria, R. Biological Activity and Chemical Constituents of Red and Brown Algae from the Persian Gulf. *Iran. J. Pharm. Res.* **2013**, *12*, 339–348.
5. Chakraborty, K.; Joseph, D.; Praveen, N.K. Antioxidant activities and phenolic contents of three red seaweeds (Division: *Rhodophyta*) harvested from the Gulf of Mannar of Peninsular India. *J. Food Sci. Technol.* **2015**, *52*, 19–24. [CrossRef]
6. Suganthy, N.; Karutha Pandian, S.; Pandima Devi, K. Neuroprotective effect of seaweeds inhabiting south Indian coastal area (Hare island, Gulf of Mannar Marine Biosphere Reserve): Cholinesterase inhibitory effect of *Hypnea valentiae* and *Ulva reticulata*. *Neurosci. Lett.* **2010**, *468*, 216–219. [CrossRef]
7. Bitencourt, F.S.; Figueiredo, J.G.; Mota, M.R.L.; Bezerra, C.C.R.; Silvestre, P.P.; Vale, M.R.; Nascimento, K.S.; Sampaio, A.H.; Nagano, C.S.; Saker-Sampaio, S.; et al. Antinociceptive and anti-inflammatory effects of a mucin-binding agglutinin isolated from the red marine alga *Hypnea cervicornis*. *Naunyn. Schmiedebergs. Arch. Pharmacol.* **2008**, *377*, 139–148. [CrossRef]
8. Selim, S.; Amin, A.; Hassan, S.; Hagazey, M. Antibacterial, cytotoxicity and anticoagulant activities from *Hypnea esperi* and *Caulerpa prolifera* marine algae. *Pak. J. Pharm. Sci.* **2015**, *28*, 525–530.
9. Hayee-Memon, A.; Shameel, M.; Usmanghani, K.; Ahmad, V.U. Fatty acid composition of three species of *Hypnea (Gigartinales, Rhodophyta)* from Karachi coast. *Pak. J. Mar. Sci.* **1992**, *1*, 7–10.
10. Moses, B.J.; Trivedi, G.K.; Mathur, H.H. Diketosteroid from marine red alga *Hypnea musciformis*. *Phytochemistry* **1989**, *28*, 3237–3239. [CrossRef]
11. Moses, B.J.; Sridharan, R.; Mathur, H.H.; Trivedi, G.K. A dihydroxyketosteroid from the marine red alga *Hypnea musciformis*. *Phytochemistry* **1990**, *29*, 3965–3967. [CrossRef]
12. Xu, X.; Xiao, C.; Jian-Hua, L.; Guang-Min, Y.; Yan-Ming, L.; Long-Mei, Z. A novel Lanostanoid lactone from the Alga *Hypnea Cemcornis*. *Chin. J. Chem.* **2001**, *19*, 702–704. [CrossRef]
13. Nasir, M.; Saeidnial, S.; Mashinchian-Moradi, A.; Gohari, A.R. Sterols from the red algae, *Gracilaria salicornia* and *Hypnea flagelliformis*, from Persian Gulf. *Pharm. Mag.* **2011**, *7*, 97–100. [CrossRef]
14. Ragasa, C.Y.; Ebajo, V.D., Jr.; Lazaro-Llanos, N.; Brkljača, R.; Urban, S. Chemical constituents of *Hypnea nidulans* Setchell. *Der. Pharm. Chem.* **2015**, *7*, 473–478.

15. Chakraborty, K.; Joseeph, D.; Joy, M.; Raola, V.K. Characterization of substituted aryl meroterpenoids from red seaweed *Hypnea musciformis* as potential antioxidants. *Food Chem.* **2016**, *212*, 778–788. [CrossRef]
16. Ogretmen, B.; Hannun, Y. Biologically active sphingolipids in cancer pathogenesis and treatment. *Nat. Rev. Cancer* **2004**, *4*, 604–616. [CrossRef]
17. Ogretmen, B. Sphingolipid metabolism in cancer signalling and therapy. *Nat. Rev. Cancer* **2018**, *18*, 33–50. [CrossRef]
18. Truman, J.P.; Garcia-Barros, M.; Obeid, L.M.; Hannun, Y.A. Evolving concepts in cancer therapy through targeting sphingolipid metabolism. *Biochim. Biophys. Acta* **2014**, *184*, 1174–1188. [CrossRef]
19. Struckhoff, A.P.; Bittman, R.; Burow, M.E.; Clejan, S.; Elliott, S.; Hammond, T.; Tang, Y.; Beckman, B.S. Novel ceramide analogs as potential chemotherapeutic agents in breast cancer. *J. Pharm. Exp. Ther.* **2004**, *309*, 523–532. [CrossRef]
20. Abdelhameed, R.F.A.; Habib, E.S.; Goda, M.S.; Fahim, J.R.; Hassanean, H.A.; Eltamany, E.E.; Ibrahim, A.K.; AboulMagd, A.M.; Fayez, S.; El-kader, A.M.A.; et al. Thalassosterol, a New Cytotoxic Aromatase Inhibitor Ergosterol Derivative from the Red Sea Seagrass *Thalassodendron ciliatum*. *Mar. Drugs* **2020**, *18*, 354. [CrossRef]
21. Lopez, A.; Gerwick, W.H. Ptilodene, a novel icosanoid inhibitor of 5-lipoxygenase and Na^+/K^+ ATPase from the red marine algae *Ptilota filicina* J. Agardh. *Tetrahedron Lett.* **1988**, *29*, 1505–1506. [CrossRef]
22. Graber, M.A.; Gerwick, W.H.; Cheney, D.P. The isolation and characterization of agardhilactone, a novel oxylipin from the marine red alga *Agardhiella subulate*. *Tetrahedron Lett.* **1996**, *37*, 4635–4638. [CrossRef]
23. Joshi, P.; Misra, L.; Siddique, A.A.; Srivastava, M.; Kumar, S.; Darokar, M.P. Epoxide group relationship with cytotoxicity in withanolide derivatives from *Somnifera*. *Steroids* **2014**, *79*, 19–27. [CrossRef]
24. Dang, H.T.; Lee, H.J.; Yoo, E.S.; Hong, J.; Choi, J.S.; Jung, J.H. The occurrence of 15-keto-prostaglandins in the red alga *Gracilaria verrucosa*. *Arch. Pharm. Res.* **2010**, *33*, 1325–1329. [CrossRef]
25. Howell, A.R.; Ndakala, A.J. The Preparation and Biological Significance of Phytosphingosine. *Curr. Org. Chem.* **2002**, *6*, 365–391. [CrossRef]
26. Fischer, C.L. Antimicrobial Activity of Host-Derived Lipids. *Antibiotics* **2020**, *9*, 75. [CrossRef]
27. Qu, Y.; Zhangb, H.; Liua, J. Isolation and Structure of a New Ceramide from the Basidiomycete *Hygrophorus eburnesus*. *Z. Naturforsch.* **2004**, *59*, 241–244. [CrossRef]
28. Yaoita, Y.; Kohata, R.; Kakuda, R.; Machida, K.; Kikuchi, M. Ceramide Constituents from Five Mushrooms. *Chem. Pharm. Bull.* **2002**, *50*, 681–684. [CrossRef]
29. Zhang, Y.; Wang, S.; Li, X.; Cui, C.; Feng, C.; Wang, B. New Sphingolipids with a previously unreported 9-methyl-C20- sphingosine moiety from a marine algous endophytic fungus *Aspergillus niger* EN-13. *Lipids* **2007**, *42*, 759–764. [CrossRef]
30. Gao, J.-M.; Yang, X.; Wang, C.-Y.; Liu, J.-K. Armillaramide, a new sphingolipid from the fungus *Armillaria Mellea*. *Fitoterapia* **2001**, *72*, 858–864. [CrossRef]
31. Dahm, F.; Bielawska, A.; Nocito, A.; Georgiev, P.; Szulc, Z.M.; Bielawski, J.; Jochum, W.; Dindo, D.; Hannun, Y.A.; Clavien, P.-A. Mitochondrially targeted ceramide LCL-30 inhibits colorectal cancer in mice. *Br. J. Cancer* **2008**, *98*, 98–105. [CrossRef]
32. Elkhayat, E.S.; Mohamed, G.A.; Ibrahim, S.R.M. Activity and Structure Elucidation of Ceramides. *Curr. Bioact. Compd.* **2012**, *8*, 370–409. [CrossRef]
33. Eltamany, E.E.; Ibrahim, A.K.; Radwan, M.M.; ElSohly, M.A.; Hassanean, H.A.; Ahmed, S.A. Cytotoxic ceramides from the Red Sea sponge *Spheciospongia vagabunda*. *Med. Chem. Res.* **2015**, *24*, 3467–3473. [CrossRef]
34. Abdelhameed, R.F.; Ibrahim, A.K.; Temraz, T.A.; Yamada, K.; Ahmed, S.A. Chemical and Biological Investigation of the Red Sea Sponge *Echinoclathria* species. *J. Pharm. Sci. Res.* **2017**, *9*, 1324–1328.
35. Che, J.; Huang, Y.; Xu, C.; Zhang, P. Increased ceramide production sensitizes breast cancer cell response to chemotherapy. *Cancer Chemother Pharm.* **2017**, *79*, 933–941. [CrossRef]
36. Pacheco, B.S.; Ziemann dos Santos, M.A.; Schultze, E.; Martins, R.M.; Lund, R.G.; Seixas, F.K.; Colepicolo, P.; Collares, T.; Paula, F.R.; Pereira De Pereira, C.M. Cytotoxic activity of fatty acids from antarctic macroalgae on the growth of human breast cancer cells. *Front. Bioeng. Biotechnol.* **2018**, *6*, 185. [CrossRef]
37. Guardiola-Serrano, F.; Beteta-Göbel, R.; Rodríguez-Lorca, R.; Ibarguren, M.; López, D.J.; Terés, S.; Alonso-Sande, M.; Higuera, M.; Torres, M.; Busquets, X.; et al. The triacylglycerol, hydroxytriolein, inhibits triple negative mammary breast cancer cell proliferation through a mechanism dependent on dihydroceramide and Akt. *Oncotarget* **2019**, *10*, 2486–2507. [CrossRef]
38. Zajkowska, M.; Lubowicka, E.; Malinowski, P.; Szmitkowski, M.; Ławicki, S. Plasma Levels of VEGF-A, VEGF B, and VEGFR-1 and Applicability of These Parameters as Tumor Markers in Diagnosis of Breast Cancer. *Acta Biochim. Pol.* **2018**, *65*, 621–628. [CrossRef]
39. Yang, X.; Zhang, Y.; Hosaka, K.; Andersson, P.; Wang, J.; Tholander, F.; Cao, Z.; Morikawa, H.; Tegnér, J.; Yang, Y.; et al. VEGF-B promotes cancer metastasis through a VEGF-A-independent mechanism and serves as a marker of poor prognosis for cancer patients. *Proc. Natl. Acad. Sci. USA* **2015**, *112*, 2900–2909. [CrossRef]
40. Lautenschlaeger, T.; George, A.; Klimowicz, A.C.; Efstathiou, J.A.; Wu, C.L.; Sandler, H.; Shipley, W.U.; Tester, W.J.; Hagan, M.P.; Magliocco, A.M.; et al. Bladder preservation therapy for muscle-invading bladder cancers on Radiation Therapy Oncology Group trials 8802, 8903, 9506, and 9706: Vascular endothelial growth factor B overexpression predicts for increased distant metastasis and shorter survival. *Oncologist* **2013**, *18*, 685–686. [CrossRef]

41. Kanda, M.; Nomoto, S.; Nishikawa, Y.; Sugimoto, H.; Kanazumi, N.; Takeda, S.; Nakao, A. Correlations of the expression of vascular endothelial growth factor B and its isoforms in hepatocellular carcinoma with clinico-pathological parameters. *J. Surg. Oncol.* **2008**, *98*, 190–196. [CrossRef]
42. Zhu, M.; Zhang, Q.; Wang, X.; Kang, L.; Yang, Y.; Liu, Y.; Yang, L.; Li, J.; Yang, L.; Liu, J.; et al. Metformin potentiates anti-tumor effect of resveratrol on pancreatic cancer by down-regulation of VEGF-B signaling pathway. *Oncotarget* **2016**, *7*, 84190–84200. [CrossRef]
43. Koch, S.; Claesson-Welsh, L. Signal transduction by vascular endothelial growth factor receptors. *Cold Spring Harb. Perspect. Med.* **2012**, *2*, 6502. [CrossRef]
44. Lal, N.; Puri, K.; Rodrigues, B. Vascular Endothelial Growth Factor B and Its Signaling. *Front. Cardiovasc. Med.* **2018**, *5*, 39. [CrossRef] [PubMed]
45. Wang, X.; Lin, Y. Tumor necrosis factor and cancer, buddies or foes? *Acta Pharm. Sin.* **2008**, *29*, 1275–1288. [CrossRef]
46. Montfort, A.; Colacios, C.; Levade, T.; Andrieu-Abadie, N.; Meyer, N.; Ségui, B. The TNF Paradox in Cancer Progression and Immunotherapy. *Front. Immunol.* **2019**, *10*, 1818. [CrossRef] [PubMed]
47. Cruceriu, D.; Baldasici, O.; Balacescu, O.; Berindan-Neagoe, I. The dual role of tumor necrosis factor-alpha (TNF-α) in breast cancer: Molecular insights and therapeutic approaches. *Cell Oncol.* **2020**, *43*, 1–18. [CrossRef] [PubMed]
48. Sheen-Chen, S.M.; Chen, W.J.; Eng, H.L.; Chou, F.F. Serum concentration of tumor necrosis factor in patients with breast cancer. *Breast Cancer Res. Treat.* **1997**, *43*, 211–215. [CrossRef]
49. Hamed, E.A.; Zakhary, M.M.; Maximous, D.W. Apoptosis, angiogenesis, inflammation, and oxidative stress: Basic interactions in patients with early and metastatic breast cancer. *J. Cancer Res. Clin. Oncol.* **2012**, *138*, 999–1009. [CrossRef]
50. Dai, J.G.; Wu, Y.F.; Li, M. Changes of serum tumor markers, immunoglobulins, TNF-α and hsCRP levels in patients with breast cancer and its clinical significance. *J. Hainan Med. Univ.* **2017**, *23*, 89–92. [CrossRef]
51. Ma, Y.; Ren, Y.; Dai, Z.J.; Wu, C.J.; Ji, Y.H.; Xu, J. IL-6, IL-8 and TNF-alpha levels correlate with disease stage in breast cancer patients. *Adv. Clin. Exp. Med.* **2017**, *26*, 421–426. [CrossRef] [PubMed]
52. Yamamoto, M.; Taguchi, Y.; Ito-Kureha, T.; Semba, K.; Yamaguchi, N.; Inoue, J. NF-kappaB non-cell-autonomously regulates cancer stem cell populations in the basal-like breast cancer subtype. *Nat. Commun.* **2013**, *4*, 2299. [CrossRef] [PubMed]
53. Mosalam, E.M.; Zidan, A.A.; Mehanna, E.T.; Mesbah, N.M.; Abo-Elmatty, D.M. Thymoquinone and Pentoxifylline Enhance the Chemotherapeutic Effect of Cisplatin by Targeting Notch Signaling Pathway in Mice. *Life Sci.* **2020**, *244*, 117299. [CrossRef] [PubMed]
54. Filippou, P.S.; Karagiannis, G.S.; Constantinidou, A. Midkine (MDK) growth factor: A key player in cancer progression and a promising therapeutic target. *Oncogene* **2020**, *39*, 2040–2054. [CrossRef] [PubMed]
55. You, Z.; Dong, Y.; Kong, X.; Beckett, L.A.; Gandour-Edwards, R.; Melamed, J. Midkine is a NF-kappaB-inducible gene that supports prostate cancer cell survival. *BMC Med. Genom.* **2008**, *1*, 6. [CrossRef] [PubMed]
56. Zhang, L.; Song, X.; Shao, Y.; Wu, C.; Jiang, J. Prognostic value of Midkine expression in patients with solid tumors: A systematic review and meta-analysis. *Oncotarget* **2018**, *9*, 24821–24829. [CrossRef]
57. Ibusuki, M.; Fujimori, H.; Yamamoto, Y.; Ota, K.; Ueda, M.; Shinriki, S.; Taketomi, M.; Sakuma, S.; Shinohara, M.; Iwase, H.; et al. Midkine in plasma as a novel breast cancer marker. *Cancer Sci.* **2009**, *100*, 1735–1739. [CrossRef]
58. Hadida, W.M.; Ali, D.A.; Elnemr, A.A.; Nofal, H.E. The Role of Midkine in Breast Cancer. *Egypt. J. Hosp. Med.* **2019**, *77*, 5089–5095. [CrossRef]
59. Muramaki, M.; Miyake, H.; Hara, I.; Kamidono, S. Introduction of midkine gene into human bladder cancer cells enhances their malignant phenotype but increases their sensitivity to antiangiogenic therapy. *Clin. Cancer Res.* **2003**, *9*, 5152–5160. [PubMed]
60. Lautz, T.; Lasch, M.; Borgolte, J.; Troidl, K.; Pagel, J.I.; Caballero-Martinez, A.; Kleinert, E.C.; Walzog, B.; Deindl, E. Midkine controls arteriogenesis by regulating the bioavailability of vascular endothelial growth factor A and the expression of nitric oxide synthase 1 and 3. *EBioMedicine* **2018**, *27*, 237–246. [CrossRef] [PubMed]
61. Reynolds, P.; Mucenski, M.; Le Cras, T.; Nichols, W.; Whitsett, J. Midkine is regulated by hypoxia and causes pulmonary vascular remodeling. *J. Biol. Chem.* **2004**, *279*, 37124–37132. [CrossRef]
62. Jono, H.; Ando, Y. Midkine: A novel prognostic biomarker for cancer. *Cancers* **2010**, *2*, 624–641. [CrossRef]
63. Muramatsu, T.; Kadomatsu, K. Midkine: An emerging target of drug development for treatment of multiple diseases. *Br. J. Pharmacol.* **2014**, *171*, 811–813. [CrossRef]
64. Kishida, S.; Kadomatsu, K. Involvement of midkine in neuroblastoma tumourigenesis. *Br. J. Pharmacol.* **2014**, *171*, 896–904. [CrossRef]
65. Kwon, Y.B.; Kim, C.D.; Kim, B.J.; Kim, M.; Park, C.S.; Yoon, T.; Seo, Y.; Suhr, K.; Park, J.; Lee, J. Anti-angiogenic effect of tetraacetyl-phytosphingosine. *Exp. Dermatol.* **2007**, *16*, 311–317. [CrossRef]
66. Jeffries, K.A.; Krupenko, N.I. Ceramide Signaling and p53 Pathways. *Adv. Cancer Res.* **2018**, *140*, 191–215. [CrossRef]
67. Lane, D.P. Cancer. p53, guardian of the genome. *Nature* **1992**, *358*, 15–16. [CrossRef]
68. Valente, L.J.; Gray, D.H.; Michalak, E.M.; Pinon-Hofbauer, J.; Egle, A.; Scott, C.L.; Janic, A.; Strasser, A. p53 efficiently suppresses tumor development in the complete absence of its cell-cycle inhibitory and proapoptotic effectors p21, Puma, and Noxa. *Cell Rep.* **2013**, *3*, 1339–1345. [CrossRef] [PubMed]
69. Carroll, B.; Donaldson, J.C.; Obeid, L. Sphingolipids in the DNA damage response. *Adv. Biol. Regul.* **2015**, *58*, 38–52. [CrossRef] [PubMed]

70. Hoeferlin, L.A.; Fekry, B.; Ogretmen, B.; Krupenko, S.A.; Krupenko, N.I. Folate stress induces apoptosis via p53-dependent de novo ceramide synthesis and up-regulation of ceramide synthase 6. *J. Biol. Chem.* **2013**, *288*, 12880–12890. [CrossRef] [PubMed]
71. Chen, M.B.; Jiang, Q.; Liu, Y.Y.; Zhang, Y.; He, B.S.; Wei, M.X.; Lu, J.W.; Ji, W.; Lu, P.H. C6 ceramide dramatically increases vincristine sensitivity both in vivo and in vitro, involving AMP-activated protein kinase-p53 signaling. *Carcinogenesis* **2015**, *36*, 1061–1070. [CrossRef] [PubMed]
72. Kim, S.S.; Chae, H.S.; Bach, J.H.; Lee, M.W.; Kim, K.Y.; Lee, W.B.; Jung, Y.M.; Bonventre, J.V.; Suh, Y.H. P53 mediates ceramide-induced apoptosis in SKN-SH cells. *Oncogene* **2002**, *21*, 2020–2028. [CrossRef] [PubMed]
73. Park, M.; Kang, J.A.; Choi, J.; Kang, C.; Kim, T.; Bae, S.; Kang, S.; Kim, S.; Choi, W.; Cho, C.; et al. Phytosphingosine Induces Apoptotic Cell Death via Caspase 8 Activation and Bax Translocation in Human Cancer Cells. *Clin. Cancer Res.* **2003**, *9*, 878–885. [PubMed]
74. Cuvillier, O. Sphingosine in apoptosis signaling. *Biochim. Biophys. Acta* **2002**, *1585*, 153–162. [CrossRef]
75. Ponnapakam, A.P.; Liu, J.; Bhinge, K.N.; Drew, B.A.; Wang, T.L.; Antoon, J.W.; Nguyen, T.T.; Dupart, P.S.; Wang, Y.; Zhao, M.; et al. 3-Ketone-4,6-diene ceramide analogs exclusively induce apoptosis in chemo-resistant cancer cells. *Bioorg. Med. Chem.* **2014**, *22*, 1412–1420. [CrossRef]
76. Dany, M.; Ogretmen, B. Ceramide induced mitophagy and tumor suppression. *Biochim. Biophys. Acta* **2015**, *1853*, 2834–2845. [CrossRef]
77. Green, D.R.; Kroemer, G. The pathophysiology of mitochondrial cell death. *Science* **2004**, *305*, 626–629. [CrossRef]
78. Colombini, M. Ceramide channels and their role in mitochondria-mediated apoptosis. *Biochim. Biophys. Acta* **2010**, *1797*, 1239–1244. [CrossRef]
79. Saelens, X.; Festjens, N.; Vande Walle, L.; Gurp, M.; van Loo, G.; Vandenabeele, P. Toxic proteins released from mitochondria in cell death. *Oncogene* **2004**, *23*, 2861–2874. [CrossRef]
80. Yang, J.; Liu, X.; Bhalla, K.; Kim, C.N.; Ibrado, A.M.; Cai, J.; Peng, T.I.; Jones, D.P.; Wang, X. Prevention of apoptosis by Bcl-2: Release of cytochrome c from mitochondria blocked. *Science* **1997**, *275*, 1129–1132. [CrossRef]
81. Dai, Q.; Liu, J.; Chen, J.; Durrant, D.; McIntyre, T.M.; Lee, R.M. Mitochondrial ceramide increases in UV-irradiated HeLa cells and is mainly derived from hydrolysis of sphingomyelin. *Oncogene* **2004**, *23*, 3650–3658. [CrossRef] [PubMed]
82. Allouche, M.; Bettaieb, A.; Vindis, C.; Rousse, A.; Grignon, C.; Laurent, G. Influence of Bcl-2 overexpression on the ceramide pathway in daunorubicin-induced apoptosis of leukemic cells. *Oncogene* **1997**, *14*, 1837–1845. [CrossRef]
83. Tchikov, V.; Bertsch, U.; Fritsch, J.; Edelmann, B.; Schutze, S. Subcellular compartmentalization of TNF receptor-1 and CD95 signaling pathways. *Eur. J. Cell Biol.* **2011**, *90*, 467–475. [CrossRef]
84. Grassme, H.; Cremesti, A.; Kolesnick, R.; Gulbins, E. Ceramide-mediated clustering is required for CD95-DISC formation. *Oncogene* **2003**, *22*, 5457–5470. [CrossRef] [PubMed]
85. Heffernan-Stroud, L.A.; Obeid, L.M. p53 and regulation of bioactive sphingolipids. *Adv. Enzym. Regul.* **2011**, *51*, 219–228. [CrossRef]
86. Lavin, M.F.; Gueven, N. The complexity of p53 stabilization and activation. *Cell Death Differ.* **2006**, *13*, 941–950. [CrossRef] [PubMed]
87. Fekry, B.; Jeffries, K.A.; Esmaeilniakooshkghazi, A.; Szulc, Z.M.; Knagge, K.J.; Kirchner, D.R.; Horita, D.A.; Krupenko, S.A.; Krupenko, N.I. C16-ceramide is a natural regulatory ligand of p53 in cellular stress response. *Nat. Commun.* **2018**, *9*, 4149. [CrossRef]
88. Mishra, S.; Tamta, A.K.; Sarikhani, M.; Desingu, P.A.; Kizkekra, S.M.; Pandit, A.S.; Kumar, S.; Khan, D.; Raghavan, S.C.; Sundaresan, N.R. Subcutaneous Ehrlich Ascites Carcinoma mice model for studying cancer-induced cardiomyopathy. *Sci. Rep.* **2018**, *8*, 5599. [CrossRef]
89. Lazarus, A.A. Behaviour rehearsal vs. non-directive therapy vs. advice in effecting behaviour change. *Behav. Res. Ther.* **1966**, *4*, 209–212. [CrossRef]
90. Saleh, A.H.; Abdelwaly, A.; Darwish, K.M.; Eissa, A.A.H.M.; Chittiboyina, A.; Helal, M.A. Deciphering the molecular basis of the kappa opioid receptor selectivity: A Molecular Dynamics study. *J. Mol. Graph. Model.* **2021**, *106*, 07940. [CrossRef]

MDPI

Article

Comparative Study of *Sargassum fusiforme* Polysaccharides in Regulating Cecal and Fecal Microbiota of High-Fat Diet-Fed Mice

Bin Wei [1,2,†], Qiao-Li Xu [2,†], Bo Zhang [2], Tao-Shun Zhou [2], Song-Ze Ke [2], Si-Jia Wang [2,3], Bin Wu [4], Xue-Wei Xu [1,*] and Hong Wang [2,*]

1 Key Laboratory of Marine Ecosystem and Biogeochemistry, State Oceanic Administration & Second Institute of Oceanography, Ministry of Natural Resources, Hangzhou 310012, China; binwei@zjut.edu.cn
2 College of Pharmaceutical Science & Collaborative Innovation Center of Yangtze River Delta Region Green Pharmaceuticals, Zhejiang University of Technology, Hangzhou 310014, China; liujiahui0829@163.com (Q.-L.X.); zhangbo_1632021@163.com (B.Z.); zhoutaoshun@yeah.net (T.-S.Z.); kesongzeke@163.com (S.-Z.K.); new8090@hotmail.com (S.-J.W.)
3 Center for Human Nutrition, David Geffen School of Medicine, University of California, Rehabilitation Building 32-21, 1000 Veteran Avenue, Los Angeles, CA 90024, USA
4 Zhoushan Campus, Ocean College, Zhejiang University, Zhoushan 316021, China; wubin@zju.edu.cn
* Correspondence: xuxw@sio.org.cn (X.-W.X.); hongw@zjut.edu.cn (H.W.); Tel.: +86-571-8832-0622 (H.W.)
† These authors contributed equally to this work.

Abstract: Seaweed polysaccharides represent a kind of novel gut microbiota regulator. The advantages and disadvantages of using cecal and fecal microbiota to represent gut microbiota have been discussed, but the regulatory effects of seaweed polysaccharides on cecal and fecal microbiota, which would benefit the study of seaweed polysaccharide-based gut microbiota regulator, have not been compared. Here, the effects of two *Sargassum fusiforme* polysaccharides prepared by water extraction (SfW) and acid extraction (SfA) on the cecal and fecal microbiota of high-fat diet (HFD) fed mice were investigated by 16S rRNA gene sequencing. The results indicated that 16 weeks of HFD dramatically impaired the homeostasis of both the cecal and fecal microbiota, including the dominant phyla Bacteroidetes and Actinobacteria, and genera *Coriobacteriaceae*, *S24-7*, and *Ruminococcus*, but did not affect the relative abundance of Firmicutes, *Clostridiales*, *Oscillospira*, and *Ruminococcaceae* in cecal microbiota and the Simpson's index of fecal microbiota. Co-treatments with SfW and SfA exacerbated body weight gain and partially reversed HFD-induced alterations of *Clostridiales* and *Ruminococcaceae*. Moreover, the administration of SfW and SfA also altered the abundance of genes encoding monosaccharide-transporting ATPase, α-galactosidase, β-fructofuranosidase, and β-glucosidase with the latter showing more significant potency. Our findings revealed the difference of cecal and fecal microbiota in HFD-fed mice and demonstrated that SfW and SfA could more significantly regulate the cecal microbiota and lay important foundations for the study of seaweed polysaccharide-based gut microbiota regulators.

Keywords: *Sargassum fusiforme* polysaccharides; high-fat diet; cecal microbiota; fecal microbiota; 16S rRNA gene sequencing

Citation: Wei, B.; Xu, Q.-L.; Zhang, B.; Zhou, T.-S.; Ke, S.-Z.; Wang, S.-J.; Wu, B.; Xu, X.-W.; Wang, H. Comparative Study of *Sargassum fusiforme* Polysaccharides in Regulating Cecal and Fecal Microbiota of High-Fat Diet-Fed Mice. *Mar. Drugs* **2021**, *19*, 364. https://doi.org/10.3390/md19070364

Academic Editor: Marc Diederich

Received: 25 May 2021
Accepted: 21 June 2021
Published: 24 June 2021

1. Introduction

Gut microbiota is a population of microorganisms that colonizes the intestines. This not only protects against pathogens, provides nutrients, and maintains the integrity of the mucosal barrier, but also plays an important role in numerous diseases, such as inflammatory bowel disease, obesity, diabetes mellitus, metabolic syndrome, atherosclerosis, non-alcoholic fatty liver disease, etc. [1–4]. Owing to sampling difficulties, most studies chose to characterize the gut microbiota composition by sequencing the fecal samples, other than the cecal contents [5,6]. The differences of cecal microbiota and fecal microbiota in mice and human volunteers have been compared by several research groups [7–9]. Guo et al. reported that the relative abundance of an unidentified genus from the S24-7 family (*S24-7*) in cecal microbiota was much higher than that in fecal microbiota, and oral administration of a marine carotenoid, fucoxanthin, significantly increased the abundance of *S24-7* in cecal microbiota but decreased the abundance of the genus in fecal microbiota [7]. Stanley et al. found that fecal and cecal microbiotas showed qualitative similarities but quantitative differences [8]. Marteau et al. compared the bacterial compositions within the human cecal and fecal microbiota and found that the abundance of *Bifidobacteria*, *Bacteroides*, *Clostridium coccoides* group, and *Clostridium leptum* subgroup were significantly lower in the cecum [9]. Therefore, comparative studies on the cecal and fecal microbiota in a specific situation are necessary to better understand the characteristics of gut microbiota.

High-fat diet (HFD) feeding has been widely used as a model for studying metabolic syndrome and gut microbiota dysbiosis [10,11]. Seaweed polysaccharides represent a kind of promising natural product that alleviates HFD-induced metabolic syndrome and promotes the healthy growth of gut bacteria [12]. In recent years, polysaccharides prepared from *Sargassum fusiforme*, a well-known edible alga, have attracted extensive research interest due to their potential biomedical application [13–17]. For example, our recent study indicated five polysaccharides prepared from *S. fusiforme* could selectively regulate the relative abundance of *Oscillospira* and *Clostridiales* in cecal microbiota of HFD-fed mice [16]. Cheng and colleagues prepared an *S. fusiforme* polysaccharide that could decrease the relative abundances of the diabetes-related fecal microbiota [17]. However, the regulatory effects of *S. fusiforme* polysaccharides on cecal microbiota and fecal microbiota, which would be helpful for the study of seaweed polysaccharide-based gut microbiota regulators, have not been compared.

In this study, two polysaccharides were prepared from *S. fusiforme* by water extraction (SfW) and acid extraction (SfA), and their chemical structures were characterized according to our recent report [16]. Then, the effects of 16 weeks of SfW and SfA administration on the cecal and fecal microbiota of HFD-fed mice were investigated.

2. Results

2.1. Chemical Structures of *Sargassum fusiforme* Polysaccharides

The physicochemical properties of the two *S. fusiforme* polysaccharides prepared by water extraction (SfW) and acid extraction (SfA) are shown in Table 1. The results indicated that the chemical structures of SfW and SfA were quite similar. For example, they had comparable contents of total sugar (70.0% vs. 62.4%) and sulfate group (28.5% vs. 31.3%), and their average molecular weights were also very close. Moreover, the monosaccharide compositions of the two polysaccharides were very similar. Both of them were mainly composed of glucose, fucose, and galactose with small amounts of mannose, glucuronic acid, and xylose, but the detailed molar ratio had a slight difference. For example, the content of xylose in SfW was lower than that in SfA (0.14 vs. 0.03), while glucose was more abundant in SfA (1.05 vs. 1.13).

Table 1. Physicochemical properties of *Sargassum fusiforme* polysaccharides.

Samples	Total Sugar (%)	Sulfate (%)	Mw (kDa)	Monosaccharide (Molar Ratio) *					
				Man	Glc A	Glc	Gal	Xyl	Fuc
SfW	70.0	28.5	166/5.9	0.07	0.07	1.05	0.41	0.14	1
SfA	62.4	31.3	276/5.8	0.05	0.06	1.13	0.38	0.03	1

* Man, mannose; GlcA, glucuronic acid; Glc, glucose; Gal, galactose; Xyl, xylose; Fuc, fucose.

Here, ^{1}H NMR spectra of SfW and SfA are shown in Figure S1. The resonance signals of the two polysaccharides at 3.0–5.5 ppm were ascribed to the typical distribution of ^{1}H NMR signals of the polysaccharides [18]. The unresolved peaks at 5.3–5.5 ppm were assigned to the anomeric protons of α-L-fucopyranosyl units [19]. The resonance signals of the two polysaccharides at 3.3–4.5 ppm were apportioned to the ring protons H-2 to H-5, but the pattern was different from each other, and the chemical shifts at 1.1 and 1.4 ppm were assigned to methyl groups of fucose units [20]. In addition, due to the complex and heterogeneous structure of sulfated polysaccharides, the broadening and overlapping of ^{1}H NMR peaks makes it difficult to completely describe their structural characteristics.

2.2. Effects of SfW and SfA on HFD-Induced Metabolic Disorders

The effects of SfW and SfA on HFD-induced metabolic disorders were evaluated after 16 weeks of co-treatment with HFD and polysaccharides. As shown in Figure 1A, 16-week HFD feeding significantly increased the body weight of the mice compared to that of the blank group, and co-treatment with SfW and SfA exacerbated body weight gain. The fasting blood glucose level significantly increased after 4 weeks of HFD-feeding, and only treatment with SfW at the fourth week reversed the increase ($p < 0.05$) (Figure 1B). In OGTT, the blood glucose reached the maximal level at 30 min after the dextrose gavage, and the blood glucose level of the control group was significantly higher than that of the blank group ($p < 0.001$). The polysaccharides administration could not attenuate the HFD-induced glucose intolerance (Figure 1C). Treatments with SfW and SfA did not protect HFD-induced insulin resistance and epididymal fat weight gain (Figure 1D–F).

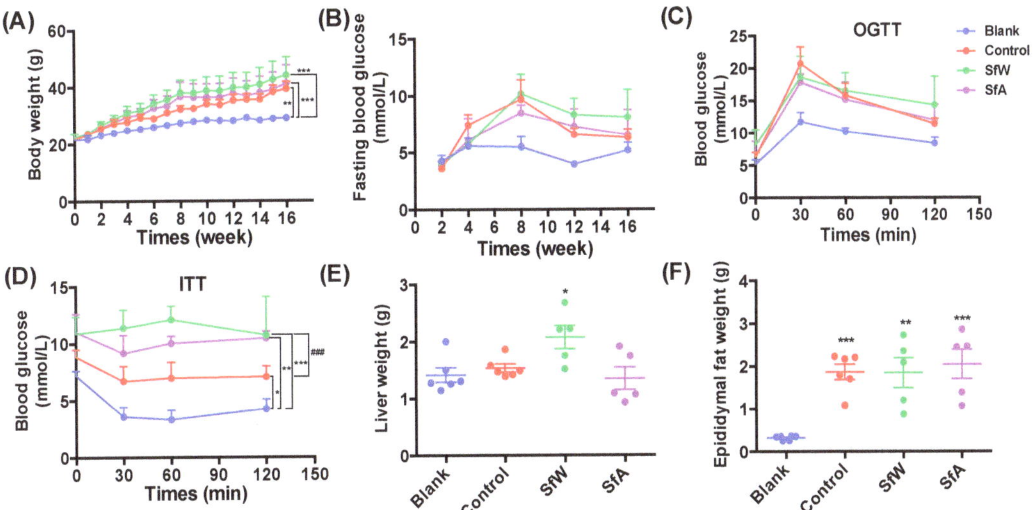

Figure 1. Effects of SfW and SfA on (**A**) body weight, (**B**) fasting blood glucose, (**C**) OGTT, (**D**) ITT, (**E**) liver weight, and (**F**) epididymal fat weight in HFD-fed mice. Values are mean ± SD (n = 5–6). * $p < 0.05$, ** $p < 0.01$, *** $p < 0.001$ vs. Blank. ### $p < 0.001$ vs. Control. Data sets in (**A–D**) were analyzed using unpaired two-tailed Student's t-test. Data sets in (**E–F**) were analyzed using one-way ANOVA followed by a Turkey's test.

The effects of 16 weeks of HFD and polysaccharide administration on gut microbiota in fecal samples and cecal contents of mice were analyzed by 16S rRNA high-throughput sequencing. As shown in Figure 2A–B, HFD significantly decreased both the Chao1 and Simpson's indices of cecal microbiota but showed negligible effect on the Simpson's index of fecal microbiota. Oral administration of SfW and SfA did not significantly improve the decrease in α-diversity of fecal and cecal microbiota. The unsupervised principal components analysis (PCA) plot at the phylum level showed that PC1 and PC2 were able to explain 56% and 41.3% of the variation, respectively, and exhibited significant distinction between the cecal and fecal microbiota of the blank group (Figure 2C). Polysaccharide administration showed no significant regulatory effect on the dysbiosis of cecal and fecal microbiota at the phylum level (Figure 2C). In detail, HFD significantly increased the relative abundance of Actinobacteria and decreased the abundance of Bacteroidetes and Verrucomicrobia in both the cecal and fecal microbiota but only enriched Firmicutes and Proteobacteria in fecal microbiota. SfW only presented a regulatory effect on Proteobacteria in fecal microbiota (Figure 2H,J).

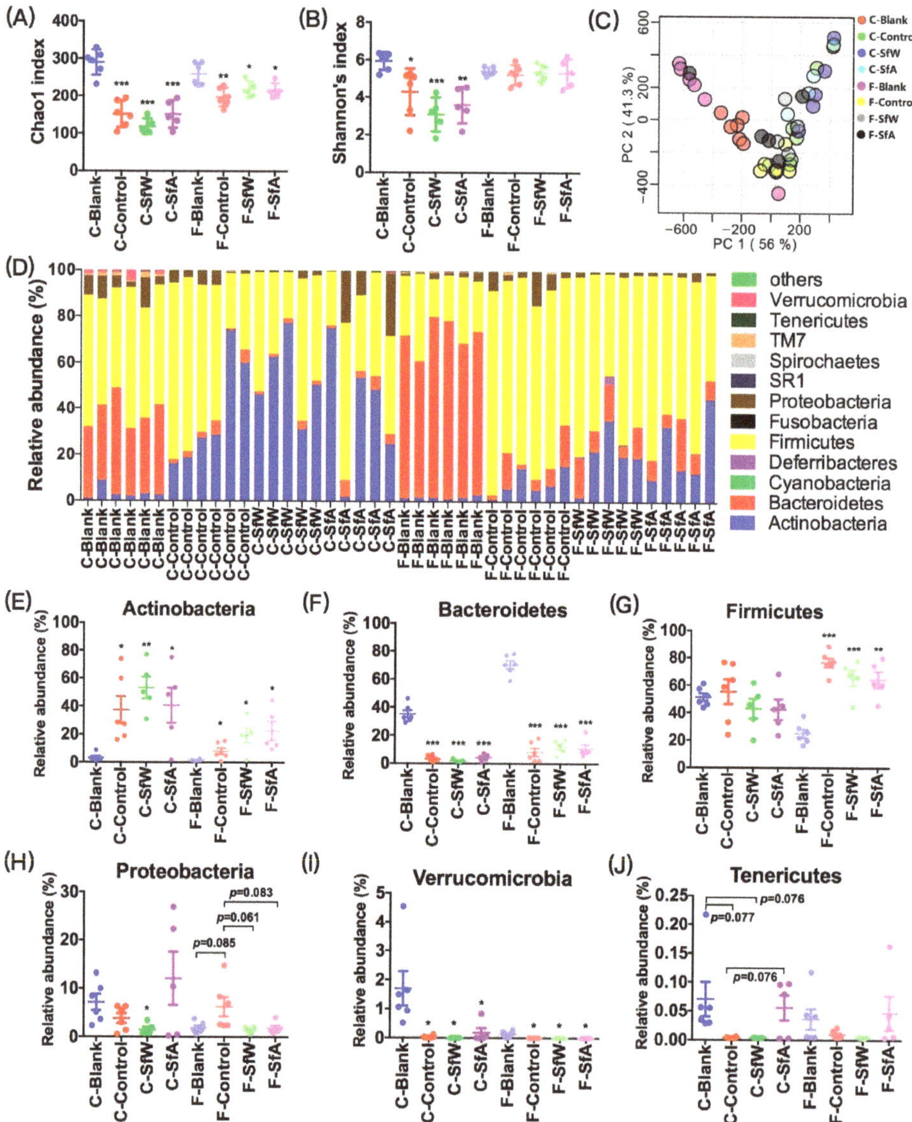

Figure 2. Effects of SfW and SfA on the (**A**) Chao1 diversity and (**B**) Shannon's diversity indices of cecal and fecal microbiota in HFD-fed mice. (**C**) PCA and (**D**) bar plots of cecal and fecal microbiota at the phylum level. The relative abundance of (**E**) Actinobacteria, (**F**) Bacteroidetes, (**G**) Firmicutes, (**H**) Proteobacteria, (**I**) Verrucomicrobia, and (**J**) Tenericutes in cecal and fecal microbiota. Values are mean ± SEM. * $p < 0.05$, ** $p < 0.01$, *** $p < 0.001$ vs. the corresponding blank group. Data sets in (**A**), (**B**), (**E–J**) were analyzed using one-way ANOVA followed by a Turkey's test. C-, cecal microbiota; F-, fecal microbiota.

The effects of SfW and SfA on the cecal and fecal microbiota in HFD-fed mice were also investigated at the genus level. As shown in Figure 3A, bacteria in these cecal and fecal samples mainly consisted of 27 genera, including *Coriobacteriaceae* and *S24-7*. However, the detailed composition of them between the cecal and fecal microbiota and between the blank and control groups were different. The far distance between the F-Blank group and F-SfW or F-SfA group in the clustering scheme suggested that HFD-induced dysbiosis of cecal microbiota could be more significantly regulated by SfW and SfA. The PCA plot at the genus level further confirmed the compositional difference between the cecal and fecal samples (Figure 3B). HFD mainly altered the relative abundance of *Coriobacteriaceae*, *S24-7*, *Ruminococcus*, *Clostridiales*, *Oscillospira*, *Ruminococcaceae*, and *Akkermansia* in both the cecal and fecal microbiota but only enriched *Oscillospira* in fecal microbiota (Figure 3C–I). Notably, oral administration of SfW and SfA could partially alleviate the increase of *Clostridiales* and *Ruminococcaceae* in fecal microbiota (Figure 3F,H).

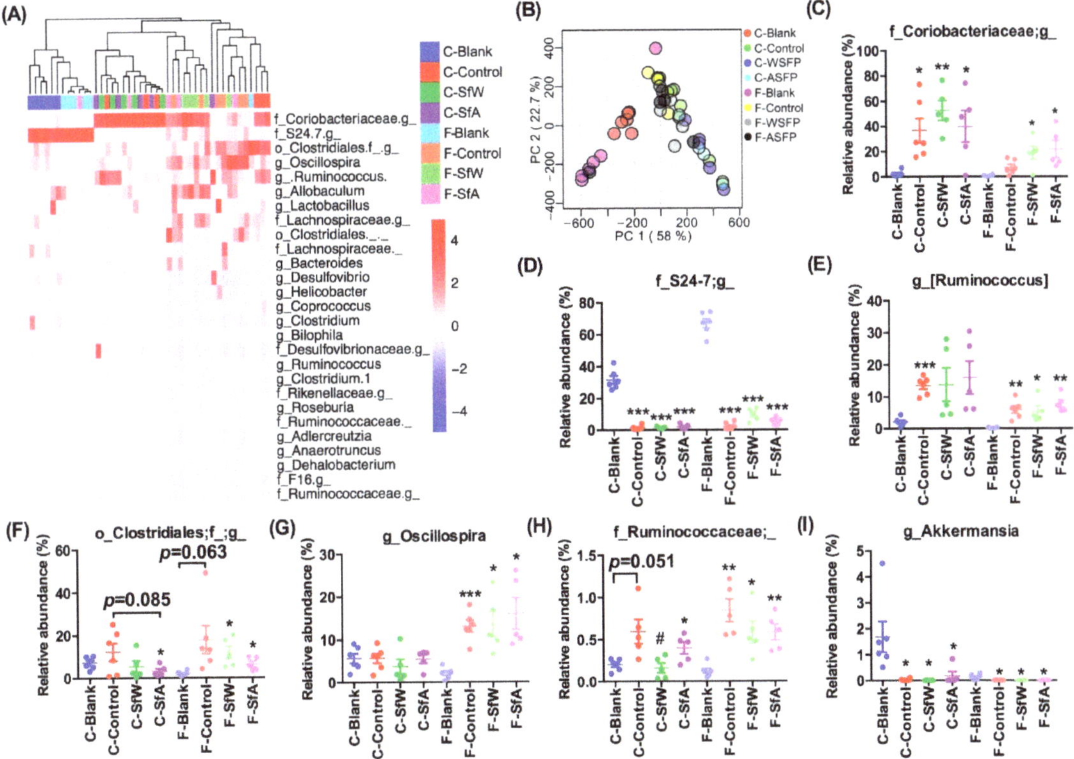

Figure 3. Effects of SfW and SfA on the cecal and fecal microbiota in HFD-fed mice. (**A**) Heatmap and (**B**) PCA plots of cecal and fecal microbiota at the genus level. The relative abundance of (**C**) *Coriobacteriaceae*, (**D**) *S24-7*, (**E**) *Ruminococcus*, (**F**) *Clostridiales*, (**G**) *Oscillospira*, (**H**) *Ruminococcaceae*, and (**I**) *Akkermansia* in cecal and fecal microbiota. Values are mean $\pm$ SEM. * $p < 0.05$, ** $p < 0.01$, *** $p < 0.001$ vs. the corresponding blank group. # $p < 0.05$ vs. the corresponding control group. Data sets in (**C**–**I**) were analyzed using one-way ANOVA followed by a Turkey's test. C-, cecal microbiota; F-, fecal microbiota.

2.3. *Effects of SfW and SfA on the Abundance of Genes Encoding Carbohydrate-Metabolizing Enzymes in Cecal and Fecal Microbiota*

The effects of *S. fusiforme* polysaccharides on the abundance of genes encoding carbohydrate-metabolizing enzymes in the cecal and fecal microbiota of HFD-fed mice were investigated using PICRUSt2 based on the 16S rRNA gene sequencing data. The results demonstrated that the HFD significantly decreased the abundance of genes encoding α-fucosidase (Figure 4A) and β-glucuronidase (Figure 4G), and increased that of monosaccharide-transporting ATPase (Figure 4B) and β-fructofur-anosidase (Figure 4E) in both the cecal and fecal microbiota, but the alterations of genes encoding α-galactosidase (Figure 4D) and β-glucosidase (Figure 4F) were only observed in fecal microbiota, and that of β-mannosidase (Figure 4H) was only presented in cecal microbiota. The administration of SfW and SfA mainly regulated the abundance of genes encoding monosaccharide-transporting ATPase, α-galactosidase, β-fructofuranosidase, and β-glucosidase with the latter showed more significant potency. For example, SfA alleviated the increase of genes encoding monosaccharide-transporting ATPase and β-glucosidase (Figure 4B–F).

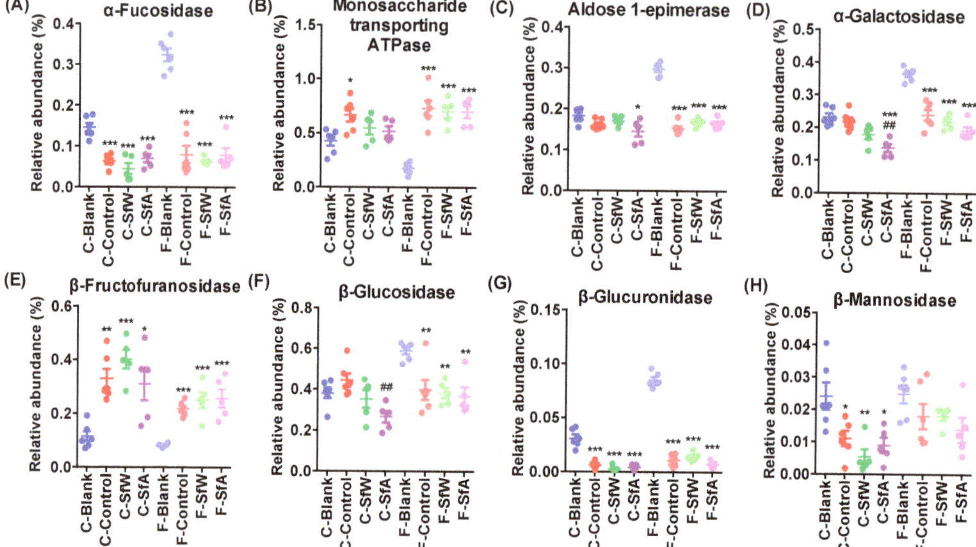

Figure 4. Effects of SfW and SfA on functional gene of cecal and fecal microbiota in HFD-fed mice. (**A**) α-Fucosidase, (**B**) monosaccharide-transporting ATPase, (**C**) aldose 1-epimerase, (**D**) α-galactosidase, (**E**) β-fructofuranosidase, (**F**) β-glucosidase, (**G**) β-glucuronidase, and (**H**) β-mannosidase. Values are mean ± SEM. * $p < 0.05$, ** $p < 0.01$, *** $p < 0.001$ vs. the corresponding blank group. ## $p < 0.01$ vs. the corresponding control group. Data sets were analyzed using one-way ANOVA followed by a Turkey's test. C-, cecal microbiota; F-, fecal microbiota.

3. Discussion

Seaweed polysaccharides have various biological activities [11,13,21–25] and also show potential to be developed as prebiotics that promote the healthy growth of gut bacteria. Previous studies have compared the similarities and differences between the cecal and fecal microbiota of animals and human volunteers [6–8]. However, the regulatory effects of polysaccharides extracted from *S. fusiforme* (a common and widely eaten seaweed) on the cecal and fecal microbiota of HFD-fed mice, which would benefit the study of seaweed polysaccharide-based gut microbiota regulators, have not been compared. Here, the impacts of 16 weeks of water and acid extracted *S. fusiforme* polysaccharides (SfW and SfA) on the cecal and fecal microbiota of HFD-fed mice were investigated. We found that the HFD significantly altered the dominant phyla Bacteroidetes and Actinobacteria, and the dominant genera *Coriobacteriaceae*, *S24-7*, and *Ruminococcus*, but did not affect the abundance of Firmicutes, *Clostridiales*, *Oscillospira*, and *Ruminococcaceae* in cecal microbiota and the Simpson's index of fecal microbiota. Co-treatments with SfW and SfA partially reversed the dysbiosis of Firmicutes, *S24-7*, *Ruminococcus*, *Clostridiales*, and *Ruminococcaceae*. The administration of SfW and SfA also altered the abundance of some genes encoding carbohydrate-metabolizing enzymes between cecal and fecal microbiota in 16-week HFD-fed mice and demonstrated that cecal microbiota was more significantly regulated by *S. fusiforme* polysaccharides.

Our recent study reported that five *S. fusiforme* polysaccharides prepared through hot-water and acid extraction showed a wide range of molecular weight (10–698.3 kDa), and the HCl-extracted polysaccharide (Sf-A) mainly consisted of glucose, fucose, and galactose [16]. Cheng et al. prepared a 205.8 kDa *S. fusiforme* polysaccharide (SFF) by acid extraction and the polysaccharide was mainly composed of fucose and galactose [17]. The chemical structures of SfW and SfA were more similar to Sf-A than SFF, which may be because they are sourced from the same supplier.

HFD has been demonstrated to adversely affect gut microbiota composition through increasing the abundance of Firmicutes and Proteobacteria and decreasing Bacteroidetes [26–29]. The present study demonstrated that 16 weeks of HFD significantly increased the relative abundance of Actinobacteria and decreased the abundance of Bacteroidetes and Verrucomicrobia in both the cecal and fecal microbiota but only enriched Firmicutes and Proteobacteria in fecal microbiota. The discrepancy between the previous studies with our findings indicates that HFD consistently decreases the abundance of Bacteroidetes in both the cecal and fecal microbiota, but the impacts on other phyla depend on the sampling position and the specific conditions of the animal models. Our recent study demonstrated that four weeks of HFD feeding significantly increased the relative abundance of *Coriobacteriaceae* and *Oscillospira* in cecal microbiota [11], but in the present study, the abundance of *Oscillospira* in cecal microbiota was not enriched by HFD, which may be explained by the prolonged HFD treatment. These findings further deepen our understanding about the impact of HFD on gut microbiota composition.

It is well known that microbiota composition varies in different parts of the gastrointestinal tract, and there are significant differences in the quantity and quality of microorganisms in the cecum contents and feces [8]. According to a recent report, the cecal and fecal microbiota exhibits different taxonomic structures, functional activities, and metabolic pathways [30]. Guo et al. also reported that the abundance of Firmicutes in the cecum of mice fed a normal diet was much higher than that in the feces, and *S24-7* was much lower in the cecal contents, which was highly consistent with the present study [9]. Here, we comprehensively compared the similarities and differences between the cecal and fecal microbiota in 16-week HFD-fed mice, including the abundance of genes encoding carbohydrate-metabolizing enzymes in cecal and fecal microbiota. HFD showed a more significant influence on the α-diversity of cecal microbiota than that of fecal microbiota, suggesting that cecal microbiota may be more suitable to represent the gut microbiota composition. Some bacteria (Actinobacteria and *Coriobacteriaceae*) were significantly enriched in the cecal contents, while others (Firmicutes, *Oscillospira*, and *Ruminococcaceae*) were enriched in the fecal samples, which may be associated with the different function of the cecum and colon. For example, *Coriobacteriaceae*, a family within the phylum Actinobacteria, are strictly anaerobic bacteria and contribute to the metabolism of bile salts, steroids, and dietary polyphenols [31]. Ariangela et al. reported that the TNBS colitis severity was most closely correlated with the composition of colonic mucus microbiome, but not fecal or cecal microbiome [32], suggesting that the choice of sampling site depends on the experimental design. Interestingly, SfA was mainly composed of glucose, fucose, and galactose, and administration of SfA significantly decreased the abundance of genes encoding α-galactosidase and β-glucosidase, which may be ascribed to the interactions between the polysaccharide and gut microbiota.

Previous studies reported the effects of several *S. fusiforme* polysaccharides on gut microbiota composition [11,17,33]. Chen et al. reported that 12 months of oral administration of an *S. fusiforme* polysaccharide decreased the abundance of the phyla Firmicutes, Proteobacteria, and the genera *Lactobacillus* and *Helicobacter* in small intestinal microbiota [33]. Cheng and colleagues found that 6 weeks of oral administration of an *S. fusiforme* polysaccharide significantly decreased the relative abundances of several diabetes-related microbiota in fecal samples, including *Bilophila*, *Oscillibacter*, and *Mucispirillum* [17]. Our previous study reported that 4 weeks of *S. fusiforme* polysaccharide treatment significantly altered the relative abundance of the phylum Bacteroidetes, and the genera *Oscillospira*, *Mucispirillum*, and *Clostridiales* in cecal microbiota [11]. In the present study, 16 weeks of SfW administration partially reversed HFD-induced alterations of *Clostridiales* and *Ruminococcaceae* in both the cecal and fecal microbiota. The differences in the regulatory effects of *S. fusiforme* polysaccharides on gut micobiota may be explained by the structural difference of polysaccharides, or difference in the animal model, and sampling position. For example, the microbiota composition in the intestinal contents from the ICR mice was investigated by Chen et al., while in the present study, we determined the microbiota composition in fecal and cecal samples from the ICR mice.

4. Materials and Methods

4.1. Materials

The brown algae *S. fusiforme* was sourced from Qingdao, China, on August, 2018. Standards of monosaccharides and 1-phenyl-3-methyl-5-pyrazolone (PMP) were purchased from Aladdin Chemistry Co., Ltd. (Shanghai, China). Dextran standards were purchased from American Polymer Standards Corporation (Mentor, OH, USA). Other reagents and solvents were of analytical grade.

4.2. Preparation of *Sargassum fusiforme* Polysaccharides

As previously reported, *S. fusiforme* polysaccharides were prepared with slight modifications [11]. Briefly, dry *S. fusiforme* was cut into pieces and pre-treated with 95% ethanol to remove the pigment. Polysaccharides were extracted from the residual materials by hot water for 1 h and 0.1 M HCl at 60 °C for 1 h, respectively. The hot water-extracted polysaccharide was further treated with 0.05 M $MgCl_2$ to eliminate the alginate, and ultra-filtered to obtain the SfW. The HCl-extracted crude polysaccharide was dialyzed and precipitated using ethanol to obtain the SfA.

4.3. Structural Analysis of *Sargassum fusiforme* Polysaccharides

The chemical analysis of SfW and SfA was based on previous studies [11]. The total sugar content was measured by the phenol-sulfuric acid method using D-glucose as the standard [34]. The content of sulfate was analyzed with the $BaCl_2$-gelation method using Na_2SO_4 as the standard [35]. The molecular weight analysis was conducted using High-Performance Size Exclusion Chromatography using a Waters 2487 HPLC system with a refractive index detector 2414 (Waters, Milford, MA, USA). The chromatography conditions refer to previous studies [11]. The molar ratio of monosaccharide and fucose content was determined by the PMP derivatization method with minor modification.

SfW and SfA fraction (100 mg) were dissolved in 0.55 mL of D_2O (99.9%) followed by centrifugation and lyophilization, and the process was repeated three times. Finally, the resulting polysaccharide was dissolved with 0.55 mL of D_2O. ^1H NMR spectra were recorded on AVANCE III NMR 600 MHz spectrometer (Bruker Inc., Billerica, MA, USA) at 25 °C.

4.4. Animal Experiments

Twenty-two male C57 mice (Specific-pathogen-free grade, six-week-old) were purchased from SPF (Beijing, China) Biotechnology Co., Ltd. (Beijing, China), and kept at the Animal Center of Zhejiang, University of Technology. All mice were randomly divided into four groups (the blank and control groups: $n = 6$; the SfW and SfA groups: $n = 5$) and fed a normal diet for one week to stabilize all the metabolic conditions. Each group was housed in one standard cage under a condition of 22 ± 1 °C, a humidity of 55 ± 5%, and a 12 h light/dark cycle. From the beginning of the experiment, mice in the control, SfW, and SfA groups were fed with an HFD (TP23300, 60 kcal% fat, Trophic Animal Feed High-Tech Co., Ltd, Nantong, China), and mice in the blank group were still fed a standard lab chow diet. The mice in the SfW and SfA groups had free access to 1 mg/mL of SfW and SfA in drinking water, respectively, for a period of 16 weeks, and the other two groups were treated with sterile water.

During the experiments, the body weight was measured weekly, and fasting blood glucose was measured monthly via tail vein using a glucometer (Johnson and Johnson, New Brunswick, NJ, USA) according to the instruction after fasting for 16 h. A sugar tolerance test (OGTT) and insulin tolerance test (ITT) were measured at week 15 and 16, respectively, and the feces were collected before the mice were dissected. All mice were sacrificed from asphyxiation by carbon dioxide. The liver and epididymal fat were harvested and their weight was measured. The cecum contents were collected and stored at −80 °C. The protocol was approved by the Animal Ethics Committee of the Zhejiang University of Technology, China. All efforts were made to minimize the suffering of the mice.

4.5. Gut Microbiota Analysis by 16S rRNA Gene Sequencing

Fecal samples and cecal contents were used for gut microbiota analysis by sequencing the 16S rRNA genes. The DNA extraction, PCR amplification, sequencing, and data analysis were conducted according to our previous study [11].

4.6. Statistical Analysis

The significance of the differences between the two groups was assessed using the unpaired two-tailed Student's *t*-test (Figure 1). Data sets that involved more than two groups were assessed by one-way ANOVA followed by a Turkey's test (Figures 1–4). *p* values and the significance level are indicated in the associated figure legends for each figure. Statistical analysis was performed with SPSS statistics software (Version 19.0).

5. Conclusions

In conclusion, we compared the microbiota composition in the cecal and fecal microbiota of 16-week HFD-fed mice and revealed that HFD dramatically altered the abundance of Bacteroidetes and Actinobacteria, and *Coriobacteriaceae*, *S24-7*, and *Ruminococcus* in both the cecal and fecal microbiota but did not affect the relative abundance of Firmicutes, *Clostridiales*, *Oscillospira*, *Ruminococcaceae* in cecal microbiota and the Simpson's index of fecal microbiota. Co-treatments with SfW and SfA exacerbate body weight gain and partially reverse HFD-induced alterations of *Clostridiales* and *Ruminococcaceae* and alter the abundance of genes encoding monosaccharide-transporting ATPase, α-galactosidase, β-fructofuranosidase, and β-glucosidase. Our findings provide important insights for the study of seaweed polysaccharide-based gut microbiota regulators.

Supplementary Materials: The following are available online at https://www.mdpi.com/1660-3397/19/7/364/s1. Figure S1 ^{1}H-NMR spectra of *Sargassum fusiforme* polysaccharides SfW and SfA.

Author Contributions: X.-W.X. and H.W. designed the study; B.W.(Bin Wei), Q.-L.X., B.Z., T.-S.Z., and S.-Z.K. performed the animal study; S.-J.W. characterized the structures of polysaccharides; B.W.(Bin Wei) and Q.-L.X. wrote the manuscript; B.W.(Bin Wu) revised the manuscript; X.-W.X. and H.W. supervised the study. All authors have read and agreed to the published version of the manuscript.

Funding: This work was financially supported by the National Key Research and Development Program (No. 2017YFE0103100) and the programs of the National Natural Science Foundation of China (No. 81903534, No. 81773628, and No. 81741165) and the Scientific Research Fund of Second Institute of Oceanography, MNR (No. JB2002).

Institutional Review Board Statement: The study was conducted according to the guidelines of the Declaration of Helsinki, and approved by the Ethics Committee of the Zhejiang University of Technology, China (No. 20190219043).

Data Availability Statement: The data in this study are available on request from the corresponding author.

Conflicts of Interest: There are no conflict to declare.

References

1. Valdes, A.M.; Walter, L.; Segal, E.; Spector, T.D. Role of the gut microbiota in nutrition and health. *BMJ Brit. Med. J.* **2018**, *361*, j2179. [CrossRef] [PubMed]
2. Chassaing, B.; Koren, O.; Goodrich, J.K.; Poole, A.C.; Srinivasan, S.; Ley, R.E.; Gewirtz, A.T. Dietary emulsifiers impact the mouse gut microbiota promoting colitis and metabolic syndrome. *Nature* **2015**, *519*, U92–U192. [CrossRef] [PubMed]
3. Stephens, R.W.; Arhire, L.; Covasa, M. Gut microbiota: From microorganisms to metabolic organ influencing obesity. *Obesity* **2018**, *26*, 801–809. [CrossRef] [PubMed]
4. Hills, R.D.; Pontefract, B.A.; Mishcon, H.R.; Black, C.A.; Sutton, S.C.; Theberge, C.R. Gut microbiome: Profound implications for diet and disease. *Nutrients* **2019**, *11*, 1613. [CrossRef]
5. Yu, F.; Han, W.; Zhan, G.F.; Li, S.; Jiang, X.H.; Wang, L.; Xiang, S.K.; Zhu, B.; Yang, L.; Luo, A.L.; et al. Abnormal gut microbiota composition contributes to the development of type 2 diabetes mellitus in db/db mice. *Aging* **2019**, *11*, 10454–10467. [CrossRef]
6. Nguyen, S.G.; Kim, J.; Guevarra, R.B.; Lee, J.H.; Kim, E.; Kim, S.I.; Unno, T. *Laminarin* favorably modulates gut microbiota in mice fed a high-fat diet. *Food Funct.* **2016**, *7*, 4193–4201. [CrossRef]
7. Marteau, P.; Pochart, P.; Dore, J.; Bera-Maillet, C.; Bernalier, A.; Corthier, G. Comparative study of bacterial groups within the human cecal and fecal microbiota. *Appl. Environ. Microb.* **2001**, *67*, 4939–4942. [CrossRef]
8. Stanley, D.; Geier, M.S.; Chen, H.; Hughes, R.J.; Moore, R.J. Comparison of fecal and cecal microbiotas reveals qualitative similarities but quantitative differences. *BMC Microbiol.* **2015**, *15*, 1–11. [CrossRef]
9. Guo, B.B.; Yang, B.; Pang, X.Y.; Chen, T.P.; Chen, F.; Cheng, K. Fucoxanthin modulates cecal and fecal microbiota differently based on diet. *Food Funct.* **2019**, *10*, 5644–5655. [CrossRef]
10. Shang, Q.S.; Song, G.R.; Zhang, M.F.; Shi, J.J.; Xu, C.Y.; Hao, J.J.; Li, G.Y.; Yu, G.L. Dietary fucoidan improves metabolic syndrome in association with increased *Akkermansia* population in the gut microbiota of high-fat diet-fed mice. *J. Funct. Foods* **2017**, *28*, 138–146. [CrossRef]
11. Li, S.; Li, J.H.; Mao, G.Z.; Wu, T.T.; Hu, Y.Q.; Ye, X.Q.; Tian, D.; Linhardt, R.J.; Chen, S.G. A fucoidan from sea cucumber Pearsonothuria graeffei with well-repeated structure alleviates gut microbiota dysbiosis and metabolic syndromes in HFD-fed mice. *Food Funct.* **2018**, *9*, 1039. [CrossRef]
12. You, L.; Gong, Y.; Li, L.; Hu, X.; Brennan, C.; Kulikouskaya, V. Beneficial effects of three brown seaweed polysaccharides on gut microbiota and their structural characteristics: An overview. *Int. J. Food Sci. Technol.* **2020**, *55*, 1199–1206. [CrossRef]
13. Zhang, R.; Zhang, X.; Tang, Y.; Mao, J. Composition, isolation, purification and biological activities of *Sargassum fusiforme* polysaccharides: A review. *Carbohydr. Polym.* **2020**, *228*, 115381. [CrossRef]
14. Kong, Q.; Zhang, R.; You, L.; Ma, Y.; Liao, L.; Pedisić, S. In vitro fermentation characteristics of polysaccharide from *Sargassum fusiforme* and its modulation effects on gut microbiota. *Food Chem. Toxicol.* **2021**, *151*, 112145. [CrossRef]
15. Zhang, Y.; Zuo, J.; Yan, L.; Cheng, Y.; Li, Q.; Wu, S.; Chen, L.; Thring, R.W.; Yang, Y.; Gao, Y.; et al. *Sargassum fusiforme* fucoidan alleviates high-fat diet-induced obesity and insulin resistance associated with the improvement of hepatic oxidative stress and gut microbiota profile. *J. Agric. Food Chem.* **2020**, *68*, 10626–10638. [CrossRef]
16. Wei, B.; Zhong, Q.W.; Ke, S.Z.; Zhou, T.S.; Xu, Q.L.; Wang, S.J.; Chen, J.W.; Zhang, H.W.; Jin, W.H.; Wang, H. *Sargassum fusiforme* polysaccharides prevent high-fat diet-induced early fasting hypoglycemia and regulate the gut microbiota composition. *Mar. Drugs* **2020**, *18*, 444. [CrossRef]
17. Cheng, Y.; Sibusiso, L.; Hou, L.F.; Jiang, H.J.; Chen, P.C.; Zhang, X.; Wu, M.J.; Tong, H.B. *Sargassum fusiforme* fucoidan modifies the gut microbiota during alleviation of streptozotocin-induced hyperglycemia in mice. *Int. J. Biol. Macromol.* **2019**, *131*, 1162–1170. [CrossRef]
18. Palanisamy, S.; Vinosha, M.; Marudhupandi, T.; Rajasekar, P.; Prabhu, N.M. Isolation of fucoidan from *Sargassum polycystum* brown algae: Structural characterization, in vitro antioxidant and anticancer activity. *Int. J. Biol. Macromol.* **2017**, *102*, 405–412. [CrossRef]
19. Somasundaram, S.N.; Shanmugam, S.; Subramanian, B.; Jaganathan, R. Cytotoxic effect of fucoidan extracted from *Sargassum cinereum* on colon cancer cell line HCT-15. *Int. J. Biol. Macromol.* **2016**, *91*, 1215–1223. [CrossRef]
20. Synytsya, A.; Kim, W.J.; Kim, S.M.; Pohl, R.; Synytsya, A.; Kvasnička, F.; Čopíková, J.; Park, Y. Structure and antitumour activity of fucoidan isolated from sporophyll of Korean brown seaweed *Undaria pinnatifida*. *Carbohydr. Polym.* **2010**, *81*, 41–48. [CrossRef]
21. Zhong, Q.W.; Zhou, T.S.; Qiu, W.H.; Wang, Y.K.; Xu, Q.L.; Ke, S.Z.; Wang, S.J.; Jin, W.H.; Chen, J.W.; Zhang, H.W.; et al. Characterization and hypoglycemic effects of sulfated polysaccharides derived from brown seaweed *Undaria pinnatifida*. *Food Chem.* **2021**, *341*, 128148. [CrossRef]
22. Zhong, Q.W.; Wei, B.; Wang, S.J.; Ke, S.Z.; Chen, J.W.; Zhang, H.W.; Wang, H. The antioxidant activity of polysaccharides derived from marine organisms: An overview. *Mar. Drugs* **2019**, *17*, 672. [CrossRef]
23. Kim, M.J.; Jeon, J.; Lee, J.S. Fucoidan prevents high-fat diet-induced obesity in animals by suppression of fat accumulation. *Phytother. Res.* **2014**, *28*, 137–143. [CrossRef]
24. Fitton, J.H.; Stringer, D.N.; Karpiniec, S.S. Therapies from fucoidan: An update. *Mar. Drugs* **2015**, *13*, 5920–5946. [CrossRef]
25. Fitton, J.H. Therapies from fucoidan; multifunctional marine polymers. *Mar. Drugs* **2011**, *9*, 1731–1760. [CrossRef]
26. Li, B.; Lu, F.; Wei, X.J.; Zhao, R.X. Fucoidan: Structure and bioactivity. *Molecules* **2008**, *13*, 1671–1695. [CrossRef]

27. Agus, A.; Clément, K.; Sokol, H. Gut microbiota-derived metabolites as central regulators in metabolic disorders. *Gut* **2020**, *70*, 1174–1182. [CrossRef]
28. Ding, N.; Zhang, X.; Zhang, X.D.; Jing, J.; Zhao, A.Z. Impairment of spermatogenesis and sperm motility by the high-fat diet-induced dysbiosis of gut microbes. *Gut* **2020**, *69*, 1608–1619. [CrossRef] [PubMed]
29. Shin, N.R.; Lee, J.C.; Lee, H.Y.; Kim, M.S.; Whon, T.W.; Lee, M.S.; Bae, J.W. An increase in the *Akkermansia spp.* population induced by metformin treatment improves glucose homeostasis in diet-induced obese mice. *Gut* **2014**, *63*, 727–735. [CrossRef] [PubMed]
30. Alessandro, T.; Valeria, M.; Cristina, F.; Antonio, P.; Marcello, A.; Massimo, D.; Michael, S.; Sergio, U. Metaproteogenomics reveals taxonomic and functional changes between cecal and fecal microbiota in mouse. *Front. Microbiol.* **2017**, *8*, 391.
31. Shahinozzaman, M.; Raychaudhuri, S.; Fan, S.; Obanda, D.N. Kale attenuates inflammation and modulates gut microbial composition and function in C57BL/6J mice with diet-induced obesity. *Microorganisms* **2021**, *9*, 238. [CrossRef]
32. Kozik, A.J.; Nakatsu, C.H.; Chun, H.; Jones-Hall, Y.L. Comparison of the fecal, cecal, and mucus microbiome in male and female mice after TNBS-induced colitis. *PLoS ONE* **2019**, *14*, e0225079. [CrossRef]
33. Chen, P.; Yang, S.; Hu, C.; Zhao, Z.; Wu, M. *Sargassum fusiforme* polysaccharide rejuvenates the small intestine in mice through altering its physiology and gut microbiota composition. *Curr. Mol. Med.* **2017**, *17*, 350–358.
34. Lin, F.M.; Pomeranz, Y. Effect of borate on colorimetric determinations of carbohydrates by the phenol-sulfuric acid method. *Anal. Biochem.* **1968**, *24*, 128–131. [CrossRef]
35. Dodgson, K.; Price, R. A note on the determination of the ester sulphate content of sulphated polysaccharides. *Biochem. J.* **1962**, *84*, 106–110. [CrossRef]

MDPI
St. Alban-Anlage 66
4052 Basel
Switzerland
Tel. +41 61 683 77 34
Fax +41 61 302 89 18
www.mdpi.com

Marine Drugs Editorial Office
E-mail: marinedrugs@mdpi.com
www.mdpi.com/journal/marinedrugs

www.ingramcontent.com/pod-product-compliance
Lightning Source LLC
LaVergne TN
LVHW070853170726
843515LV00005B/637

* 9 7 8 3 0 3 6 5 7 0 0 8 2 *